STUDENT'S SOLUTIONS MANUAL

DIACRITECH

ELEMENTARY STATISTICS: PICTURING THE WORLD

SEVENTH EDITION

Ron Larson

Pennsylvania State University
The Behrend College

Betsy Farber

Bucks County Community College

Reproduced by Pearson from electronic files supplied by the author.

Publishing as Pearson, 330 Hudson Street, NY NY 10013

ISBN-13: 978-0-13-468361-4
ISBN-10: 0-13-468361-7

CONTENTS

CHAPTER 1

Introduction to Statistics

1.1 AN OVERVIEW OF STATISTICS

1.1 TRY IT YOURSELF SOLUTIONS

1. The population consists of the responses of all ninth to twelfth graders in the United States. The sample consists of the responses of the 1501 ninth to twelfth graders in the survey. The sample data set consists of 1215 ninth to twelfth graders who said leaders today are more concerned with their own agenda than with achieving the overall goals of the organization they serve and 286 ninth to twelfth graders who did not say that.

2a. Population parameter, because the total spent on employees' salaries, $5,150,694, is based on the entire company.

b. Sample statistic, because 43% is based on a subset of the population.

3a. The population consists of the responses of all U.S. adults, and the sample consists of the responses of the 1000 U.S. adults in the study.

b. The part of this study that represents the descriptive branch of statistics involves the statement "three out of four adults will consult with their physician or pharmacist and only 8% visit a medication-specific website [when they have a question about their medication]."

c. A possible inference drawn from the study is that most adults consult with their physician or pharmacist when they have a question about their medication.

1.1 EXERCISE SOLUTIONS

1. A sample is a subset of a population.

3. A parameter is a numerical description of a population characteristic. A statistic is a numerical description of a sample characteristic.

5. False. A statistic is a numerical measure that describes a sample characteristic.

7. True

9. False. A population is the collection of *all* outcomes, responses, measurements, or counts that are of interest.

11. Population, because it is a collection of the salaries of each member of a Major League Baseball team.

13. Sample, because the collection of the 300 people is a subset of the population of 13,000 people in the auditorium.

15. Sample, because the collection of the 10 patients is a subset of the population of 50 patients at the clinic.

17. Population, because it is a collection of all the gamers' scores in the tournament.

19. Population, because it is a collection of all the U.S. senators' political parties.

21. Population: Parties of registered voters.
Sample: Parties of registered voters who respond to a survey

23. Population: Ages of adults in the United States who own automobiles.
Sample: Ages of adults in the United States who own Honda automobiles

25. Population: Collections of the responses of all U.S. adults.
Sample: Collection of the responses of the 1020 U.S. adults surveyed
Sample data set: 42% of adults who said they trust their political leaders and 58% who said they did not

27. Population: Collection of the influenza immunization status of all adults in the United States.
Sample: Collection of the influenza immunization status of the 3301 U.S. adults surveyed
Sample data set: 39% of U.S. adults who received an influenza vaccine and 61% who did not

29. Population: Collection of the average hourly billing rates of all U.S. law firms.
Sample: Collection of the average hourly billing rates for partners of the 159 U.S. law firms surveyed
Sample data set: The average hourly billing rate for partners of 159 U.S. law firms is $604

31. Population: Collection of all U.S. adults.
Sample: Collection of the responses of those suffering with chronic pain of the 1029 U.S. adults surveyed
Sample data set: 23% of respondents suffering with chronic pain who were diagnosed with a sleeping disorder and 77% who were not

33. Population: Collection of all companies listed in the Standard & Poor's 500.
Sample: Collection of the responses of the 54 Standard & Poor's 500 companies surveyed
Sample data set: Starting salaries of the 54 companies surveyed

35. Sample statistic. The value $72,000 is a numerical description of a sample of average salaries.

37. Population Parameter. The 62 surviving passengers out of 97 total passengers is a numerical description of all of the passengers of the Hindenburg that survived.

39. Sample statistic. The value 7% is a numerical description of a sample of computer users.

41. Sample statistic. The value 80% is a numerical description of a sample of U.S. adults.

43. The statement "23% of those suffering with chronic pain had been diagnosed with a sleep disorder" is an example of descriptive statistics. Using inferential statistics, you may conclude that an association exists between chronic pain and sleep disorders.

45. Answers will vary.

47. The inference may incorrectly imply that exercise increases a person's cognitive ability. The study shows a slower decline in cognitive ability, not an increase.

49. (a) The sample is the results on the standardized test by the participants in the study.

(b) The population is the collection of all the results of the standardized test.

(c) The statement "the closer that participants were to an optimal sleep duration target, the better they performed on a standardized test" is an example of descriptive statistics.

(d) Individuals who obtain optimal sleep will be more likely to perform better on a standardized test then they would without optimal sleep.

1.2 DATA CLASSIFICATION

1.2 TRY IT YOURSELF SOLUTIONS

1. The city names are nonnumerical entries, so these are qualitative data. The city populations are numerical entries, so these are quantitative data.

2. (1) Ordinal, because the data can be put in order.

(2) Nominal, because no mathematical computations can be made.

3. (1) Interval, because the data can be ordered and meaningful differences can be calculated, but it does not make sense to write a ratio using the temperatures.

(2) Ratio, because the data can be ordered, meaningful differences can be calculated, the data can be written as a ratio, and the data set contains an inherent zero.

1.2 EXERCISE SOLUTIONS

1. Nominal and ordinal

3. False. Data at the ordinal level can be qualitative or quantitative.

5. False. More types of calculations can be performed with data at the interval level than with data at the nominal level.

7. Quantitative, because dog weights are numerical measurements.

9. Qualitative, because hair colors are attributes.

11. Quantitative, because infant heights are numerical measurements.

13. Qualitative, because the poll responses are attributes.

15. Interval. Data can be ordered and meaningful differences can be calculated, but it does not make sense to say one year is a multiple of another.

17. Nominal. No mathematical computations can be made and data are categorized using numbers.

19. Ordinal. Data can be arranged in order, but the differences between data entries are not meaningful.

21. Horizontal: Nominal; Vertical: Ratio

23. Horizontal: Nominal; Vertical: Ratio

25. (a) Interval (b) Nominal (c) Ratio (d) Ordinal

27. Qualitative. Ordinal. Data can be arranged in order, but differences between data entries are not meaningful.

29. Qualitative. Nominal. No mathematical computations can be made and data are categorized by region.

31. Qualitative. Ordinal. Data can be arranged in order, but differences between data entries are not meaningful.

33. An inherent zero is a zero that implies "none." Answers will vary.

1.3 DATA COLLECTION AND EXPERIMENTAL DESIGN

1.3 TRY IT YOURSELF SOLUTIONS

1. This is an observational study.

2. There is no way to tell why the people quit smoking. They could have quit smoking as a result of either chewing the gum or watching the DVD. The gum and the DVD could be confounding variables. To improve the study, two experiments could be done, one using the gum and the other using the DVD. Or just conduct one experiment using either the gum or the DVD.

3. Sample answer: Assign numbers 1 to 79 to the employees of the company. Use the table of random numbers and obtain 63, 7, 40, 19, and 26. The employees assigned these numbers will make up the sample.

4. (1) The sample was selected by using the students in a randomly chosen class. This is cluster sampling.

(2) The sample was selected by numbering each student in the school, randomly choosing a starting number, and selecting students at regular intervals from the starting number. This is systematic sampling.

1.3 EXERCISE SOLUTIONS

1. In an experiment, a treatment is applied to part of a population and responses are observed. In an observational study, a researcher measures characteristics of interest of a part of a population but does not change existing conditions.

3. In a random sample, every member of the population has an equal chance of being selected. In a simple random sample, every possible sample of the same size has an equal chance of being selected.

5. False. A placebo is a fake treatment.

7. False. Using stratified sampling guarantees that members of each group within a population will be sampled.

9. False. To select a systematic sample, a population is ordered in some way and then members of the population are selected at regular intervals.

11. Observational study. The study does not apply a treatment to the adults.

13. Experiment. The study applies a treatment (different photographs) to the subjects.

15. Answers will vary. *Sample answer*: Starting at the left-most number in row 6:
28/70/35/17/09/94/45/64/83/96/73/78/
The numbers would be 28,70,35,17,9,94,45,64,83,96,73,78.

17. Answers will vary.

19. (a) The experimental units are the 500 females ages 25 to 45 years old who suffer from migraine headaches. The treatment is the new drug used to treat migraine headaches.

(b) A problem with the design is that the sample is not representative of the entire population because only females ages 25 to 45 were used. To increase validity, use a stratified sample.

(c) For the experiment to be double-blind, neither the subjects nor the company would know whether the subjects are receiving the drug or the placebo.

21. Answers will vary. *Sample answer*: Number the volunteers from 1 to 18. Using the random number table in Appendix B, starting with the left-most number in row 16:
29/55/31/84/32/**13**/63/00/55/29/**02**/79/**18**/**10**/**17**/49/02/77/90/31/50/91/20/93/99
23/50/**12**/26/42/63/**08**/10/81/91/89/42/**06**/78/00/55/13/75/47/**07**/
Treatment group: Maria, Adam, Bridget, Carlos, Susan, Rick, Dan, Mary, and Connie.
Control group: Jake, Mike, Lucy, Ron, Steve, Vanessa, Kate, Pete, and Judy.

23. Simple random sampling is used because each employee has an equal chance of being contacted, and all samples of 300 people have an equal chance of being selected. A possible source of bias is that the random sample may contain a much greater percentage of employees from one department than from others.

25. Cluster sampling is used because the disaster area is divided into grids, and 30 grids are then entirely selected. A possible source of bias is that certain grids may have been much more severely damaged than others.

27. Stratified sampling is used because a sample is taken from each one-acre subplot (stratum).

29. Census, because it is relatively easy to obtain the ages of the 115 residents

31. The question is biased because it already suggests that eating whole-grain foods improves your health. The question might be rewritten as "How does eating whole-grain foods affect your health?"

33. The survey question is unbiased because it does not imply how much exercise is good or bad.

35. The households sampled represent various locations, ethnic groups, and income brackets. Each of these variables is considered a stratum. Stratified sampling ensures that each segment of the population is represented.

37. Answers will vary.

39. Open Question
Advantage: Allows respondent to express some depth and shades of meaning in the answer. Allows for new solutions to be introduced.
Disadvantage: Not easily quantified and difficult to compare surveys.

Closed Question
Advantage: Easy to analyze results.
Disadvantage: May not provide appropriate alternatives and may influence the opinion of the respondent.

CHAPTER 1 REVIEW EXERCISE SOLUTIONS

1. Population: Collection of the responses of all U.S. adults.
Sample: Collection of the responses of the 4787 U.S. adults who were sampled
Sample data set: 15% of adults who use ride-hailing applications and 85% who do not

3. Population: Collection of the responses of all U.S. adults.
Sample: Collection of the responses of the 2223 U.S. adults who were sampled
Sample data set: 62% of adults who would encourage a child to pursue a career as a video game developer or designer and 38% who would not

5. Population parameter. The value $22.7 million is a numerical description of the total infrastructure-strengthening investments.

7. Population Parameter. The 10 students minoring in physics is a numerical description of all math majors at a university.

9. The statement "62% would encourage a child to pursue a career as a video game developer or designer" is an example of descriptive statistics. An inference drawn from the sample is that a majority of people encourage children to pursue a career as a video game developer or designer.

11. Quantitative, because ages are numerical measurements.

13. Quantitative, because revenues are numerical measures.

15. Interval. The data can be ordered and meaningful differences can be calculated, but it does not make sense to say that 84 degrees is 1.05 times as hot as 80 degrees.

17. Nominal. The data are qualitative and cannot be arranged in a meaningful order.

19. Experiment. The study applies a treatment (drug to treat hypertension in patients with obstructive sleep apnea) to the subjects.

21. *Sample answer:* The subjects could be split into male and female and then be randomly assigned to each of the five treatment groups.

23. Simple random sampling is used because random telephone numbers were generated and called. A potential source of bias is that telephone sampling only samples individuals who have telephones, who are available, and who are willing to respond.

25. Cluster sampling is used because each district is considered a cluster and every pregnant woman in a selected district is surveyed. A potential source of bias is that the selected districts may not be representative of the entire area.

27. Stratified sampling is used because the population is divided by grade level and then 25 students are randomly selected from each grade level.

29. Answers will vary. Sample answer: Sampling, because the population of students at the university is too large for their favorite spring break destinations to be easily recorded. Random sampling would be advised because it would be easy to select students randomly and then record their favorite spring break destination.

CHAPTER 1 QUIZ SOLUTIONS

1. Population: Collection of the school performance of all Korean adolescents.
Sample: Collection of the school performance of the 359,264 Korean adolescents in the study

2. (a) Sample statistic. The value 52% is a numerical description of a sample of U.S. adults.

(b) Population Parameter . The 90% of members that approved the contract of the new president is a numerical description of all Board of Trustees members.

(c) Sample statistic. The value 25% is a numerical description of a sample of small business owners.

3. (a) Qualitative, because debit card personal identification numbers are labels and it does not make sense to find differences between numbers.

(b) Quantitative, because final scores are numerical measurements.

4. (a) Ordinal, because badge numbers can be ordered and often indicate seniority of service, but no meaningful mathematical computation can be performed.

(b) Ratio, because horsepower of one car can be expressed as a multiple of another.

(c) Ordinal, because data can be arranged in order, but the differences between data entries make no sense.

(d) Interval, because meaningful differences between years can be calculated, but a zero entry is not an inherent zero.

5. (a) Observational study. The study does not attempt to influence the responses of the subjects and there is no treatment.

(b) Experiment. The study applies a treatment (multivitamin) to the subjects.

6. Randomized block design

7. (a) Convenience sampling is used because all the people sampled are in one convenient location.

(b) Systematic sampling is used because every tenth machine part is sampled.

(c) Stratified sampling is used because the population is first stratified and then a sample is collected from each stratum.

8. Convenience sampling. People at campgrounds may be strongly against air pollution because they are at an outdoor location.

CHAPTER 2

Descriptive Statistics

2.1 FREQUENCY DISTRIBUTIONS AND THEIR GRAPHS

2.1 TRY IT YOURSELF SOLUTIONS

1. The number of classes is 6.

Min = 14, Max = 55, $\text{Class width} = \dfrac{\text{Range}}{\text{Number of classes}} = \dfrac{55-14}{6} = 6.83 \Rightarrow 7$

The minimum data entry is a convenient lower limit for the first class. Then add the class width to get the lower limits of the other classes. The upper limits are one less than the lower limit of the next class.

Lower limit	Upper limit
14	20
21	27
28	34
35	41
42	48
49	55

Make a tally mark for each entry in the appropriate class. The number of tally marks for a class is the frequency of that class.

Class	Frequency, f
14-20	8
21-7	15
28-34	14
35-41	7
42-48	4
49-55	3

2. Find each midpoint, relative frequency, and cumulative frequency.

$$\text{Midpoint} = \frac{(\text{Lower class limit}) + (\text{Upper class limit})}{2}$$

$$\text{Relative frequency} = \frac{\text{Class frequency}}{\text{Sample size}} = \frac{f}{n}$$

The cumulative frequency of a class is the sum of the frequencies of that class and all previous classes.

Class	f	Midpoint	Relative frequency	Cumulative frequency
14–20	8	$\frac{14+20}{2}=17$	$\frac{8}{51}\approx 0.1569$	8
21–27	15	$\frac{21+27}{2}=24$	$\frac{15}{51}\approx 0.2941$	$8+15=23$
28–34	14	$\frac{28+34}{2}=31$	$\frac{14}{51}\approx 0.2745$	$23+14=37$
35–41	7	$\frac{35+41}{2}=38$	$\frac{7}{51}\approx 0.1373$	$37+7=44$
42–48	4	$\frac{42+48}{2}=45$	$\frac{4}{51}\approx 0.0784$	$44+4=48$
49–55	3	$\frac{49+55}{2}=52$	$\frac{3}{51}\approx 0.0588$	$48+3=51$
	$\Sigma f=51$		$\Sigma\frac{f}{n}=1$	

Sample answer: The most common range of points scored by winning teams is 21 to 27. About 14% of the winning teams scored more than 41 points.

3. Find the class boundaries. Because the data entries are integers, subtract 0.5 from each lower limit to find the lower class boundaries and add 0.5 to each upper limit to find the upper class boundaries.

Class	Class Boundaries	Frequency, F
14–20	13.5–20.5	8
21–27	20.5–27.5	15
28–34	27.5–34.5	14
35–41	34.5–41.5	7
42–48	41.5–48.5	4
49–55	48.5–55.5	3

Use class midpoints for the horizontal scale and frequency for the vertical scale. (Class boundaries can also be used for the horizontal scale.)

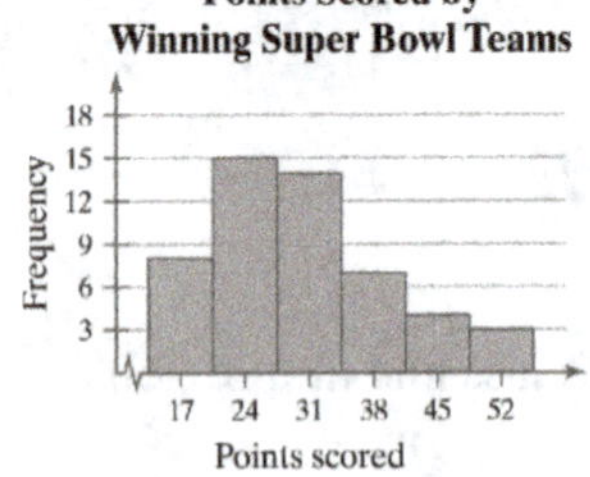

Sample answer: The most common range of points scored by winning teams is 21 to 27. About 14% of the winning teams scored more than 41 points.

4. To construct the frequency polygon, use the same horizontal and vertical scales that were used in the histogram labeled with the class midpoints in Try It Yourself 3. Then plot the points that represent the midpoint and frequency of each class and connect the points with line segments. Extend the left side and right side to one class width before the first class midpoint and after the last class midpoint, respectively.

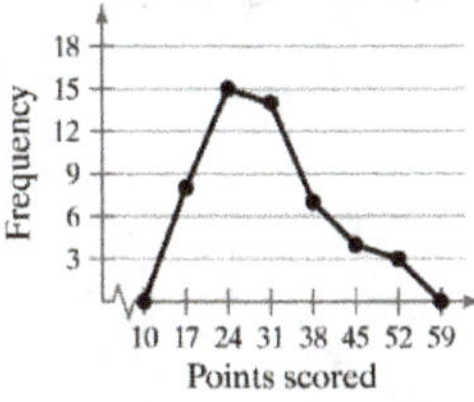

Sample answer: The frequency of points scored increases up to 24 points and then decreases.

5. Notice the shape of the relative frequency histogram is the same as the shape of the frequency histogram constructed in Try It Yourself 3. The only difference is that the vertical scale measures the relative frequencies.

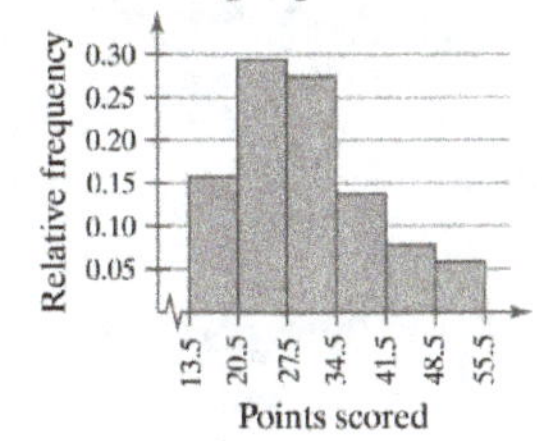

6. Use upper class boundaries for the horizontal scale and cumulative frequency for the vertical scale.

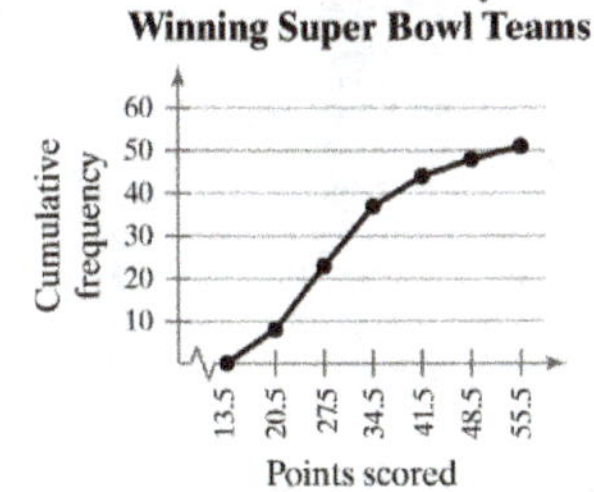

Sample answer: The greatest increase in cumulative frequency occurs between 20.5 and 27.5.

7.

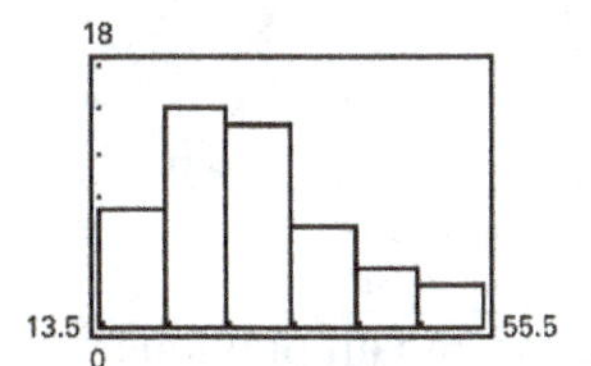

2.1 EXERCISE SOLUTIONS

1. Organizing the data into a frequency distribution may make patterns within the data more evident. Sometimes it is easier to identify patterns of a data set by looking at a graph of the frequency distribution.

3. Class limits determine which numbers can belong to that class.
Class boundaries are the numbers that separate classes without forming gaps between them.

5. The sum of the relative frequencies must be 1 or 100% because it is the sum of all portions or percentages of the data.

7. False. Class width is the difference between the lower (or upper limits) of consecutive classes.

9. False. An ogive is a graph that displays cumulative frequencies.

11. $\text{Class width} = \dfrac{\text{Range}}{\text{Number of classes}} = \dfrac{64-9}{7} \approx 7.9 \Rightarrow 8$

Lower class limits: 9, 17, 25, 33, 41, 49, 57
Upper class limits: 16, 24, 32, 40, 48, 56, 64

13. $\text{Class width} = \dfrac{\text{Range}}{\text{Number of classes}} = \dfrac{135-17}{8} = 14.75 \Rightarrow 15$

Lower class limits: 17, 32, 47, 62, 77, 92, 107, 122
Upper class limits: 31, 46, 61, 76, 91, 106, 121, 136

15. (a) $\text{Class width} = 11 - 0 = 11$

(b) and (c)

$\text{Midpoint} = \dfrac{(\text{Lower class limit}) + (\text{Upper class limit})}{2}$

Find the class boundaries. Because the data entries are integers, subtract 0.5 from each lower limit to find the lower class boundaries and add 0.5 to each upper limit to find the upper class boundaries.

Class	Midpoint	Class boundaries
0 – 10	5	−0.5 – 10.5
11 – 21	16	10.5 – 21.5
22 – 32	27	21.5 – 32.5
33 – 43	38	32.5 – 43.5
44 – 54	49	43.5 – 54.5
55 – 65	60	54.5 – 65.5
66 – 76	71	65.5 – 76.5

17. $\text{Relative frequency} = \dfrac{\text{Class frequency}}{\text{Sample size}} = \dfrac{f}{n}$

The cumulative frequency of a class is the sum of the frequencies of that class and all previous classes.

Class	Frequency, f	Midpoint	Relative frequency	Cumulative frequency
0 – 10	188	5	0.15	188
11 – 21	372	16	0.30	560
22 – 32	264	27	0.22	824
33 – 43	205	38	0.17	1029
44 – 54	83	49	0.07	1112
55 – 65	76	60	0.06	1188
66 – 76	32	71	0.03	1220
	$\sum f = 1220$		$\sum \frac{f}{n} = 1$	

19. (a) Number of classes: 7

(b) Greatest frequency: about 300
Least frequency: about 10

(c) Class width: 10

(d) *Sample answer:* About half of the employee salaries are between $50,000 and $69,000.

21. Identify the highest point and its respective class. Class with greatest frequency: 506 – 510
Identify the lowest point (not including the points on the horizontal axis) and its respective class. Class with least frequency: 474 – 478

23. (a) Identify the tallest bar and its respective class. Class with greatest relative frequency: 35 – 36 centimeters
Identify the shortest bar and its respective class. Class with least relative frequency: 39 – 40 centimeters

(b) Greatest relative frequency ≈ 0.25
Least relative frequency ≈ 0.01

(c) *Sample answer*: From the graph, 0.25 or 25% of females have a fibula length between 35 and 36 centimeters.

25. (a) Locate the cumulative frequency of the highest (right-most) point. The number in the sample is 75.

(b) Locate the neighboring points where the pitch between them is the steepest. The greatest increase in frequency is from 158.5 – 201.5 pounds.

27. (a) Locate 201.5 on the horizontal axis and find the corresponding cumulative frequency at the point on the ogive: 47

(b) Locate 68 on the vertical axis and find the corresponding weight at the point on the ogive: 287.5 pounds

(c) Subtract the cumulative frequency for each weight: $62 - 22 = 40$

(d) Subtract the cumulative frequency for bears weighing 330.5 pounds from the number in the sample: $75 - 69 = 6$

29. $\text{Class width} = \dfrac{\text{Range}}{\text{Number of classes}} = \dfrac{39-0}{5} = 7.8 \Rightarrow 8$

Class	Frequency, f	Midpoint	Relative frequency	Cumulative frequency
0 –7	8	3.5	0.33	8
8–15	7	11.5	0.29	15
16–23	3	19.5	0.13	18
24–31	3	27.5	0.13	21
32–39	3	35.5	0.13	24
	$\sum f = 24$		$\sum \frac{f}{n} \approx 1$	

Class with greatest frequency: 0 – 7
Classes with least frequency: 16 – 23, 24 – 31, 32 – 39

31. $\text{Class width} = \dfrac{\text{Range}}{\text{Number of classes}} = \dfrac{7119-1000}{6} \approx 1019.8 \Rightarrow 1020$

Class	Frequency, f	Mid-point	Relative frequency	Cumulative frequency
1000 –2019	11	1509.5	0.52	11
2020 –3039	3	2529.5	0.14	14
3040– 4059	2	3549.5	0.10	16
4060– 5079	3	4569.5	0.14	19
5080– 6099	1	5589.5	0.05	20
6100– 7119	1	6609.5	0.05	21
	$\sum f = 21$		$\sum \frac{f}{n} = 1$	

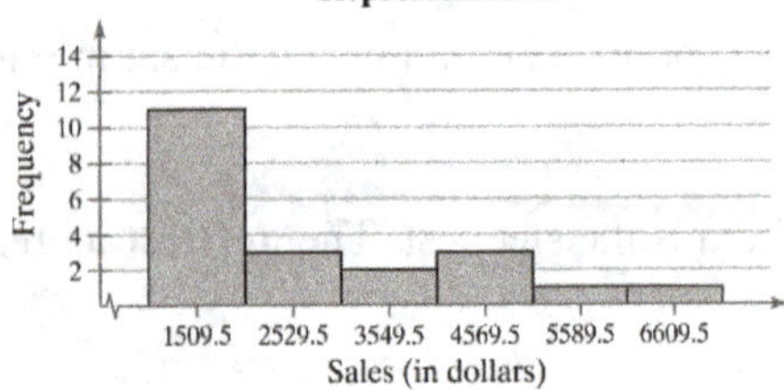

Sample answer: The graph shows that most of the sales representatives at the company sold from \$1000 to \$2019.

33. Class width $= \dfrac{\text{Range}}{\text{Number of classes}} = \dfrac{514-291}{8} = 27.875 \Rightarrow 28$

Class	Frequency, f	Midpoint	Relative frequency	Cumulative frequency
291-318	5	304.5	0.1667	5
319-346	4	332.5	0.1333	9
347-374	3	360.5	0.1000	12
375-402	5	388.5	0.1667	17
403-430	6	416.5	0.2000	23
431-458	4	444.5	0.1333	27
459-486	1	472.5	0.0333	28
487-514	2	500.5	0.0667	30
	$\sum f = 30$		$\sum \frac{f}{n} = 1$	

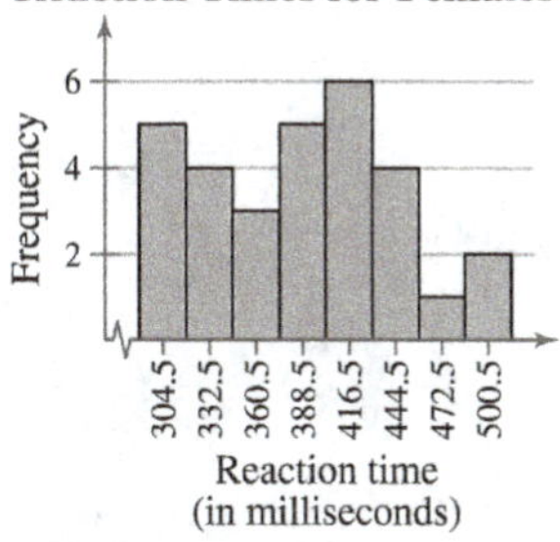

Sample answer: The graph shows that the most frequent reaction times were between 403 and 430 milliseconds.

35. Class width $= \dfrac{\text{Range}}{\text{Number of classes}} = \dfrac{70-42}{6} \approx 4.7 \Rightarrow 5$

Class	Frequency, f	Midpoint	Relative frequency	Cumulative frequency
42–46	4	44	0.0889	4
47–51	11	49	0.2444	15
52–56	14	54	0.3111	29
57–61	9	59	0.2000	38
62–66	4	64	0.0889	42
67–71	3	69	0.0667	45
	$\sum f = 45$		$\sum \frac{f}{n} = 1$	

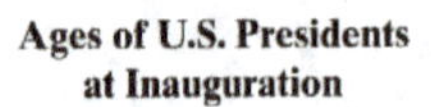

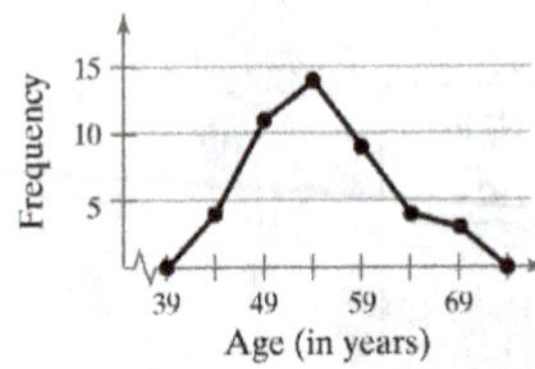

Sample answer: The graph shows that the number of U.S. presidents who were 52 or older at inauguration was twice as many as those who were 51 and younger.

37. Class width $= \dfrac{\text{Range}}{\text{Number of classes}} = \dfrac{10-1}{5} = 1.8 \Rightarrow 2$

Class	Frequency, f	Midpoint	Relative frequency	Cumulative frequency
1–2	7	1.5	0.19	7
3–4	8	3.5	0.22	15
5–6	10	5.5	0.28	25
7–8	2	7.5	0.06	27
9–10	9	9.5	0.25	36
	$\sum f = 36$		$\sum \frac{f}{n} \approx 1$	

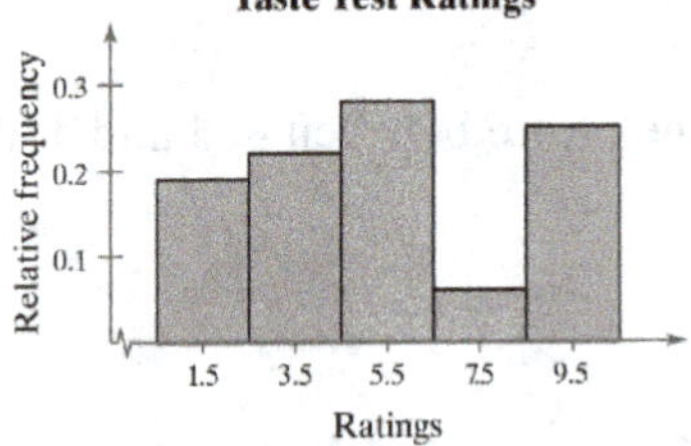

Class with greatest relative frequency: 5 – 6
Class with least relative frequency: 7 – 8

39. Class width $= \dfrac{\text{Range}}{\text{Number of classes}} = \dfrac{75-60}{5} = 3$

Notice that using a class width of 3 is not wide enough to include all the data with 5 classes. Therefore, use a class width of 4.

Class	Frequency, f	Midpoin t	Relative frequency	Cumulative frequency
60–63	6	61.5	0.2143	6
64–67	7	65.5	0.2500	13
68–71	9	69.5	0.3214	22
72–75	6	73.5	0.2143	28
76–79	0	77.5	0.0000	28
	$\sum f = 28$		$\sum \frac{f}{n} \approx 1$	

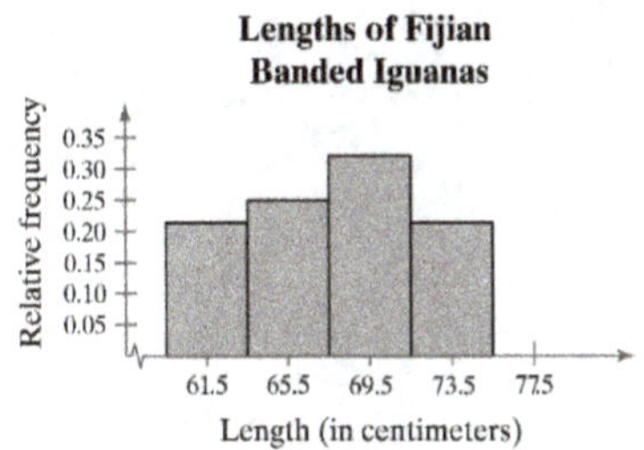

Class with greatest relative frequency: 68 – 71
Class with least relative frequency: 76 – 79

41. Class width $= \dfrac{\text{Range}}{\text{Number of classes}} = \dfrac{75-52}{6} \approx 3.8 \Rightarrow 4$

Class	Frequency, f	Relative frequency	Cumulative frequency
52–55	6	0.1714	6
56–59	4	0.1143	10
60–63	6	0.1714	16
64–67	10	0.2857	26
68–71	5	0.1429	31
72–75	4	0.1143	35
	$\sum f = 35$		$\sum \frac{f}{n} \approx 1$

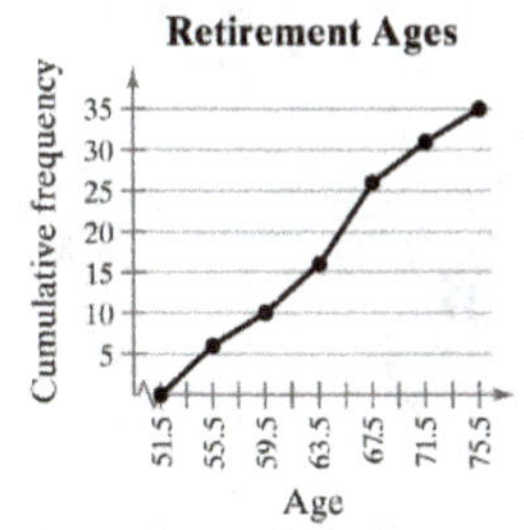

Location of the greatest increase in frequency: 64 – 67

43. (a) Class width $= \dfrac{\text{Range}}{\text{Number of classes}} = \dfrac{120-65}{6} \approx 9.2 \Rightarrow 10$

Class	Frequency, f	Midpoint	Relative frequency	Cumulative frequency
65-74	4	69.5	0.1667	4
75-84	7	79.5	0.2917	11
85-94	4	89.5	0.1667	15
95-104	5	99.5	0.2083	20
105-114	3	109.5	0.1250	23
115-124	1	119.5	0.0417	24
	$\sum f = 24$		$\sum \frac{f}{N} \approx 1$	

(b)

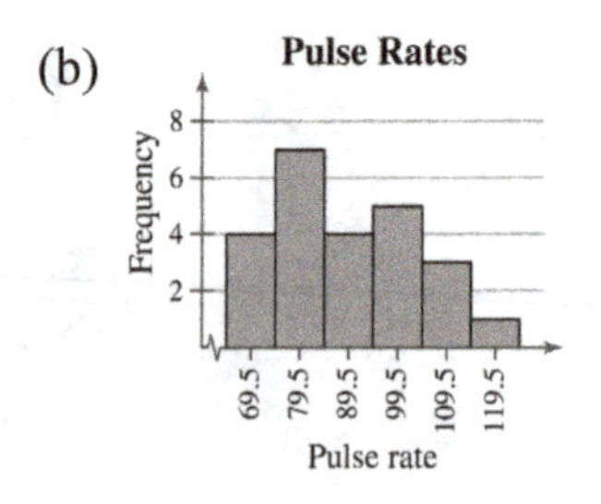

(c)

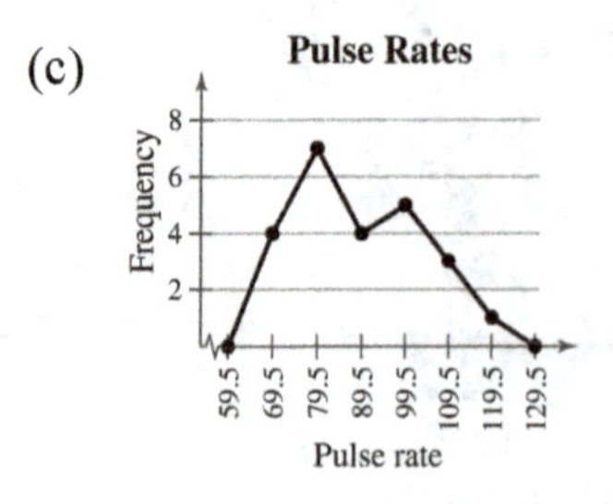

(d)

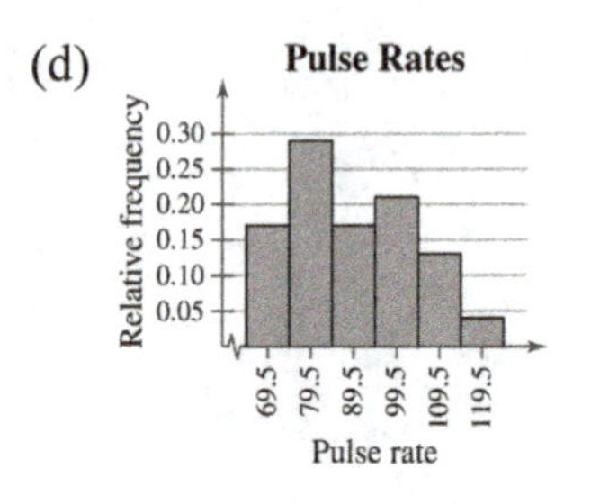

(e)

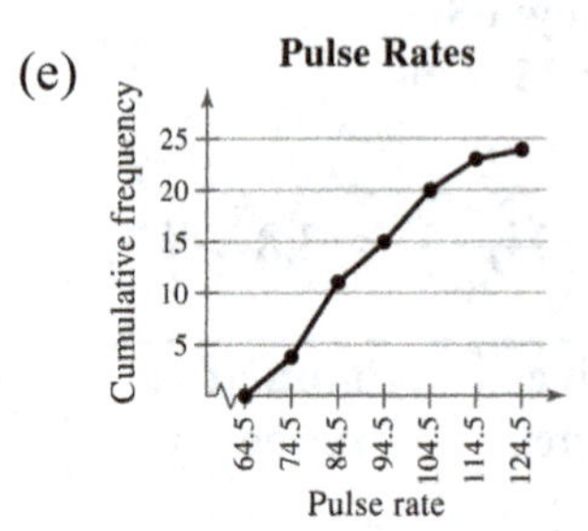

45. (a) Class width $= \dfrac{\text{Range}}{\text{Number of classes}} = \dfrac{104-61}{8} = 5.375 \Rightarrow 6$

Class	Frequency, f	Midpoint	Relative frequency
61-66	1	63.5	0.033
67-72	3	69.5	0.100
73-78	6	75.5	0.200
79-84	10	81.5	0.333
85-90	5	87.5	0.167
91-96	2	93.5	0.067
97-102	2	99.5	0.067
103-108	1	105.5	0.033
	$\sum f = 30$		$\sum \frac{f}{n} = 1$

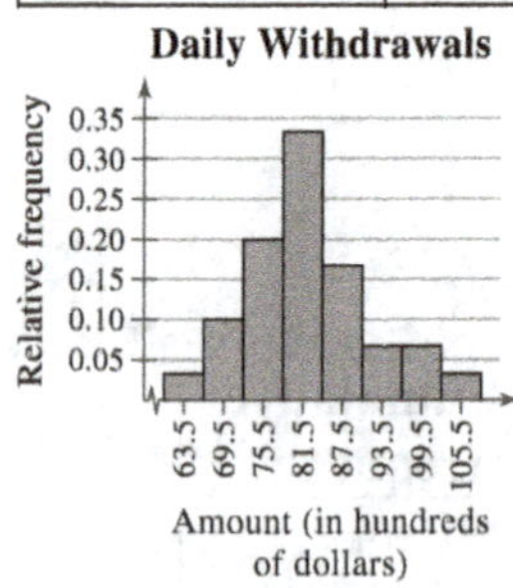

(b) 16.7%, because the sum of the relative frequencies for the last three classes is 0.167.

(c) \$9700, because the sum of the relative frequencies for the last two classes is 0.10.

47.

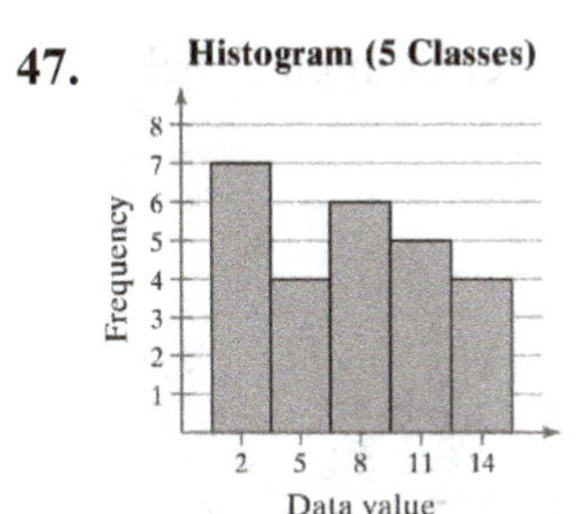

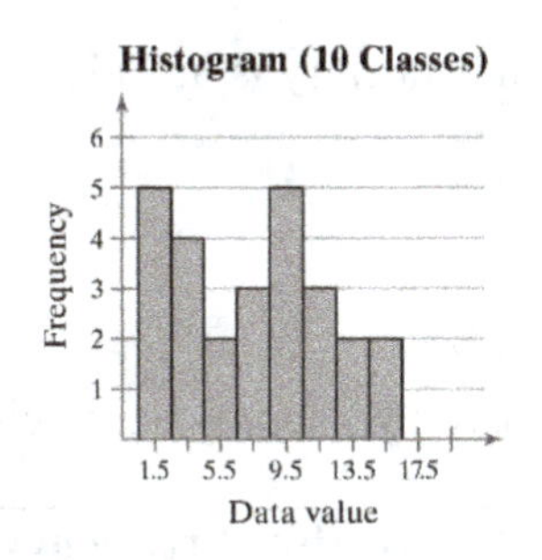

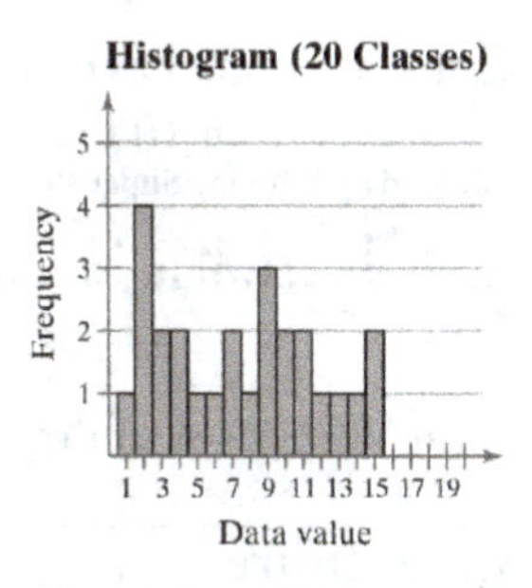

In general, a greater number of classes better preserves the actual values of the data set but is not as helpful for observing general trends and making conclusions. In choosing the number of classes, an important consideration is the size of the data set. For instance, you would not want to use 20 classes if your data set contained 20 entries. In this particular example, as the number of classes increases, the histogram shows more fluctuation. The histograms with 10 and 20 classes have classes with zero frequencies. Not much is gained by using more than five classes. Therefore, it appears that five classes would be best.

2.2 MORE GRAPHS AND DISPLAYS

2.2 TRY IT YOURSELF SOLUTIONS

1. Because the data entries go from a low of 14 to a high of 55, use stem values from 1 to 5. List the stems to the left of a vertical line. For each data entry, list a leaf to the right of its stem.

Stem	Leaves	
1	4 6 6 6 7	Key: 1 | 4 = 14
2	0 0 0 1 1 1 3 3 4 4 4 4 6 7 7 7 7 7 8 9	
3	0 1 1 1 1 2 2 3 4 4 4 4 5 5 5 7 8 8 9	
4	2 3 6 8 9	
5	2 5	

Sample answer: Most of the winning teams scored between 20 and 39 points.

2. Use the leaves 0, 1, 2, 3, and 4 in the first stem row and the leaves 5, 6, 7, 8, and 9 in the second stem row.

Stem	Leaves	
1	4	Key: 1 | 4 = 14
1	6 6 6 7	
2	0 0 0 1 1 1 3 3 4 4 4 4	
2	6 7 7 7 7 7 8 9	
3	0 1 1 1 1 2 2 3 4 4 4 4	
3	5 5 5 7 8 8 9	
4	2 3	
4	6 8 9	
5	2	
5	5	

Sample answer: Most of the winning teams scored from 20 to 35 points.

3. Choose the horizontal axis so that each data entry is included in the dot plot. For example, label the horizontal axis from 10 to 55.

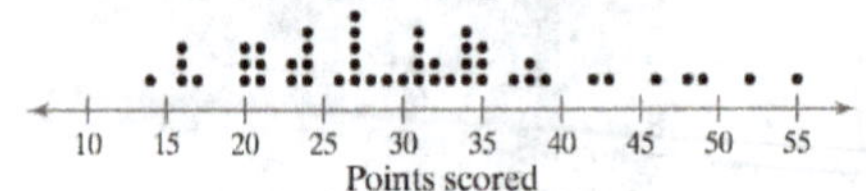

Sample answer: Most of the points scored by winning teams cluster between 20 and 40.

4.

Type of Degree	f	Relative Frequency	Angle
Associate's	455	0.235	85°
Bachelor's	1051	0.542	195°
Master's	330	0.170	61°
Doctoral	104	0.054	19°
	$\sum f = 1940$	$\sum \frac{f}{n} \approx 1$	$\sum = 360°$

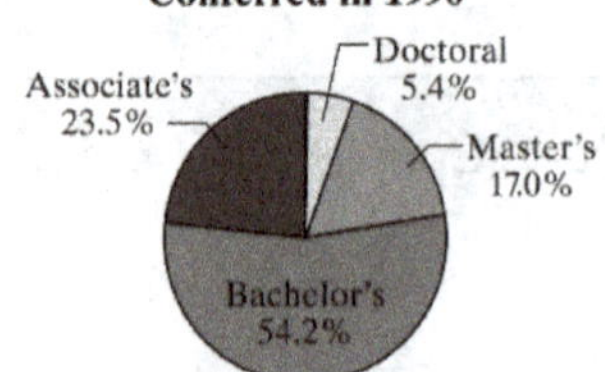

From 1990 to 2014, as percentages of the total degrees conferred, associate's degrees increased by 2.9%, bachelor's degrees decreased by 5.1%, master's degrees increased by 2.8%, and doctoral degrees decreased by 0.7%.

5.

Cause	Frequency, f
Auto dealers (used cars)	16,281
Insurance companies	8384
Mortgage brokers	3634
Collection agencies	19,277
Travel agencies and bureaus	6985

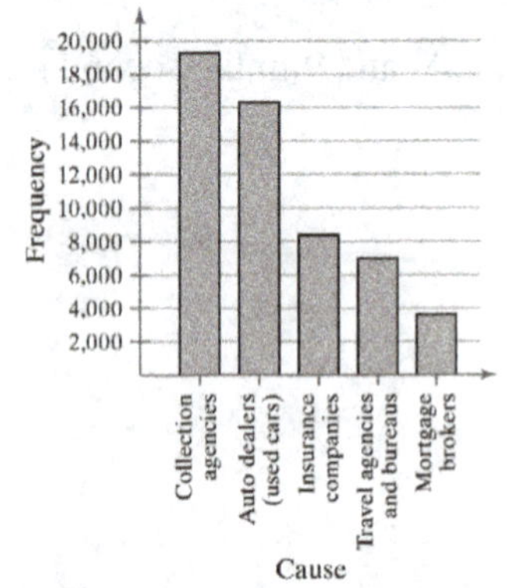

Collection agencies are the greatest cause of complaints.

6.

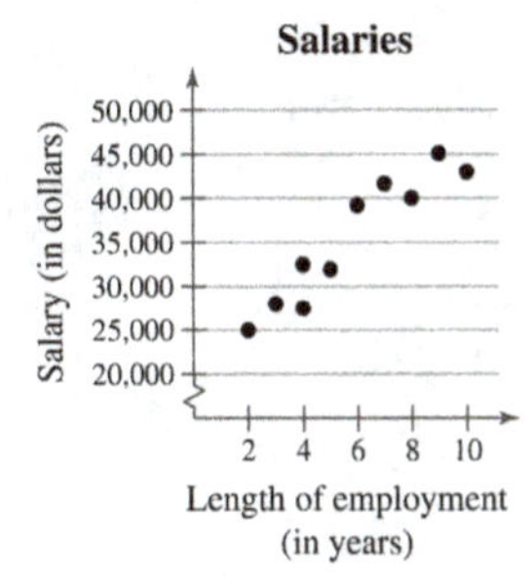

It appears that the longer an employee is with the company, the greater the employee's salary.

7. Let the horizontal axis represent the years and let the vertical axis represent the number of burglaries (in millions).

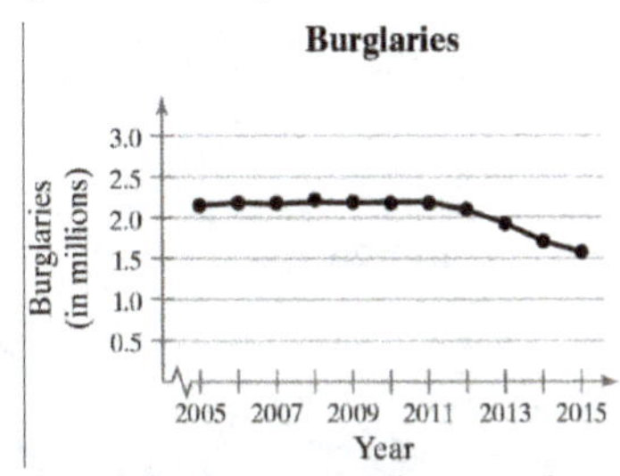

Sample answer: The number of burglaries remained about the same until 2012 and then decreased through 2015.

2.2 EXERCISE SOLUTIONS

1. Quantitative: stem-and-leaf plot, dot plot, histogram, time series chart, scatter plot.
Qualitative: pie chart, Pareto chart

3. Both the stem-and-leaf plot and the dot plot allow you to see how data are distributed, determine specific data entries, and identify unusual data values.

5. b **7.** a

9. 27, 32, 41, 43, 43, 44, 47, 47, 48, 50, 51, 51, 52, 53, 53, 53, 54, 54, 54, 54, 55, 56, 56, 58, 59, 68, 68, 68, 73, 78, 78, 85
Max: 85 Min: 27

11. 13, 13, 14, 14, 14, 15, 15, 15, 15, 15, 16, 17, 17, 18, 19
Max: 19 Min: 13

13. *Sample answer*: Facebook has the most users, and Pinterest has the least. Tumblr and Instagram have about the same number of users.

15. *Sample answer*: The Texter is the least popular driver. The Left-Lane Hog is tolerated more than the Tailgater. The Speedster and the Drifter have the same popularity.

17. Exam Scores Key: 6|7 = 67

```
6|7 8
7|3 5 5 6 9
8|0 0 2 3 5 5 7 7 8
9|0 1 1 1 2 4 5 5
```

Sample answer: Most grades for the biology midterm were in the 80s and 90s.

19. Ice Thickness (in centimeters) Key: 4|3 = 4.3

```
4|3 9
5|1 8 8 8 9
6|4 8 9 9 9
7|0 0 2 2 2 5
8|0 1
```

Sample answer: Most of the ice had a thickness of 5.8 centimeters to 7.2 centimeters.

21. Incomes (in millions) of Highest Paid Athletes

Key: 3 | 3 = 33

```
3 | 3 4 4 4
3 | 5 6 7 7 8 8 8
4 | 1 2 3 4 4
4 | 5 5 5 6
5 | 0 3 3 3
5 | 6 6
6 |
6 | 8
7 |
7 | 7
8 | 1
8 | 8
```

Sample answer: Most of the highest-paid athletes have an income of \$33 million to \$56 million.

23.

105 115 125 135 145 155 165 175 185 195 205
Systolic blood pressure (in mmHg)

Sample answer: Systolic blood pressure tends to be from 120 to 150 millimeters of mercury.

25.

Balance Owed	f	Relative Frequency	Angle
\$1 to \$10,000	16.7	0.379	136°
\$10,001 to \$25,000	12.4	0.281	101°
\$25,001 to \$50,000	8.3	0.188	68°
\$50,001 +	6.7	0.152	55°
	$\sum f = 44.1$	$\sum \frac{f}{n} = 1$	$\sum = 360°$

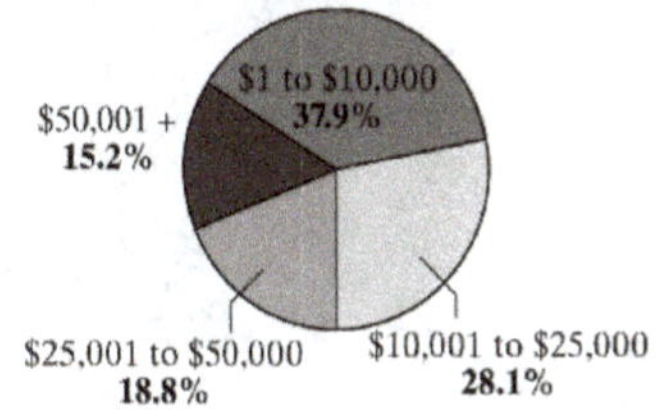

Sample answer: The majority of student loan borrowers owe $25,000 or less.

27.

Country	Medals
Germany	42
Great Britain	67
United States	121
Russia	56
China	70

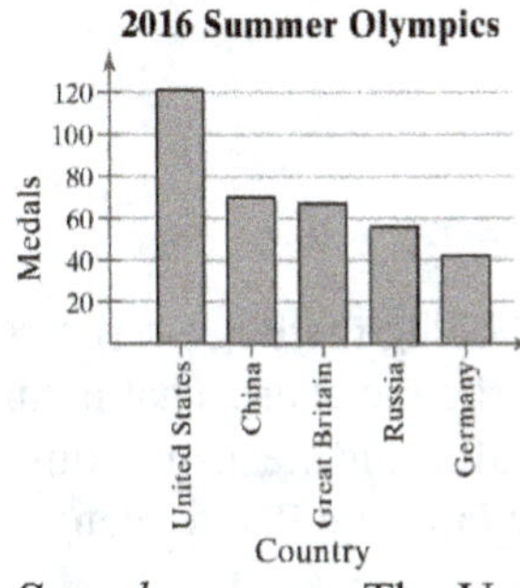

Sample answer: The United States won the most medals out of the five countries and Germany won the least.

29. Let the horizontal axis represent hours and let the vertical axis represent the hourly wage (in dollars).

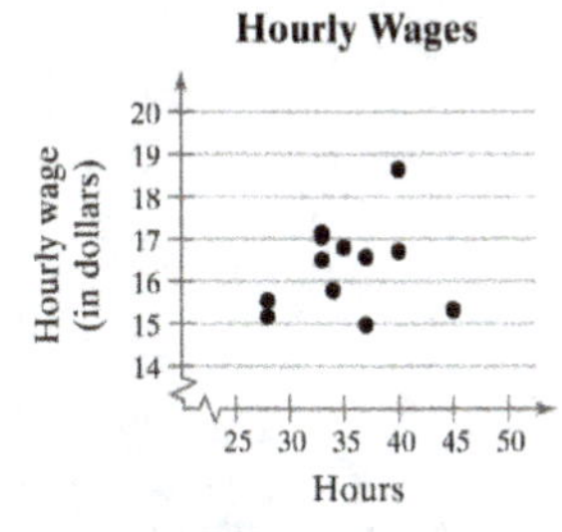

Sample answer: It appears that there is no relation between hourly wages and hours worked.

31. Let the horizontal axis represent the years and let the vertical axis represent the number of degrees (in thousands).

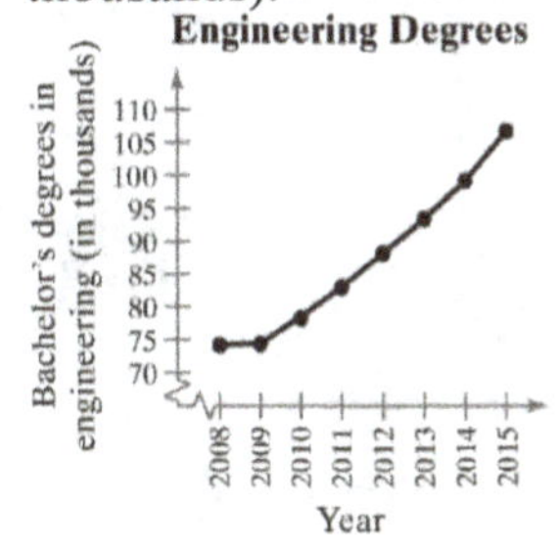

Sample answer: The number of bachelor's degrees in engineering conferred in the U.S. has increased from 2008 to 2015.

33. Heights (in inches)

7	2 2 4	Key: 7 | 2 = 72
7	5 5 5 5 6	
8	1 1 2 2 2 4	
8		

The dot plot helps you see that the data are clustered from 72 to 76 and 81 to 84, with 75 being the most frequent value. The stem-and-leaf plot helps you see that most values are 75 or greater.

35. **Favorite Season of U.S. Adults Ages 18 and Older**

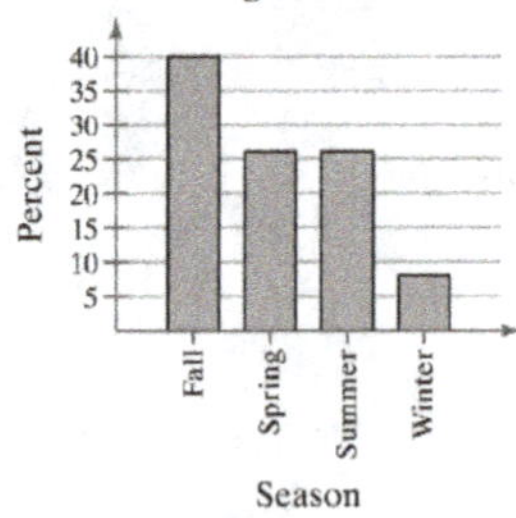

The pie chart helps you to see the percentages as parts of a whole, with fall being the largest. It also shows that while fall is the largest percentage, it makes up less than half of the pie chart. That means that a majority of U.S. adults ages 18 and older prefer a season other than fall. This means it would not be a fair statement to say that most U.S. adults ages 18 and older prefer fall. The Pareto chart helps you to see the rankings of the seasons.

37. (a) The graph is misleading because the large gap from 0 to 90 makes it appear that the sales for the 3rd quarter are disproportionately larger than the other quarters.

(b)

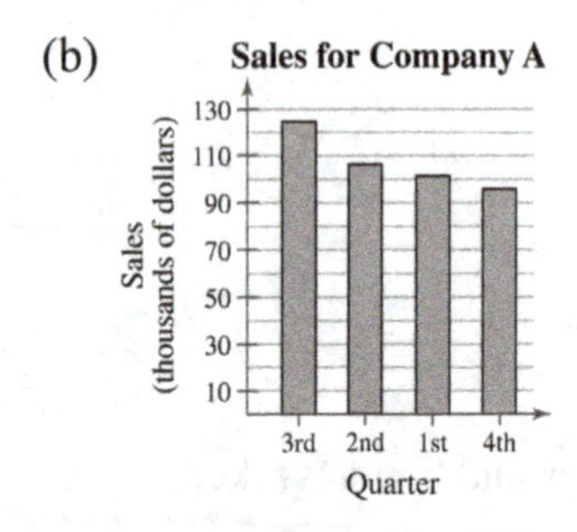

39. (a) The graph is misleading because the angle makes it appear as though the 3rd quarter had a larger percent of sales than the others, when the 1st and 3rd quarters have the same percent.

(b)

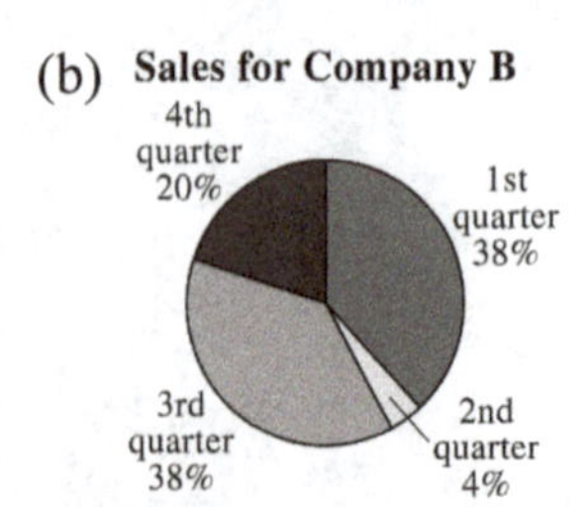

41. (a) At Law Firm A, the lowest salary was \$90,000 and the highest salary was \$203,000. At Law Firm B, the lowest salary was \$90,000 and the highest salary was \$190,000. There are 30 lawyers at Law Firm A and 32 lawyers at Law Firm B.

(b) At Law Firm A, the salaries tend to be clustered at the far ends of the distribution range. At Law Firm B, the salaries are spread out.

43. (a) Key: 2 | 6 = 26

Stem	Leaves
2	6
3	1
4	0 4 4 5 6 7 9 9
0	4 4 5 6 7 9 9
5	5 5 5
6	3 4 4
7	0 1 2 2

(b)

(c)

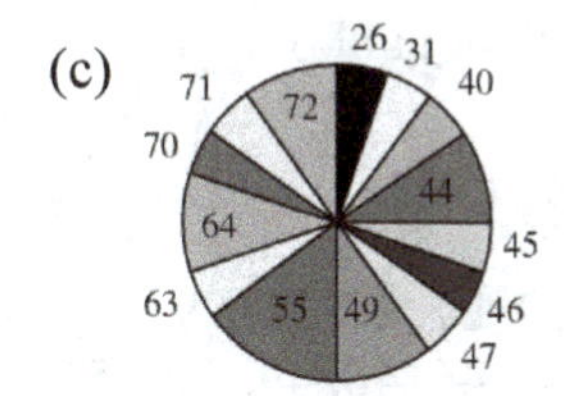

(d)

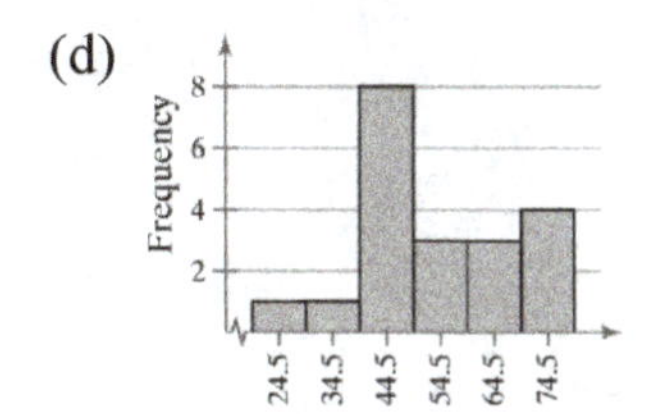

(e)

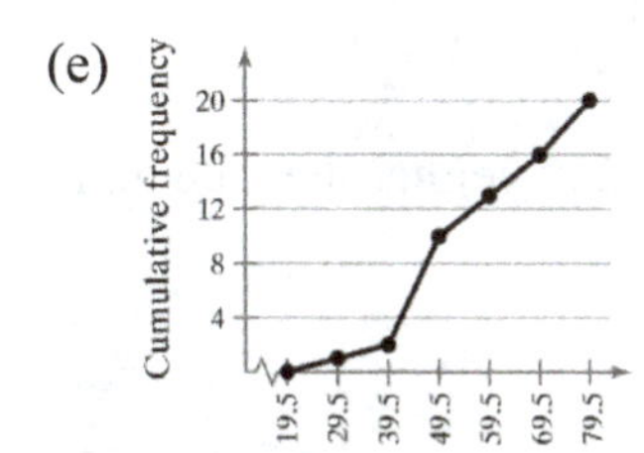

Sample answer: The stem-and-leaf plot, dot plot, frequency histogram, and ogive display the data best because the data is quantitative.

2.3 MEASURES OF CENTRAL TENDENCY

2.3 TRY IT YOURSELF SOLUTIONS

1. $\sum x = 35 + 33 + 16 + 23 + 16 + \ldots + 28 + 24 + 34 = 1541$

$$\bar{x} = \frac{\sum x}{n} = \frac{1541}{51} \approx 30.2$$

The mean points scored by the 51 winning teams is about 30.2.

2. Order the data from smallest to largest.
Note: The stem-and-leaf plot from Try It Yourself 1 in Section 2.2 may be helpful to in ordering the data.
14, 16, 16, 16, 17, 20, 20, 20, 21, 21, 21, 23, 23, 24, 24, 24, 24, 26, 27, 27, 27, 27, 27, 28, 29, 30, 31, 31, 31, 31, 32, 32, 33, 34, 34, 34, 34, 35, 35, 35, 37, 38, 38, 39, 42, 43, 46, 48, 49, 52, 55
Because there are 51 entries (an odd number), the median is the middle, or 26th entry. So, the median is 30 points.

3. Order the data from smallest to largest.
17, 20, 21, 21, 24, 24, 27, 28, 29, 31, 31, 32, 34, 34, 43, 48
Because there are an even number of entries, the median is the mean of the two middle entries.

$$\text{median} = \frac{28+29}{2} = \frac{57}{2} = 28.5$$

The median points scored by the winning teams in the Super Bowls for the National Football League's 2001 through 2016 seasons is 28.5 points.

4. Look at the ordered data from Try It Yourself 1
14, 16, 16, 16, 17, 20, 20, 20, 21, 21, 21, 23, 23, 24, 24, 24, 24, 26, 27, 27, 27, 27, 27, 28, 29, 30, 31, 31, 31, 31, 32, 32, 33, 34, 34, 34, 34, 35, 35, 35, 37, 38, 38, 39, 42, 43, 46, 48, 49, 52, 55
The mode is the data entry that occurs with the greatest frequency.
The entry 27 occurs the most, so the mode is 27.

5. "some" occurs with the greatest frequency (578). The mode is "some".

6. $\bar{x} = \frac{\sum x}{n} = \frac{410}{19} \approx 21.6$

median = 21
mode = 20

The mean in Example 6 ($\bar{x} \approx 23.8$) was heavily influenced by the entry 65. Neither the median nor the mode was affected as much by the entry 65.

7.

Final Grade	Points, x	Weight, w	$x \cdot w$
C	2	3	6
C	2	4	8
D	1	1	1
A	4	3	12
B	3	2	6
B	3	3	9
		$\sum w = 16$	$\sum(x \cdot w) = 42$

$$\bar{x} = \frac{\sum(x \cdot w)}{\sum w} = \frac{42}{16} \approx 2.6$$

The new weighted mean is about 2.6.

8.

Class	Midpoint, x	Frequency, f	$x \cdot f$
14-20	17	8	136
21-27	24	15	360
28-34	31	14	434
35-41	38	7	266
42-48	45	4	180
49-55	52	3	156
		$\sum f = 51 = n$	$\sum(x \cdot f) = 1532$

$$\bar{x} = \frac{\sum(x \cdot f)}{n} = \frac{1532}{51} \approx 30.0$$

This is very close to the mean found using the original data set.

2.3 EXERCISE SOLUTIONS

1. True

3. True

5. *Sample answer:* 1, 2, 2, 2, 3

7. *Sample answer:* 2, 5, 7, 9, 35

9. The shape of the distribution is skewed right because the bars have a "tail" to the right.

11. The shape of the distribution is uniform because the bars are approximately the same height.

13. (11), because the distribution values range from 1 to 12 and has (approximately) equal frequencies.

15. (12), because the distribution has a maximum value of 90 and is skewed left due to a few students scoring much lower than the majority of the students.

17. $\bar{x} = \frac{\sum x}{n} = \frac{209}{14} \approx 14.9$

12 12 13 14 14 15 $\underbrace{15\ 15}$ 16 16 16 16 17 18

$$\text{median} = \frac{15+15}{2} = 15$$

mode = 16 (occurs 4 times)

19. $\bar{x} = \frac{\sum x}{n} = \frac{6316}{7} \approx 902.3$

650 662 709 **788** 803 1242 1462

median = 788

mode = none

The mode cannot be found because no data entry is repeated.

The mean does not represent the center of the data because it is influenced by the outliers of 1242 and 1462.

21. $\bar{x}=\frac{\sum x}{n}=\frac{697}{14}\approx 49.8$

45 47 48 48 48 49 $\underbrace{50\ 51}$ 51 51 51 51 52 55

$$\text{median}=\frac{50+51}{2}=50.5$$

mode = 51 (occurs 5 times)

23. $\bar{x}=\frac{\sum x}{n}=\frac{119}{16}\approx 7.4$

1 2 2 3 3 5 6 $\underbrace{6\ 6}$ 8 10 10 10 11 17 19

$$\text{median}=\frac{6+6}{2}=6$$

mode = 6, 10 (both occur 3 times)

25. $\bar{x}=\frac{\sum x}{n}=\frac{100}{7}=14.3$

6, 7, 8, 9, 13, 15, 42

The median is the middle value, 9.

mode = none

The mode cannot be found because no data entry is repeated.

The mean does not represent the center of the data set because it is influenced by the outlier of 42.

27. $\bar{x}$ is not possible (nominal data)

median = not possible (nominal data)

mode = "Search and buy online"

The mean and median cannot be found because the data are at the nominal level of measurement.

29. $\bar{x}$ is not possible (nominal data)

median is not possible (nominal data)

mode = "Junior"

The mean and median cannot be found because the data are at the nominal level of measurement.

31. $\bar{x}=\frac{\sum x}{n}=\frac{817}{28}\approx 29.2$

5 8 10 11 13 16 21 23 23 23 26 27 27 $\underbrace{30\ 31}$ 32 34 34 34 35 37 38 43 44 45 46 49 52

$$\text{median}=\frac{30+31}{2}=30.5$$

mode = 23, 34 (both occur 3 times each)

33. $\bar{x}=\frac{\sum x}{n}=\frac{292}{15}\approx 19.5$

5 8 10 15 15 15 17 **20** 21 22 22 25 28 32 37

↖ median = 20

mode = 15 (occurs 3 times)

35. Cluster around 275 – 425

37. Mode, because the data are at the nominal level of measurement.

39. Mean, because the distribution is symmetric and there are no outliers.

41.

Source	Score, x	Weight, w	$x \cdot w$
Homework	85	0.05	4.25
Quizzes	80	0.35	28
Project/Speech	100	0.35	35
Final exam	93	0.25	23.25
		$\sum w = 1$	$\sum (x \cdot w) = 90.5$

$$\bar{x} = \frac{\Sigma(x \cdot w)}{\Sigma w} = \frac{90.5}{1} = 90.5$$

43.

Balance, x	Days, w	$x \cdot w$
\$523	24	12,552
\$2415	2	4830
\$250	4	1000
	$\sum w = 30$	$\sum (x \cdot w) = 18,382$

$$\bar{x} = \frac{\Sigma(x \cdot w)}{\Sigma w} = \frac{18,382}{30} \approx \$612.73$$

45.

Source	Score, x	Weight, w	$x \cdot w$
Engineering	85	9	765
Business	81	13	1053
Math	90	5	450
		$\sum w = 27$	$\sum (x \cdot w) = 2268$

$$\bar{x} = \frac{\Sigma(x \cdot w)}{\Sigma w} = \frac{2268}{27} = 84$$

47.

Source	Score, x	Weight, w	$x \cdot w$
Homework	85	0.05	4.25
Quizzes	80	0.35	28
Project/Speech	100	0.35	35
Final exam	85	0.25	21.25
		$\sum w = 1$	$\sum (x \cdot w) = 88.5$

$$\bar{x} = \frac{\Sigma(x \cdot w)}{\Sigma w} = \frac{88.5}{1} = 88.5$$

49.

Class	Midpoint, x	Frequency, f	$x \cdot f$
29-33	31	11	341
34-38	36	12	432
39-43	41	2	82
44-48	46	5	230
		$n = 30$	$\sum(x \cdot f) = 1085$

$$\bar{x} = \frac{\Sigma(x \cdot f)}{n} = \frac{1085}{30} \approx 36.2 \text{ miles per gallon}$$

51.

Class	Midpoint, x	Frequency, f	$x \cdot f$
0-9	4.5	78	351
10-19	14.5	97	1406.5
20-29	24.5	54	1323
30-39	34.5	63	2173.5
40-49	44.5	69	3070.5
50-59	54.5	86	4687
60-69	64.5	73	4708.5
70-79	74.5	53	3948.5
80-89	84.5	43	3633.5
90-99	94.5	15	1417.5
		$n = 631$	$\sum(x \cdot f) = 26{,}719.5$

$$\bar{x} = \frac{\Sigma(x \cdot f)}{n} = \frac{26{,}719.5}{631} \approx 42.3 \text{ years old}$$

53. $\text{Class width} = \dfrac{\text{Range}}{\text{Number of classes}} = \dfrac{297 - 127}{5} = 34 \Rightarrow 35$

Class	Frequency, f	Midpoint
127–161	7	144
162–196	6	179
197–231	3	214
232–266	3	249
267–301	1	284

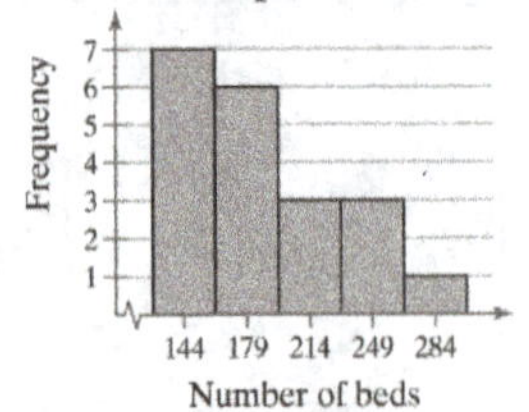

Positively skewed

55. $\text{Class width} = \dfrac{\text{Range}}{\text{Number of classes}} = \dfrac{76-62}{5} = 2.8 \Rightarrow 3$

Class	Midpoint	Frequency, f
62-64	63	3
65-67	66	7
68-70	69	9
71-73	72	8
74-76	75	3
		$\sum f = 30$

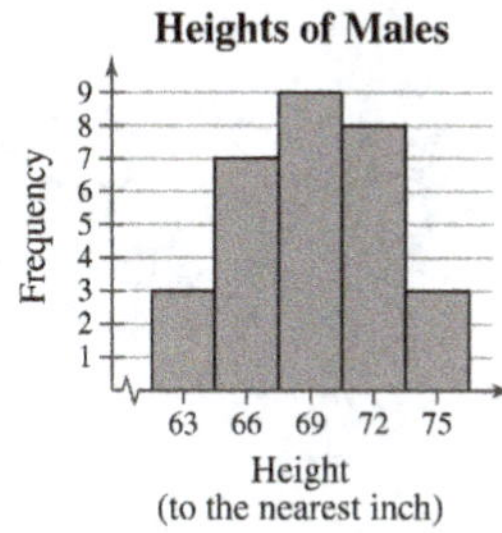

Shape: Symmetric

57. (a) $\bar{x} = \dfrac{\sum x}{n} = \dfrac{9109}{6} \approx 1518.2$

1502 1511 $\underbrace{1516\ 1525}$ 1526 1529

$$\text{median} = \frac{1516+1525}{2} = 1520.5$$

(b) $\bar{x} = \dfrac{\sum x}{n} = \dfrac{9127}{6} = 1521.2$

1511 1516 $\underbrace{1520\ 1525}$ 1526 1529

$$\text{median} = \frac{1520+1525}{2} = 1522.5$$

(c) The mean was affected more.

59. The data are skewed right.
A = mode, because it is the data entry that occurred most often.
B = median, because the median is to the left of the mean in a skewed right distribution.
C = mean, because the mean is to the right of the median in a skewed right distribution.

61. Increase one of the three-credit B classes to an A. The three-credit class is weighted more than the two-credit classes, so it will have a greater effect on the grade point average.

63. Car A

$\bar{x} = \frac{\sum x}{n} = \frac{152}{5} = 30.4$

28 28 **30** 32 34

↖ median = 30

mode = 28 (occurs 2 times)

Car B

$\bar{x} = \frac{\sum x}{n} = \frac{151}{5} = 30.2$

29 29 **31** 31 31

↖ median = 31

mode = 31 (occurs 3 times)

Car C

$\bar{x} = \frac{\sum x}{n} = \frac{151}{5} = 30.2$

28 29 **30** 32 32

↖ median = 30

mode = 32 (occurs 2 times)

(a) Mean should be used because Car A has the highest mean of the three.

(b) Median should be used because Car B has the highest median of the three.

(c) Mode should be used because Car C has the highest mode of the three.

65. (a) $\bar{x} = \frac{\sum x}{n} = \frac{1477}{30} \approx 49.2$

11 13 22 28 36 36 36 37 37 37 38 41 43 44 $\underbrace{46\ 47}$ 51 51 51 53 61 62 63 64

72 72 74 76 85 90

$\text{median} = \frac{46+47}{2} = 46.5$

(b) Key: $3|6 = 36$

```
1|1 3
2|2 8
3|6 6 6 7 7 7 8
4|1 3 4 6 7
5|1 1 1 3
6|1 2 3 4
7|2 2 4 6
8|5
9|0
```

median

mean

(c) The distribution is positively skewed.

2.4 MEASURES OF VARIATION

2.4 TRY IT YOURSELF SOLUTIONS

1. Min = 23, or \$23,000 and Max = 58, or \$58,000,
Range = Max – Min = 58 – 23 = 35, or \$35,000
The range of the starting salaries for Corporation B is 35, or \$35,000. This is much larger than the range for Corporation A.

2. $\mu = 41.5$, or \$41,500

Salary, x	$x-\mu$	$(x-\mu)^2$
23	–18.5	342.25
29	–12.5	156.25
32	–9.5	90.25
40	–1.5	2.25
41	–0.5	0.25
41	–0.5	0.25
49	7.5	56.25
50	8.5	72.25
52	10.5	110.25
58	16.5	272.25
$\sum x = 415$	$\sum(x-\mu) = 0$	$\sum(x-\mu)^2 = 1102.5$

$$\sigma^2 = \frac{\Sigma(x-\mu)^2}{N} = \frac{1102.5}{10} \approx 110.3$$

$$\sigma = \sqrt{\sigma^2} = \sqrt{\frac{1102.5}{10}} = 10.5, \text{ or } \$10{,}500$$

The population variance is about 110.3 and the population standard deviation is 10.5, or \$10,500.

3. $\bar{x} = \frac{\Sigma x}{n} = \frac{316}{8} = 39.5$

Time, x	$x-\bar{x}$	$\left(x-\bar{x}\right)^2$
43	3.5	12.25
57	17.5	306.25
18	–21.5	462.25
45	5.5	30.25
47	7.5	56.25
33	–6.5	42.25
49	9.5	90.25
24	–15.5	240.25
$\sum x = 316$	$\sum(x-\mu) = 0$	$\sum(x-\mu)^2 = 1240$

$$SS_x = \Sigma\left(x-\bar{x}\right)^2 = 1240$$

$$s^2 = \frac{\Sigma(x-\bar{x})^2}{n-1} = \frac{1240}{7} \approx 177.1$$

$$s = \sqrt{s^2} = \sqrt{\frac{1240}{7}} \approx 13.3$$

4. Enter the data in a computer or a calculator.
$\bar{x} \approx 19.8,\quad s \approx 7.8$

5. *Sample answer:* 7, 7, 7, 7, 7, 13, 13, 13, 13, 13

Salary, x	$x-\mu$	$(x-\mu)^2$
7	–3	9
7	–3	9
7	–3	9
7	–3	9
7	–3	9
13	3	9
13	3	9
13	3	9
13	3	9
13	3	9
$\sum x = 100$	$\sum(x-\mu) = 0$	$\sum(x-\mu)^2 = 90$

$$\mu = \frac{\Sigma x}{N} = \frac{100}{10} = 10$$

$$\sigma = \sqrt{\frac{\Sigma(x-\mu)^2}{N}} = \sqrt{\frac{90}{10}} = \sqrt{9} = 3$$

6. $67.1 - 64.2 = 2.9 = 1$ standard deviation
Because 67.1 is one standard deviation above the mean height, the percent of heights between 64.2 inches and 67.1 inches is 34.13%.
Approximately 34.13% of women ages 20-29 are between 64.2 and 67.1 inches tall.

7. $39.3 - 2(23.5) = -7.7$
Because –7.7 does not make sense for an age, use 0.
$39.3 + 2(23.5) = 86.3$

$$1 - \frac{1}{k^2} = 1 - \frac{1}{(2)^2} = 1 - \frac{1}{4} = 0.75$$

At least 75% of the data lie within 2 standard deviations of the mean. At least 75% of the population of Iowa is between 0 and 86.3 years old. Because $80 < 86.3$, and age of 80 lies within two standard deviations of the mean. So, the age is not unusual.

8.

x	f	xf
0	10	0
1	19	19
2	7	14
3	7	21
4	5	20
5	1	5
6	1	6
	$n = 50$	$\sum xf = 85$

$$\bar{x} = \frac{\sum xf}{n} = \frac{85}{50} = 1.7$$

$x - \bar{x}$	$(x-\bar{x})^2$	$(x-\bar{x})^2 f$
−1.7	2.89	28.90
−0.7	0.49	9.31
0.3	0.09	0.63
1.3	1.69	11.83
2.3	5.29	26.45
3.3	10.89	10.89
4.3	18.49	18.49
		$\sum(x-\bar{x})^2 f = 106.5$

$$s = \sqrt{\frac{\sum(x-\bar{x})^2 f}{n-1}} = \sqrt{\frac{106.5}{49}} \approx 1.5$$

9.

Class	x	f	Xf
1-99	49.5	380	18,810
100-199	149.5	230	34,385
200-299	249.5	210	52,395
300-399	349.5	50	17,475
400-499	449.5	60	26,970
500+	650	70	45,500
		$n = 1000$	$\sum xf = 195{,}535$

$$\bar{x} = \frac{\sum xf}{n} = \frac{195{,}535}{1000} \approx 195.5$$

$x - \bar{x}$	$(x-\bar{x})^2$	$(x-\bar{x})^2 f$
−146.0	21,316	8,100,080
−46.0	2116	486,680
54.0	2916	612,360
154.0	23,716	1,185,800
254.0	64,516	3,870,960
454.5	206,570.25	14,459,917.5
		$\sum(x-\bar{x})^2 f = 28{,}715{,}797.5$

$$s=\sqrt{\frac{\sum\left(x-\bar{x}\right)^2 f}{n-1}}=\sqrt{\frac{28,715,797.5}{999}}\approx 169.5$$

10. Los Angeles: $\bar{x}\approx 36.88$, $s\approx 17.39$

Dallas: $\bar{x}\approx 19.8$, $s\approx 7.8$

Los Angeles: $CV=\frac{s}{\bar{x}}\cdot 100\%=\frac{17.4}{36.9}\cdot 100\%\approx 47.2\%$

Dallas: $CV=\frac{s}{\bar{x}}\cdot 100\%=\frac{7.8}{19.8}\cdot 100\%\approx 39.4\%$

The office rental rates are more variable in Los Angeles than in Dallas.

2.4 EXERCISE SOLUTIONS

1. The range is the difference between the maximum and minimum values of a data set. The advantage of the range is that it is easy to calculate. The disadvantage is that it uses only two entries from the data set.

3. The units of variance are squared. Its units are meaningless (example: dollars^2). The units of standard deviation are the same as the data.

5. When calculating the population standard deviation, you divide the sum of the squared deviations by *N*, then take the square root of that value. When calculating the sample standard deviation, you divide the sum of the squared deviations by $n-1$, then take the square root of that value.

7. Similarity: Both estimate proportions of the data contained within *k* standard deviations of the mean. Difference: The Empirical Rule assumes the distribution is approximately symmetric and bell-shaped. Chebychev's Theorem makes no such assumption.

9. Range = Max – Min = 75 – 40 = 35; Approximately 35, or $35,000

11. (a) Range = Max – Min = 38.5 – 20.7 = 17.8

(b) Range = Max – Min = 60.5 – 20.7 = 39.8

13. Range = Max – Min = 14 – 10 = 4

$$\mu=\frac{\sum x}{N}=\frac{123}{11}\approx 11.2$$

x	$x-\mu$	$(x-\mu)^2$
14	2.8	7.84
13	1.8	3.24
13	1.8	3.24
12	0.8	0.64
11	−0.2	0.04
10	−1.2	1.44
10	−1.2	1.44
10	−1.2	1.44
10	−1.2	1.44
10	−1.2	1.44
10	−1.2	1.44
$\sum x=123$	$\sum(x-\mu)\approx 0$	$\sum(x-\mu)^2=23.64$

$$\sigma^2=\frac{\Sigma(x-\mu)^2}{N}=\frac{23.64}{11}\approx 2.1$$

$$\sigma=\sqrt{\frac{\Sigma(x-\mu)^2}{N}}=\sqrt{\frac{23.64}{11}}\approx 1.5$$

15. Range = Max – Min = 23 – 17 = 6

$$\bar{x}=\frac{\Sigma x}{n}=\frac{380}{20}=19$$

x	$x-\bar{x}$	$(x-\bar{x})^2$
19	0	0
20	1	1
17	−2	4
19	0	0
17	−2	4
21	2	4
23	4	16
21	2	4
17	−2	4
17	−2	4
19	0	0
19	0	0
17	−2	4
20	1	1
23	4	16
18	−1	1
18	−1	1
18	−1	1
18	−1	1
19	0	0
$\sum x=380$	$\sum(x-\bar{x})=0$	$\sum(x-\bar{x})^2=66$

$$s^2 = \frac{\Sigma(x-\bar{x})^2}{n-1} = \frac{66}{20-1} \approx 3.5$$

$$s = \sqrt{\frac{\Sigma(x-\bar{x})^2}{n-1}} = \sqrt{\frac{66}{19}} \approx 1.9$$

17. The data set in (a) has a standard deviation of 2.4 and the data set in (b) has a standard deviation of 5 because the data in (b) have more variability.

19. Company B; An offer of $43,000 is two standard deviations from the mean of Company A's starting salaries, which makes it unlikely. The same offer is within one standard deviation of the mean of Company B's starting salaries, which makes the offer likely.

21. (a) Greatest sample standard deviation: (ii)
Data set (ii) has more entries that are farther away from the mean.
Least sample standard deviation: (iii)
Data set (iii) has more entries that are close to the mean.

(b) The three data sets have the same mean but have different standard deviations.

(c) Estimates will vary; (i) $s \approx 1.1$; (ii) $s \approx 1.3$; (iii) $s \approx 0.8$

23. (a) Greatest sample standard deviation: (i)
Data set (i) has more entries that are farther away from the mean.
Least sample standard deviation: (iii)
Data set (iii) has more entries that are close to the mean.

(b) The three data sets have the same mean, median, and mode, but have different standard deviations.

(c) Estimates will vary; (i) $s \approx 9.6$; (ii) $s \approx 9.0$; (iii) $s \approx 5.1$

25. *Sample answer:* 3,3,3,7,7,7

27. *Sample answer:* 9,9,9,9,9,9,9

29. $(63,\ 71) \rightarrow (67-1(4),\ 67+1(4)) \rightarrow (\bar{x}-s,\ \bar{x}+s)$

68% of the vehicles have speeds between 63 and 71 mph.

31. (a) $n = 75$; $68\%(75) = (0.68)(75) \approx 51$ vehicles have speeds between 63 and 71 mph.

(b) $n = 25$; $68\%(25) = (0.68)(25) \approx 17$ vehicles have speeds between 63 and 71 mph.

33. 78, 76, and 82 are unusual; 82 is very unusual because it is more than 3 standard deviations from the mean.

35. $(\bar{x}-2s,\ \bar{x}+2s) \to (0,\ 4)$ are 2 standard deviations from the mean.

$$1-\frac{1}{k^2}=1-\frac{1}{(2)^2}=1-\frac{1}{4}=0.75 \Rightarrow$$ At least 75% of the eruption times lie between 0 and 4.

If $n = 40$, at least $(0.75)(40) = 30$ households have between 0 and 4 pets.

37. $(\bar{x}-4s,\ \bar{x}+4s) \to (70,\ 94)$ are 4 standard deviations from the mean.

$$1-\frac{1}{k^2}=1-\frac{1}{4^2}=1-\frac{1}{16}=0.9375$$

At least 93.75% of the exam scores are from 70 to 94.

39.

x	f	xf	$x-\bar{x}$	$(x-\bar{x})^2$	$(x-\bar{x})^2 f$
0	3	0	−5	25	75
1	4	4	−4	16	64
2	3	6	−3	9	27
3	9	27	−2	4	36
4	3	12	−1	1	3
5	3	15	0	0	0
6	8	48	1	1	8
7	5	35	2	4	20
8	6	48	3	9	54
9	6	54	4	16	96
	$\Sigma = 50$	$\Sigma = 249$			$\Sigma = 383$

$$\bar{x}=\frac{\Sigma xf}{n}=\frac{249}{50}\approx 5.0$$

$$s=\sqrt{\frac{\Sigma(x-\bar{x})^2 f}{n-1}}=\sqrt{\frac{383}{49}}\approx 2.8$$

41.

Class	x	f	xf	$x-\bar{x}$	$(x-\bar{x})^2$	$(x-\bar{x})^2 f$
15,000–17,499	16,249.5	9	146,245.5	−4759.62	22,653,982.54	203,885,842.9
17,500–19,999	18,749.5	10	187,495	−2259.62	5,105,882.54	51,058,825.4
20,000–22,499	21,249.5	16	339,992	240.38	57,782.54	924,520.64
22,500–24,999	23,749.5	11	261,244.5	2740.38	7,509,682.54	82,606,507.94
25,000 or more	26,249.5	6	157,497	5240.38	27,461,582.54	164,769,495.20
		$n = 52$	$\Sigma xf = 1{,}092{,}4$		$\Sigma(x-\bar{x})^2 f = 503{,}245{,}192.1$	

$$\bar{x}=\frac{\Sigma xf}{n}=\frac{1{,}092{,}474}{52}\approx \$21{,}009.12$$

$$s=\sqrt{\frac{\Sigma(x-\bar{x})^2 f}{n-1}}=\sqrt{\frac{503{,}245{,}192.1}{51}}\approx \$3141.27$$

43.

x	f	xf	$x-\bar{x}$	$(x-\bar{x})^2$	$(x-\bar{x})^2 f$
1	2	2	−1.9	3.61	7.22
2	18	36	−0.9	0.81	14.58
3	24	72	0.1	0.01	0.24
4	16	64	1.1	1.21	19.36
	$n=60$	$\Sigma xf=174$		$\Sigma(x-\bar{x})^2 f=41.4$	

$$\bar{x}=\frac{\sum xf}{n}=\frac{174}{60}\approx 2.9$$

$$s=\sqrt{\frac{\sum(x-\bar{x})^2 f}{n-1}}=\sqrt{\frac{41.4}{59}}\approx 0.8$$

45. Denver: $\bar{x}=\frac{\sum x}{n}=\frac{552.3}{12}\approx 46.0$

$$s^2=\frac{\sum(x-\bar{x})^2}{n-1}=\frac{220.89}{11}\approx 20.08$$

$$s=\sqrt{\frac{\sum(x-\bar{x})^2}{n-1}}=\sqrt{\frac{220.89}{11}}\approx 4.48$$

$$CV=\frac{s}{\bar{x}}\cdot 100\%=\frac{4.48}{46.0}\cdot 100\%\approx 9.7\%$$

Los Angeles: $\bar{x}=\frac{\sum x}{n}=\frac{634.5}{12}\approx 52.9$

$$s^2=\frac{\sum(x-\bar{x})^2}{n-1}=\frac{239.97}{11}\approx 21.82$$

$$s=\sqrt{\frac{\sum(x-\bar{x})^2}{n-1}}=\sqrt{\frac{239.97}{11}}\approx 4.67$$

$$CV=\frac{s}{\bar{x}}\cdot 100\%=\frac{4.67}{52.9}\cdot 100\%\approx 8.8\%$$

Salaries for entry level architects are more variable in Denver than in Los Angeles.

47. Ages: $\mu=\frac{\sum x}{N}=\frac{491}{22}\approx 22.32$

$$\sigma^2=\frac{\sum(x-\mu)^2}{N}=\frac{194.77}{22}\approx 8.85$$

$$\sigma=\sqrt{\frac{\sum(x-\bar{x})^2}{N}}=\sqrt{\frac{194.77}{22}}\approx 2.98$$

$$CV=\frac{\sigma}{\mu}\cdot 100\%=\frac{2.98}{22.32}\cdot 100\%\approx 13.3\%$$

Heights: $\mu = \frac{\sum x}{N} = \frac{1546}{22} \approx 70.27$

$$\sigma^2 = \frac{\sum(x-\mu)^2}{N} = \frac{134.36}{22} \approx 6.11$$

$$\sigma = \sqrt{\frac{\sum(x-\bar{x})^2}{N}} = \sqrt{\frac{134.36}{22}} \approx 2.47$$

$$CV = \frac{\sigma}{\mu} \cdot 100\% = \frac{2.47}{70.27} \cdot 100\% \approx 3.5\%$$

Ages are more variable than heights for all members of the 2016 Women's U.S. Olympic swimming team.

49. Male: $\bar{x} = \frac{\sum x}{n} = \frac{8760}{8} = 1095$

$$s^2 = \frac{\sum(x-\bar{x})^2}{n-1} = \frac{359{,}400}{8-1} \approx 51{,}342.86$$

$$s = \sqrt{\frac{\sum(x-\bar{x})^2}{n-1}} = \sqrt{\frac{359{,}400}{7}} \approx 226.6$$

$$CV = \frac{s}{\bar{x}} \cdot 100\% = \frac{226.6}{1095} \cdot 100\% \approx 20.7\%$$

Female: $\bar{x} = \frac{\sum x}{n} = \frac{9120}{8} = 1140$

$$s^2 = \frac{\sum(x-\bar{x})^2}{n-1} = \frac{282{,}000}{8-1} \approx 40{,}285.71$$

$$s = \sqrt{\frac{\sum(x-\bar{x})^2}{n-1}} = \sqrt{\frac{282{,}000}{7}} = 200.7$$

$$CV = \frac{s}{\bar{x}} \cdot 100\% = \frac{200.7}{1140} \cdot 100\% \approx 17.6\%$$

SAT scores are more variable for males than for females.

51. (a) Answers will vary.

(b) Ages of students

$\sum x = 380$; $\sum x^2 = 7286$

$$s = \sqrt{\frac{\sum x^2 - \frac{(\sum x)^2}{n}}{n-1}} = \sqrt{\frac{7286 - \frac{(380)^2}{20}}{20-1}} = \sqrt{\frac{66}{19}} \approx 1.9$$

(c) The answer is the same as from Exercise 15.

53. (a) $\bar{x} \approx 42.1$; $s \approx 5.6$

(b) $\bar{x} \approx 44.3$; $s \approx 5.9$

(c) 3.5, 3, 3, 4, 4, 2.75, 4.25, 3.25, 3.25, 3.5, 3.25, 3.75, 3.5, 4.17

$\bar{x} \approx 3.5$; $s \approx 0.47$

(d) When each entry is multiplied by a constant k, the new sample mean is $k \cdot x$, and the new sample standard deviation is $k \cdot s$.

55. (a) $P = \dfrac{3(\bar{x} - \text{median})}{s} = \dfrac{3(17-19)}{2.3} \approx -2.61$; The data are skewed left.

(b) $P = \dfrac{3(\bar{x} - \text{median})}{s} = \dfrac{3(32-25)}{5.1} \approx 4.12$; The data are skewed right.

(c) $P = \dfrac{3(\bar{x} - \text{median})}{s} = \dfrac{3(9.2-9.2)}{1.8} = 0$; The data are symmetric.

(d) $P = \dfrac{3(\bar{x} - \text{median})}{s} = \dfrac{3(42-40)}{6.0} = 1$; The data are skewed right.

(e) $P = \dfrac{3(\bar{x} - \text{median})}{s} = \dfrac{3(155-175)}{20.0} = -3$; The data are skewed left.

2.5 MEASURES OF POSITION

2.5 TRY IT YOURSELF SOLUTIONS

1. Order the data from least to greatest. The median (or Q_2) is 30. This was also found in Section 2.3, Try It Yourself 2.
The first quartile is the median of the data entries to the left of Q_2 and the third quartile is the median of the data entries to the right of Q_2.
$Q_1 = 23$, $Q_2 = 30$, $Q_3 = 35$
About one-quarter of the winning scores were 23 points or less, about one-half were 30 points or less, and about three-quarters were 35 points or less.

2. Enter data
$Q_1 = 23.5$, $Q_2 = 30$, $Q_3 = 41$
About one-quarter of these universities charge tuition of $23,500 or less; about one-half charge $30,000 or less; and about three-quarters charge $41,000 or less.

3. $Q_1 = 23,\ Q_3 = 35$
$\text{IQR} = Q_3 - Q_1 = 35 - 23 = 12$
$Q_1 - 1.5(\text{IQR}) = 23 - 1.5(12) = 5$; $Q_3 + 1.5(\text{IQR}) = 35 + 1.5(12) = 53$
The score 55 is greater than $Q_3 + 1.5(\text{IQR})$. So, 55 is an outlier.

4. $\text{Min} = 14,\ Q_1 = 23,\ Q_2 = 30,\ Q_3 = 35,\ \text{Max} = 55$

Points Scored by Winning Super Bowl Teams

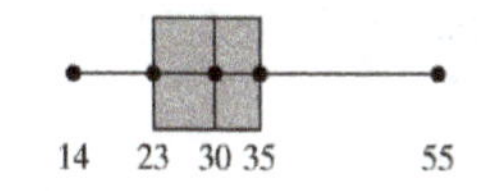

10 15 20 25 30 35 40 45 50 55
Points

About 50% of the winning scores were between 23 and 35 points. About 25% of the winning scores were less than 23 points. About 25% of the winning scores were greater than 35 points.

5. The 10th percentile is 19.5.
About 10% of the winning scores were 19.5 or less.

6. 17,18,19,20,20,23,24,26,29,29,29,30,30,34,35,36,38,39,39,43,44,44,44,45,45
7 data entries are less than 26

$$\text{Percentile of } 26 = \frac{\text{number of data entries less than } 26}{\text{total number of data entries}} \cdot 100 = \frac{7}{25} \cdot 100 = 28^{\text{th}} \text{ percentile}$$

The tuition cost of $26,000 is greater than 28% of the other tuition costs.

7. $\mu = 70,\ \sigma = 8$

$$x = 60:\ z = \frac{x - \mu}{\sigma} = \frac{60 - 70}{8} = -1.25$$

$$x = 71:\ z = \frac{x - \mu}{\sigma} = \frac{71 - 70}{8} = 0.125$$

$$x = 92:\ z = \frac{x - \mu}{\sigma} = \frac{92 - 70}{8} = 2.75$$

From the *z*-scores, the utility bill of $60 is 1.25 standard deviations below the mean, the bill of $71 is 0.125 standard deviation above the mean, and the bill of $92 is 2.75 standard deviations above the mean.

8. 5 feet = 5(12) = 60 inches

$$\text{Man: } z = \frac{x - \mu}{\sigma} = \frac{60 - 69.9}{3} = -3.3; \quad \text{Woman: } z = \frac{x - \mu}{\sigma} = \frac{60 - 64.3}{2.6} \approx -1.7$$

The *z*-score for the 5-foot-tall man is 3.3 standard deviations below the mean. This is an unusual height for a man. The *z*-score for the 5-foot-tall woman is 1.7 standard deviations below the mean. This is among the typical heights for a woman.

2.5 EXERCISE SOLUTIONS

1. The talk is longer in length than 75% of the lectures in the series.

3. The student scored higher than 89% of the students who took the Fundamentals of Engineering exam.

5. The interquartile range of a data set can be used to identify outliers because data values that are greater than $Q_3+1.5(\text{IQR})$ or less than $Q_1-1.5(\text{IQR})$ are considered outliers.

7. True

9. False. An outlier is any number above $Q_3+1.5(\text{IQR})$ or below $Q_1-1.5(\text{IQR})$.

11. (a) 51 54 56 **57** 59 60 60 **60** 60 62 63 **63** 63 65 80

Q_1 (57), Q_2 (60), Q_3 (63)

(b) $\text{IQR}=Q_3-Q_1=63-57=6$

(c) $Q_1-1.5(IQR)=57-1.5(6)=48$; $Q_3+1.5(IQR)=63+1.5(6)=72$. The date entry 80 is an outlier.

13. Min = 0, Q_1 = 2, Q_2 = 5, Q_3 = 8, Max = 10

15. (a) 24 26 27 **28** 30 32 35 **35** 36 39 39 **41** 50 51 60

Q_1 (28), Q_2 (35), Q_3 (41)

Min = 24, $Q_1=28$, $Q_2=35$, $Q_3=41$, Max = 60

(b)

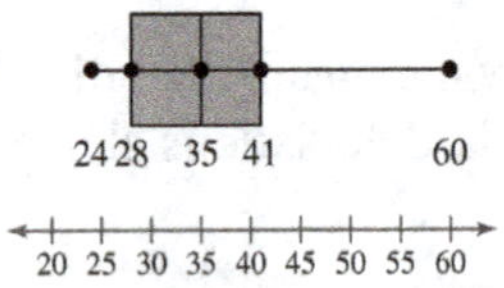

24 28 35 41 60

20 25 30 35 40 45 50 55 60

17. (a) lower half: 1 2 2 4 4 5 5 5 6 6 | upper half: 6 7 7 7 7 8 8 8 9 9

Q_1 (4.5), Q_2 (6), Q_3 (7.5)

Min = 1, $Q_1=4.5$, $Q_2=6$, $Q_3=7.5$, Max = 9

(b)

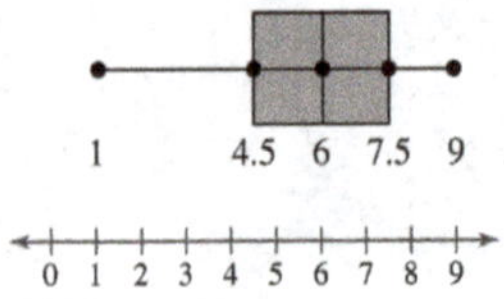

1 4.5 6 7.5 9

0 1 2 3 4 5 6 7 8 9

19. None. The Data are not skewed or symmetric.

21. Skewed left. Most of the data lie to the right in the box-and-whisker plot.

23. Min = 1, $Q_1 = 2$, $Q_2 = 3$, $Q_3 = 6.5$, Max = 8

Studying

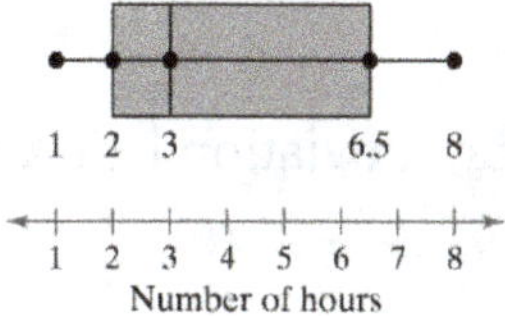

25. Min = 1, $Q_1 = 3$, $Q_2 = 7$, $Q_3 = 12$, Max = 45

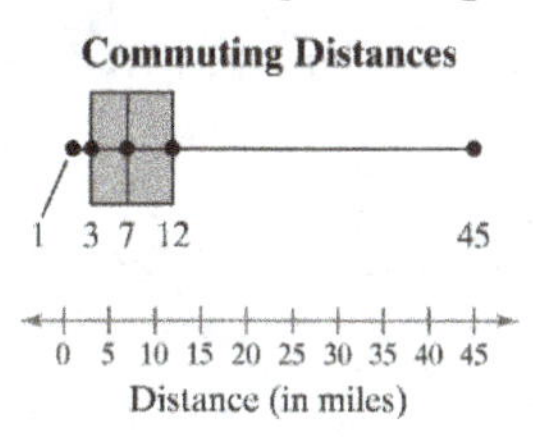

27. (a) 6.5 hours (b) About 50% (c) About 25%

29. About 158; About 70% of quantitative reasoning scores on the Graduate Record Examination are less than 158.

31. About 8th percentile; About 8% of quantitative reasoning scores on the Graduate Record Examination are less than 140.

33. $\text{Percentile of } 40 = \dfrac{\text{number of data entries less than } 40}{\text{total number of data entries}} \cdot 100 = \dfrac{3}{30} \cdot 100 = 10^{\text{th}}$ percentile

35. 75th percentile $= Q_3 = 56$; Ages over 56 are 57,57,61,61,65,66

37. **Depatment of Motor Vehicles Wait Times**

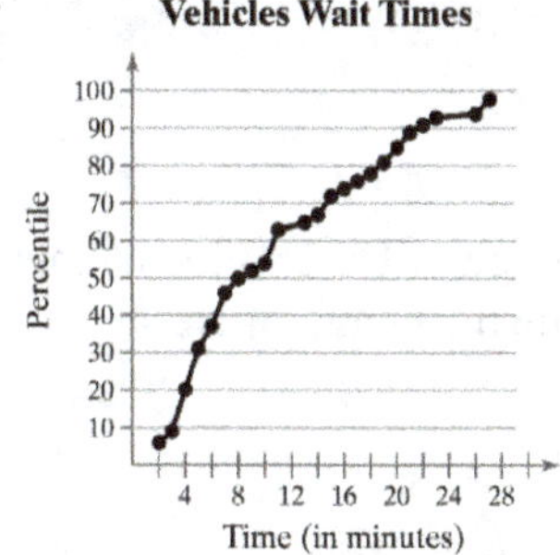

39. A wait time of 20 minutes corresponds to about the 85th percentile.

41. A $\Rightarrow z = -1.43$
B $\Rightarrow z = 0$
C $\Rightarrow z = 2.14$
The z-score 2.14 is unusual because it is so large.

43. Christopher Froome: $x=31 \Rightarrow z=\frac{x-\mu}{\sigma}=\frac{31-27.9}{3.3}\approx 0.94$

Not unusual; The z-score is 0.94, so the age of 31 is about 0.94 standard deviation above the mean.

45. Antonin Magne: $x=27 \Rightarrow z=\frac{x-\mu}{\sigma}=\frac{27-27.9}{3.3}\approx -0.27$

Not unusual; The z-score is –0.27, so the age of 27 is about 0.27 standard deviation below the mean.

47. Henri Cornet: $x=20 \Rightarrow z=\frac{x-\mu}{\sigma}=\frac{20-27.9}{3.3}\approx -2.39$

Unusual; The z-score is –2.39, so the age of 20 is about 2.39 standard deviations below the mean.

49. (a) $x=34{,}000 \Rightarrow z=\frac{x-\mu}{\sigma}=\frac{34{,}000-35{,}000}{2{,}250}\approx -0.44$

$x=37{,}000 \Rightarrow z=\frac{x-\mu}{\sigma}=\frac{37{,}000-35{,}000}{2{,}250}\approx 0.89$

$x=30{,}000 \Rightarrow z=\frac{x-\mu}{\sigma}=\frac{30{,}000-35{,}000}{2{,}250}\approx -2.22$

The tire with a life span of 30,000 miles has an unusually short life span.

(b) $x=30{,}500 \Rightarrow z=\frac{x-\mu}{\sigma}=\frac{30{,}500-35{,}000}{2{,}250}=-2 \Rightarrow$ about 2.5th percentile

$x=37{,}250 \Rightarrow z=\frac{x-\mu}{\sigma}=\frac{37{,}250-35{,}000}{2{,}250}=1 \Rightarrow$ about 84th percentile

$x=35{,}000 \Rightarrow z=\frac{x-\mu}{\sigma}=\frac{35{,}000-35{,}000}{2{,}250}=0 \Rightarrow$ about 50th percentile

51. Robert Duvall: $x=53 \Rightarrow z=\frac{x-\mu}{\sigma}=\frac{53-43.7}{8.7}\approx 1.07$

Jack Nicholson: $x=46 \Rightarrow z=\frac{x-\mu}{\sigma}=\frac{46-50.4}{13.8}\approx -0.32$

The age of Robert Duvall was about 1 standard deviation above the mean age of Best Actor winners, and the age of Jack Nicholson was less than 1 standard deviation below the mean age of Best Supporting Actor winners. Neither actor's age is unusual.

53. John Wayne: $x=62 \Rightarrow z=\frac{x-\mu}{\sigma}=\frac{62-43.7}{8.7}\approx 2.10$

Gig Young: $x=56 \Rightarrow z=\frac{x-\mu}{\sigma}=\frac{56-50.4}{13.8}\approx 0.41$

The age of John Wayne was more than 2 standard deviations above the mean age of Best Actor winners, which is unusual. The age of Gig Young was less than 1 standard deviation above the mean age of Best Supporting Actor winners, which is not unusual.

55. 1 2 3 3 5 5 7 7 8 10

(arrows under 3, 5, 7 mark Q_1, Q_2, Q_3)

$$\text{Midquartile} = \frac{Q_1 + Q_3}{2} = \frac{3+7}{2} = 5$$

57. (a) The distribution of Concert 1 is symmetric. The distribution of Concert 2 is skewed right. Concert 1 has less variation.

(b) Concert 2 is more likely to have outliers because it has more variation.

(c) Concert 1, because 68% of the data should be between ± 16.3 of the mean.

(d) No, you do not know the number of songs played at either concert or the actual lengths of the songs.

59. (a) lower half: 2 7 8 9 9 10 10 | 11 | upper half: 11 12 12 13 15 16 24

(arrows mark Q_1 at 9, Q_2 at 11, Q_3 at 13)

$Q_1 = 9$, $Q_2 = 11$, $Q_3 = 13$

$\text{IQR} = Q_3 - Q_1 = 13 - 9 = 4$

$1.5 \times \text{IQR} = 6$

$Q_1 - (1.5 \times \text{IQR}) = 9 - 6 = 3$

$Q_3 + (1.5 \times \text{IQR}) = 13 + 6 = 19$

Any values less than 6 or greater than 19 are outliers. So, 2 and 24 are outliers.

(b)

2 7 9 11 13 16 24

0 5 10 15 20 25 30

61. (a) 1, 23, 29, 35, 37, 46, **46**, 47, 49, 52, 53, 59, 83

$Q_1 = 32$, $Q_2 = 46$, $Q_3 = 52.5$

$\text{IQR} = Q_3 - Q_1 = 52.5 - 32 = 20.5$

$1.5 \times \text{IQR} = 30.75$

$Q_1 - (1.5 \times \text{IQR}) = 32 - 30.75 = 1.25$

$Q_3 + (1.5 \times \text{IQR}) = 52.5 + 30.75 = 83.25$

Any values less than 1.25 or greater than 83.25 are outliers. So, 1 is an outlier.

(b)

1 23 32 46 52.5 83

0 10 20 30 40 50 60 70 80 90

63. Answers will vary.

CHAPTER 2 REVIEW EXERCISE SOLUTIONS

1. Class width $= \dfrac{\text{Max} - \text{Min}}{\text{Number of classes}} = \dfrac{55-26}{5} = 5.8 \Rightarrow 6$

Class	Midpoint	Class boundaries	Frequency, f	Relative frequency	Cumulative frequency
26–31	28.5	25.5–31.5	5	0.25	5
32–37	34.5	31.5–37.5	4	0.20	9
38–43	40.5	37.5–43.5	6	0.30	15
44–49	46.5	43.5–49.5	3	0.15	18
50–55	52.5	49.5–55.5	2	0.10	20
			$\sum f = 20$	$\sum \frac{f}{n} = 1$	

3. Class width $= \dfrac{\text{Max} - \text{Min}}{\text{Number of classes}} = \dfrac{12.10-11.86}{7} \approx 0.03 \Rightarrow 0.04$

Class	Midpoint	Frequency, f	Relative frequency
11.86-11.89	11.875	3	0.12
11.90-11.93	11.915	5	0.20
11.94-11.97	11.955	8	0.32
11.98-12.01	11.995	7	0.28
12.02-12.05	12.035	1	0.04
12.06-12.09	12.075	0	0.00
12.10-12.13	12.115	1	0.04
		$\sum f = 25$	$\sum \frac{f}{n} = 1$

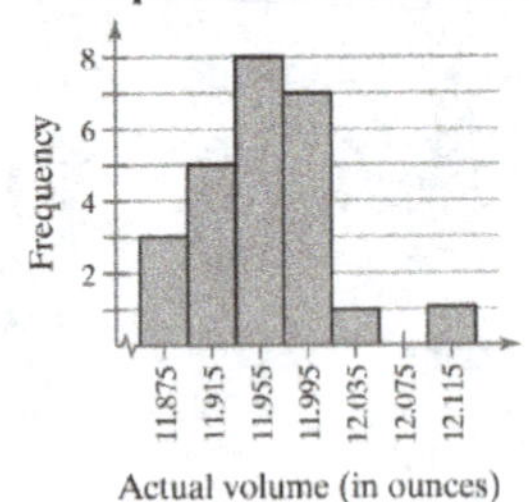

5. Class width $= \dfrac{\text{Max} - \text{Min}}{\text{Number of classes}} = \dfrac{166-79}{6} = 14.5 \Rightarrow 15$

Class	Midpoint	Frequency,
79–93	86	9
94–108	101	12
109–123	116	5
124–138	131	4
139–153	146	2
154–168	161	1
		$\sum f = 33$

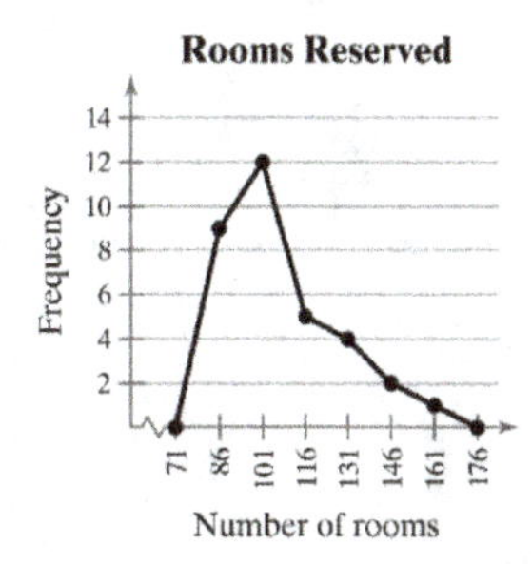

7. Because the data entries go from a low of 22 to a high of 65, use stem values from 2 to 6. List the stems to the left of a vertical line. For each data entry, list a leaf to the right of its stem.

Pollution Indices of U.S. Cities

2 | 2 3 8 8 Key: 2 | 2 = 22
3 | 2 3 6 8 8 9 9
4 | 1 1 3 6 9
5 | 0 3 4 6 7
6 | 3 5 5

Sample answer: Most U.S. cities have a pollution index from 32 to 57.

9.

Location	Frequency	Relative frequency	Degrees
Sleeping	8.8	0.3667	132°
Leisure and Sports	4.0	0.1667	60°
Working	2.3	0.0958	34°
Educational Activities	3.5	0.1458	52°
Other	5.4	0.2250	81°
	$\sum f = 24$	$\sum \frac{f}{n} = 1$	

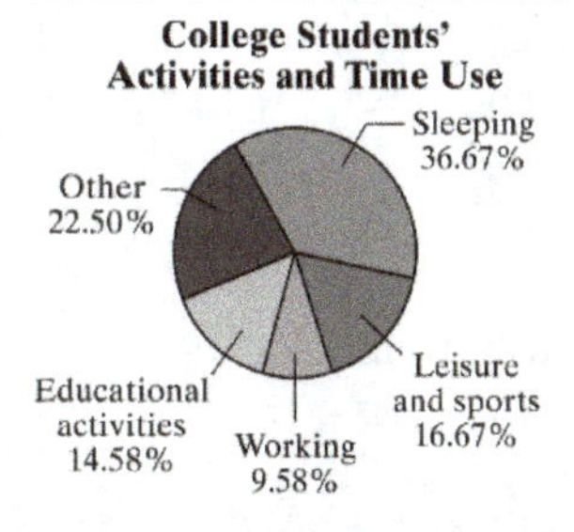

Sample answer: Full-time university and college students spend the least amount of time working.

11.

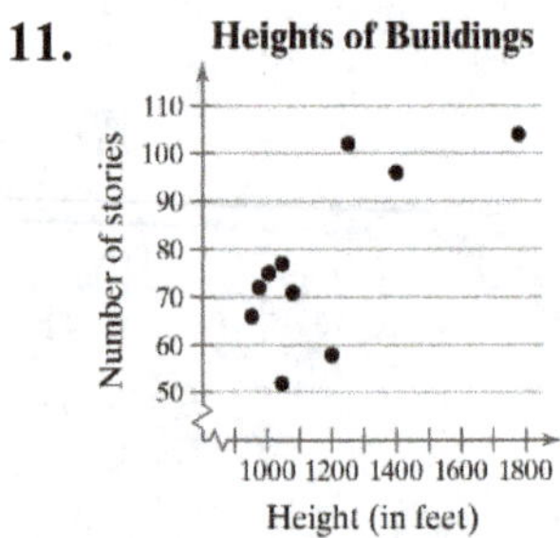

Sample answer: The number of stories appears to increase with height.

13. $\bar{x} = \frac{\sum x}{n} = \frac{295}{10} = 29.5$

21.0 24.0 26.0 28.0 **29.5** **29.5** 31.0 33.0 35.5 37.5

median = 29.5

Mode = 29.5 (occurs 2 times)

15.

Source	Score, x	Weight, w	$x \cdot w$
Test 1	78	0.15	11.7
Test 2	72	0.15	10.8
Test 3	86	0.15	12.9
Test 4	91	0.15	13.65
Test 5	87	0.15	13.05
Test 6	80	0.25	20
		$\sum w = 1$	$\sum (x \cdot w) = 82.1$

$\bar{x} = \frac{\sum (x \cdot w)}{\sum w} = \frac{82.1}{1} = 82.1$

17.

Midpoint, x	Frequency, f	$x \cdot f$
28.5	5	142.5
34.5	4	138
40.5	6	243
46.5	3	139.5
52.5	2	105
	$n = 20$	$\sum (x \cdot f) = 768$

$\bar{x} = \frac{\sum (x \cdot f)}{n} = \frac{768}{20} \approx 38.4$

19. Skewed right

21. Skewed right

23. Mean, because the mean is to the right of the median in a skewed right distribution.

25. Range = Max – Min = 15 – 1 = 14

$$\mu = \frac{\sum x}{N} = \frac{96}{14} \approx 6.9$$

x	$x-\mu$	$(x-\mu)^2$
4	−2.9	8.41
2	−4.9	24.01
9	2.1	4.41
12	5.1	26.01
15	8.1	65.61
3	−3.9	15.21
6	−0.9	0.81
8	1.1	1.21
1	−5.9	34.81
4	−2.9	8.41
14	7.1	50.41
12	5.1	26.01
3	−3.9	15.21
3	−3.9	15.21
$\sum x = 96$	$\sum(x-\mu) \approx 0$	$\sum(x-\mu)^2 = 295.74$

$$\sigma^2 = \frac{\sum(x-\mu)^2}{N} = \frac{295.74}{14} \approx 21.1$$

$$\sigma = \sqrt{\frac{\sum(x-\mu)^2}{N}} = \sqrt{\frac{295.74}{14}} \approx 4.6$$

27. Range = Max – Min = \$7439 – \$5395 = \$2044

$$\bar{x} = \frac{\sum x}{n} = \frac{100,269}{16} \approx \$6266.81$$

x	$x-\bar{x}$	$(x-\bar{x})^2$
5816	−450.81	203,230
7220	953.19	908,571
6045	−221.81	49,200
7439	1172.19	1,374,029
5612	−654.81	428,776
5395	−871.81	760,053
6341	74.19	5504
6908	641.19	411,125
6106	−160.81	25,860
5561	−705.81	498,168
7361	1094.19	1,197,252
5710	−556.81	310,037
6320	53.19	2829
5538	−728.81	531,164
6265	−1.81	3
6632	365.19	133,364
$\sum x = 100,269$	$\sum(x-\bar{x}) \approx 0$	$\sum(x-\bar{x})^2 = 6,839,165$

$$s^2 = \frac{\Sigma(x-\bar{x})^2}{n-1} = \frac{6,839,165}{15} \approx 455,944.3$$

$$s = \sqrt{\frac{\Sigma(x-\bar{x})^2}{n-1}} = \sqrt{\frac{6,839,165}{15}} \approx \$675.24$$

29 95% of the distribution lies within 2 standard deviations of the mean.

$\bar{x} - 2s = 110 - (2)(17.50) = 75$

$\bar{x} + 2s = 110 + (2)(17.50) = 145$

95% of the distribution lies between \$75 and \$145.

31. $(\bar{x} - 2s,\ \bar{x} + 2s) \rightarrow (24,\ 40)$ are 2 standard deviations from the mean.

$$1 - \frac{1}{k^2} = 1 - \frac{1}{(2)^2} = 1 - \frac{1}{4} = 0.75$$

At least (40)(0.75) = 30 customers have a mean sale between \$24 and \$40.

33. $\bar{x} = \frac{\Sigma xf}{n} = \frac{99}{40} \approx 2.5$

x	f	xf	$x-\bar{x}$	$(x-\bar{x})^2$	$(x-\bar{x})^2 f$
0	1	0	–2.5	6.25	6.25
1	8	8	–1.5	2.25	18.00
2	13	26	–0.5	0.25	3.25
3	10	30	0.5	0.25	2.50
4	5	20	1.5	2.25	11.25
5	3	15	2.5	6.25	18.75
	$n = 40$	$\sum xf = 99$			$\sum(x-\bar{x})^2 f = 60$

$$s = \sqrt{\frac{\Sigma(x-\bar{x})^2 f}{n-1}} = \sqrt{\frac{60}{39}} \approx 1.2$$

35. Freshmen: $\bar{x} = \frac{\Sigma x}{n} = \frac{23.1}{9} \approx 2.567$

x	$x-\bar{x}$	$(x-\bar{x})^2$
2.8	0.233	0.0543
1.8	−0.767	0.5833
4.0	1.433	2.0535
3.8	1.233	1.5203
2.4	−0.167	0.0279
2.0	−0.567	0.3215
0.9	−1.667	2.7789
3.6	1.033	1.0671
1.8	−0.767	0.5883
		$\sum(x-\bar{x})^2 = 9.0000$

$$s=\sqrt{\frac{\Sigma(x-\bar{x})^2}{n-1}}=\sqrt{\frac{9.0000}{8}}\approx 1.061$$

$$CV=\frac{s}{\bar{x}}\cdot 100\%=\frac{1.061}{2.567}\cdot 100\%\approx 41.3\%$$

Seniors: $\bar{x}=\frac{\sum x}{n}=\frac{26.6}{9}\approx 2.956$

x	$x-\bar{x}$	$(x-\bar{x})^2$
2.3	−0.656	0.4303
3.3	0.344	0.1183
1.8	−1.156	1.3363
4.0	1.044	1.0899
3.1	0.144	0.0207
2.7	−0.256	0.0655
3.9	0.944	0.8911
2.6	−0.356	0.1267
2.9	−0.056	0.0031
		$\sum(x-\bar{x})^2 = 4.0822$

$$s=\sqrt{\frac{\Sigma(x-\bar{x})^2}{n-1}}=\sqrt{\frac{4.0822}{8}}\approx 0.714$$

$$CV=\frac{s}{\bar{x}}\cdot 100\%=\frac{0.714}{2.956}\cdot 100\%\approx 24.2\%$$

Grade point averages are more variable for freshmen than seniors.

37. 16, 16, 22, 22, 22, 22, **25**, 25, 30, 30, 34, 34, 35, 35, 35, 41, 46, 50, 52, **56**, 58, 107, 112, 119, 124, 136
Min = 16, $Q_1 = 25$, $Q_2 = 35$, $Q_3 = 56$, Max = 136

39. 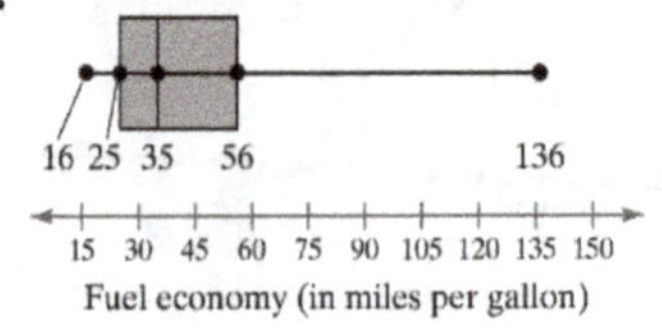

41. 21.0 24.0 <u>26.0</u> 28.0 **29.5** **29.5** 31.0 <u>33.0</u> 35.5 37.5

median = 29.5

$\text{IQR} = Q_3 - Q_1 = 33 - 26 = 7$ inches

43. The 65th percentile means that 65% had a test grade of 75 or less. So, 35% scored higher than 75.

45. $z = \dfrac{16{,}500 - 11{,}830}{2370} \approx 1.97$

Not unusual; The *z*-score is 1.97, so a towing capacity of 16,500 pounds is about 1.97 standard deviations above the mean.

47. $z = \dfrac{18{,}000 - 11{,}830}{2370} = 2.60$

Unusual; The *z*-score is 2.60, so a towing capacity of 18,000 pounds is about 2.60 standard deviations above the mean.

CHAPTER 2 QUIZ SOLUTIONS

1. (a) $\text{Class width} = \dfrac{\text{Max} - \text{Min}}{\text{Number of classes}} = \dfrac{157 - 101}{5} = 11.2 \Rightarrow 12$

Class	Midpoint	Class boundaries	Frequency, f	Relative frequency	Cumulative frequency
101-112	106.5	100.5-112.5	3	0.11	3
113-124	118.5	112.5-124.5	11	0.41	14
125-136	130.5	124.5-136.5	8	0.30	22
137-148	142.5	136.5-148.5	3	0.11	25
149-160	154.5	148.5-160.5	2	0.07	27
			$\sum f = 27$	$\sum \frac{f}{n} = 1$	

(b)

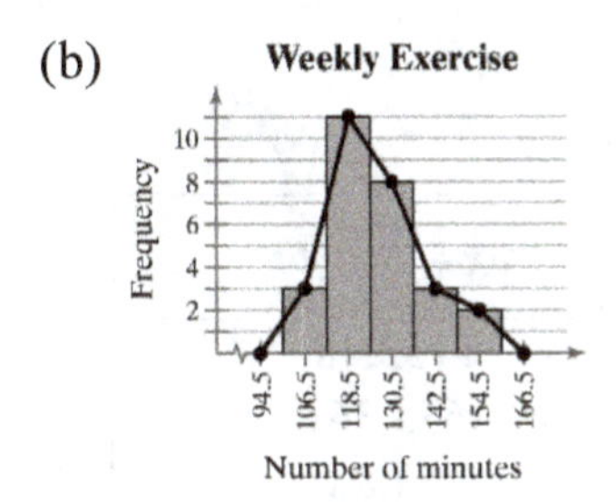

(c)

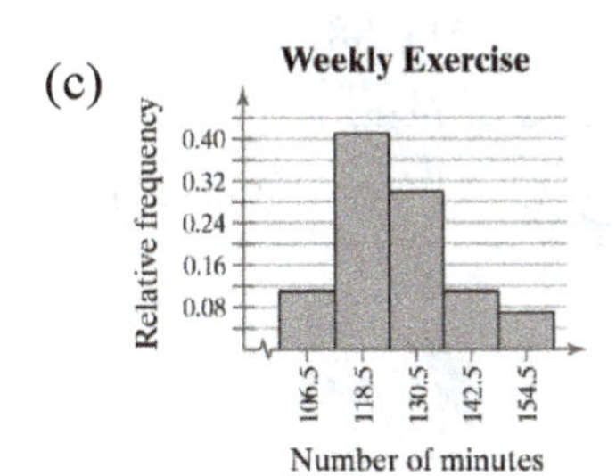

(d) Skewed right

(e)

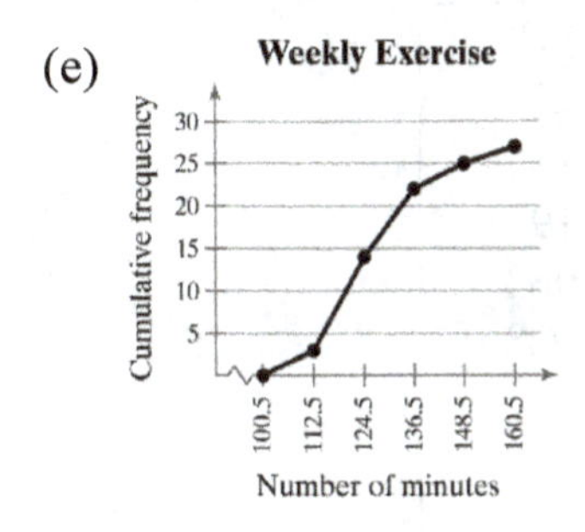

(f) **Weekly Exercise (in minutes)**

10	1 8	Key: 10 | 8 = 108
11	1 4 6 7 8 9 9	
12	0 0 3 3 4 7 7 8	
13	0 1 1 2 5 9 9	
14	2	
15	0 7	

(g) 101, 108, 111, 114, 116, 117, **118**, 119, 119, 120, 120, 123, 123, **124**, 127, 127, 128, 130, 131, 131, **132**, 135, 139, 139, 142, 150, 157

Min = 101, $Q_1 = 118$, $Q_2 = 124$, $Q_3 = 132$, Max = 157

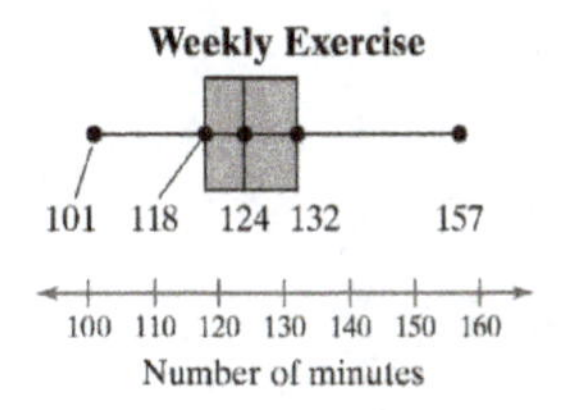

2. $\bar{x} = \dfrac{\sum xf}{n} = \dfrac{3403.5}{27} \approx 126.1$

Midpoint, x	Frequency, f	xf	$x-\bar{x}$	$\left(x-\bar{x}\right)^2$	$\left(x-\bar{x}\right)^2 f$
106.5	3	319.5	−19.6	384.16	1152.48
118.5	11	1303.5	−7.6	57.76	635.36
130.5	8	1044	4.4	19.36	154.88
142.5	3	427.5	16.4	268.96	806.88
154.5	2	309.0	28.4	806.56	1613.12
	$n=27$	$\sum xf = 3403.5$			$\sum\left(x-\bar{x}\right)^2 f = 4362.72$

$$s = \sqrt{\frac{\sum\left(x-\bar{x}\right)^2 f}{n-1}} = \sqrt{\frac{4362.72}{26}} \approx 13.0$$

3. (a)

Category	Frequency	Relative frequency	Degrees
Metals	57	0.5089	183°
Metalloids	7	0.0625	23°
Halogens	5	0.0446	16°
Noble gases	6	0.0536	19°
Rare earth elements	30	0.2679	96°
Other nonmetals	7	0.0625	23°
	$n=112$	$\sum \frac{f}{n} = 1$	

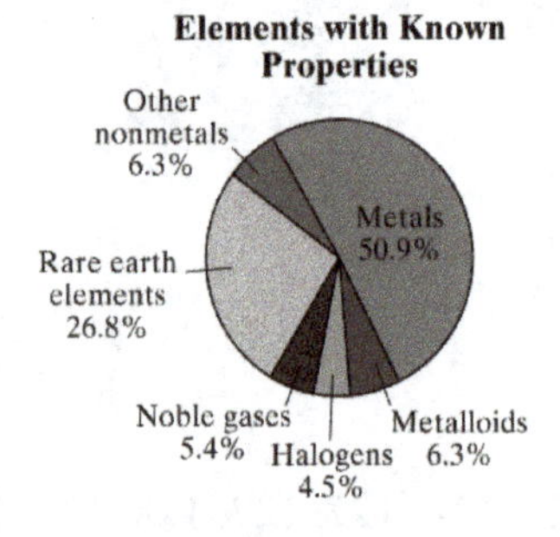

(b)

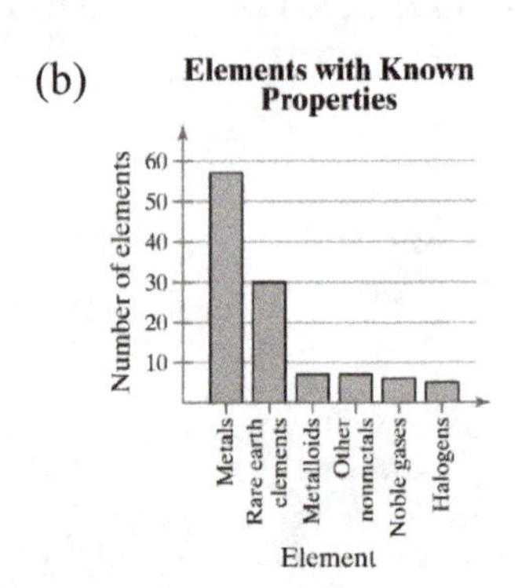

4. (a) $\bar{x} = \frac{\sum x}{n} = \frac{16,262}{16} \approx 1016.4$

718, 720, 749, 790, 860, 891, 969, **976, 1062**, 1100, 1100, 1124, 1248, 1255, 1316, 1384

$$\text{median} = \frac{976 + 1062}{2} = 1019$$

mode = 1100 (occurs twice)

The mean or median best describes a typical salary because there are no outliers.

(b) Range = Max – Min = 1384 – 718 = 666

x	$x - \bar{x}$	$(x - \bar{x})^2$
1100	83.6	6989
749	–267.4	71,503
720	–296.4	87,853
1062	45.6	2079
1384	367.6	135,130
1248	231.6	53,639
1124	107.6	11,578
891	–125.4	15,725
1255	238.6	56,930
969	–47.4	2247
976	–40.4	1632
790	–226.4	51,257
718	–298.4	89,043
860	–156.4	24,461
1316	299.6	89,760
1100	83.6	6989
		$\sum(x - \bar{x})^2 = 706,815$

$$s^2 = \frac{\sum(x - \bar{x})^2}{n - 1} = \frac{706,815}{15} \approx 47,120.9$$

$$s = \sqrt{\frac{\sum(x - \bar{x})^2}{n - 1}} = \sqrt{\frac{706,815}{15}} \approx 217.1$$

(c) $CV = \frac{s}{\bar{x}} \cdot 100\% = \frac{217.1}{1016.4} \cdot 100\% \approx 21.4\%$

5. $\bar{x} - 2s = 180,000 - 2 \cdot 15,000 = \$150,000$

$\bar{x} + 2s = 180,000 + 2 \cdot 15,000 = \$210,000$

95% of the new home prices fall between \$150,000 and \$210,000.

6. (a) $x = 225,000:\ z = \frac{x - \bar{x}}{s} = \frac{225,000 - 180,000}{15,000} = 3.0$

Unusual; The z-score is 3, so a new home price of \$225,000 is about 3 standard deviations above the mean.

(b) $x = 80{,}000: \quad z = \frac{x-\bar{x}}{s} = \frac{80{,}000 - 180{,}000}{15{,}000} \approx -6.67$

Unusual; The z-score is -6.67, so a new home price of $80,000 is about 6.67 standard deviations below the mean.

(c) $x = 200{,}000: \quad z = \frac{x-\bar{x}}{s} = \frac{200{,}000 - 180{,}000}{15{,}000} \approx 1.33$

Not unusual; The z-score is 1.33, so a new home price of $200,000 is about 1.33 standard deviations above the mean.

(d) $x = 147{,}000: \quad z = \frac{x-\bar{x}}{s} = \frac{147{,}000 - 180{,}000}{15{,}000} = -2.2$

Unusual; The z-score is -2.2, so a new home price of $147,000 is about 2.2 standard deviations below the mean.

7. 59 68 68 68 68 69 69 **71** 73 74 75 78 78 79 **81** **84** 84 86 86 86 87 87 **89** 89 91 93 94 95 95 103

$Q_1 = 71$ $\quad Q_2 = 82.5$ $\quad Q_3 = 89$

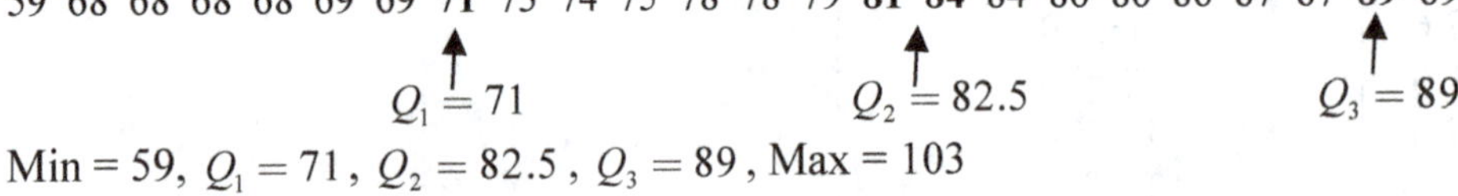

Min = 59, $Q_1 = 71$, $Q_2 = 82.5$, $Q_3 = 89$, Max = 103

Wins for Each MLB Team
59
71
82.5
89
103
55 60 65 70 75 80 85 90 95 100 105
Number of wins

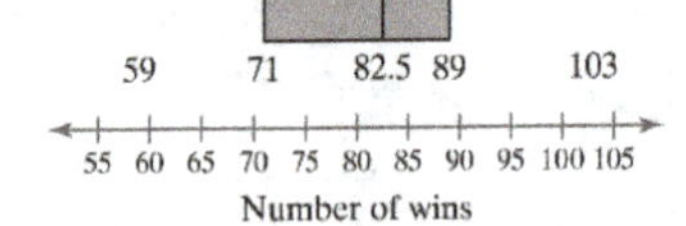

CUMULATIVE REVIEW FOR CHAPTERS 1 AND 2

1. Systematic sampling is used because every fortieth toothbrush from each assembly line is tested. It is possible for bias to enter into the sample if, for some reason, an assembly line makes a consistent error.

3.

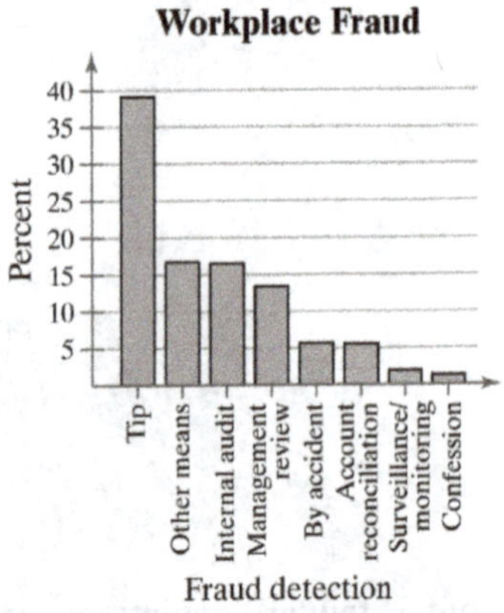

Workplace Fraud
Percent
40
35
30
25
20
15
10
5
Tip
Other means
Internal audit
Management review
By accident
Account reconciliation
Surveillance/ monitoring
Confession
Fraud detection

5. 88% is a statistic. The percent is based on a subset of the population.

7. Population: Collection of opinions of all college and university admissions directors and enrollment officers
Sample: Collection of opinions of the 339 college and university admission directors and enrollment officers surveyed

9. Experiment. The study applies a treatment (digital device) to the subjects.

11. Quantitative: The data are at the ratio level.

13. (a)

Q_1 ↘ Q_2 ↘
0 0 0 0 0 0 0 1 1 1 2 2 **2** 2 2 3 3 3 3 4 4 6 6 7 **9** **11** 11 12 15 16
16 23 23 27 31 31 32 **32** 40 44 45 46 47 48 50 55 67 87 90 99
↗ Q_3

Min = 0, $Q_1 = 2$, $Q_2 = 10$, $Q_3 = 32$, Max = 99

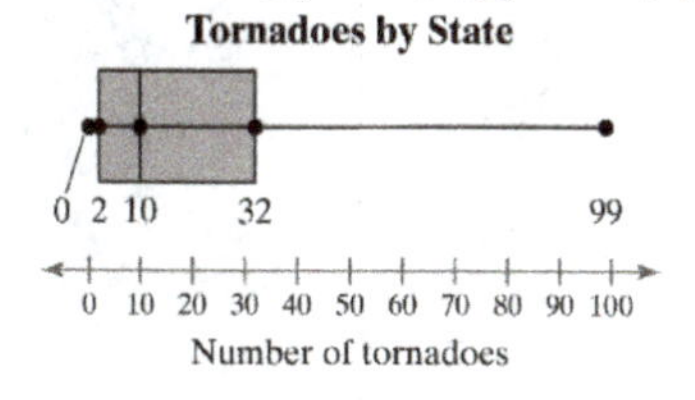

(b) The distribution of the number of tornadoes is skewed right.

15. (a) $\bar{x} = \frac{49.4}{9} \approx 5.49$

3.4 3.9 4.2 4.6 (5.4) 6.5 6.8 7.1 7.5
median = 5.4

mode = none
Both the mean and median accurately describe a typical American alligator tail length. (Answers will vary.)

(b) Range – Max – Min – 7.5 – 3.4 = 4.1

x	$x-\bar{x}$	$(x-\bar{x})^2$
3.4	–2.09	4.3681
3.9	–1.59	2.5281
4.2	–1.29	1.6641
4.6	–0.89	0.7921
5.4	–0.09	0.0081
6.5	1.01	1.0201
6.8	1.31	1.7161
7.1	1.61	2.5921
7.5	2.01	4.0401
		$\sum(x-\bar{x})^2 = 18.7289$

$$s^2 = \frac{\Sigma(x-\bar{x})^2}{n-1} = \frac{18.7289}{8} \approx 2.34$$

$$s = \sqrt{\frac{\Sigma(x-\bar{x})^2}{n-1}} = \sqrt{\frac{18.7289}{8}} \approx 1.53$$

17. Class width $= \dfrac{\text{Max} - \text{Min}}{\text{Number of classes}} = \dfrac{64-0}{8} = 8 \Rightarrow 9$

Class limits	Midpoint	Class boundaries	Frequency	Relative frequency	Cumulative frequency
0-8	4	–0.5-8.5	20	0.500	20
9-17	13	8.5-17.5	7	0.175	27
18-26	22	17.5-26.5	6	0.150	33
27-35	31	26.5-35.5	1	0.025	34
36-44	40	35.5-44.5	2	0.050	36
45-53	49	44.5-53.5	1	0.025	37
54-62	58	53.5-62.5	2	0.050	39
63-71	67	62.5-71.5	1	0.025	40
			$\sum f = 40$	$\sum \frac{f}{n} = 1$	

19.

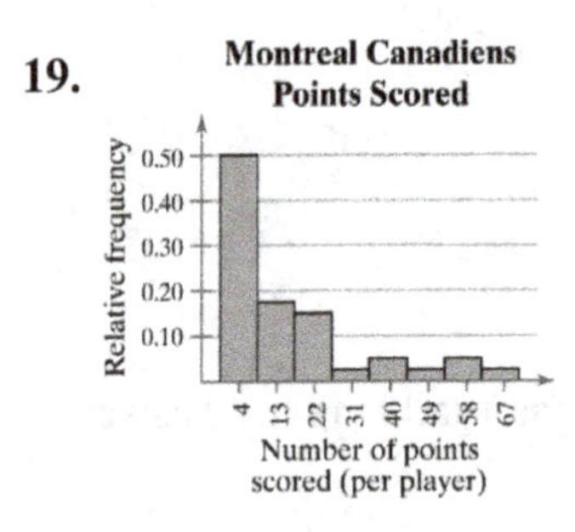

Class with greatest frequency: 0 – 8
Classes with least frequency: 27 – 35, 45 – 53, and 63 – 71

CHAPTER 3

Probability

3.1 BASIC CONCEPTS OF PROBABILITY AND COUNTING

3.1 TRY IT YOURSELF SOLUTIONS

1. (1)

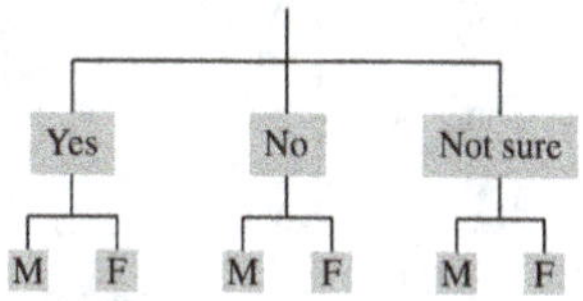

Yes No Not sure

M F M F M F

6 outcomes

Let Y = Yes, N = No, NS = Not sure, M = Male and F = Female.

Sample space = {YM, YF, NM, NF, NSM, NSF}

(2)

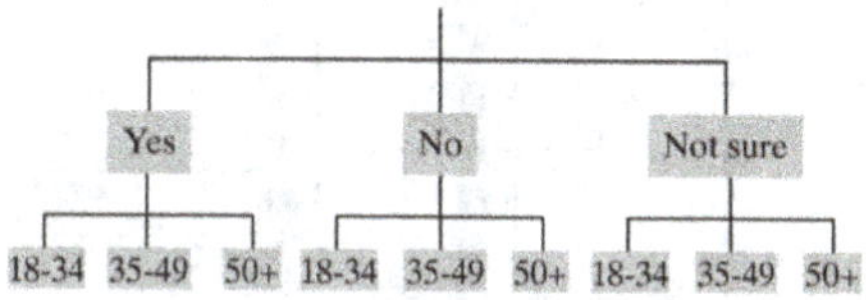

Yes No Not sure

18-34 35-49 50+ 18-34 35-49 50+ 18-34 35-49 50+

9 outcomes

Let Y = Yes, N = No, NS = Not sure,

50+ = 50 and older.

Sample space = {Y18–34, Y35–49, Y50+, N18–34, N35–49, N50+, NS18–34, NS35–49, NS50+}

(3)

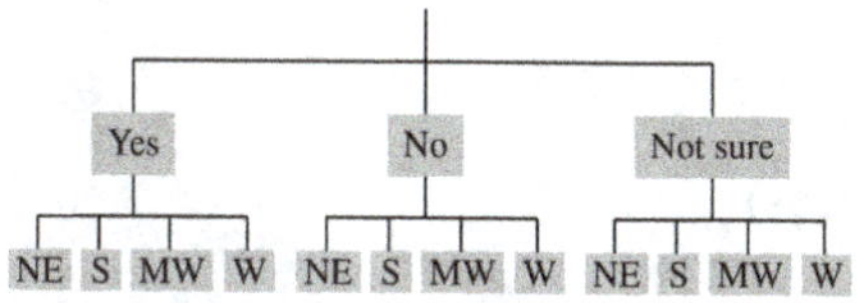

Yes No Not sure

NE S MW W NE S MW W NE S MW W

12 outcomes

Let Y = Yes, N = No, NS = Not sure, NE = Northeast, S = South, MW = Midwest, and W = West

Sample space = {YNE, YS, YMW, YW, NNE, NS, NMW, NW, NSNE, NSS, NSMW, NSW}

2. (1) Event *C* has six outcomes: choosing the ages 18, 19, 20, 21, 22, and 23.
The event is not a simple event because it consists of more than a single outcome.

(2) Event *D* has one outcome: choosing the age 20.
The event is a simple event because it consists of a single outcome.

3. Manufacturer: 4
Size: 2
Color: 5

(4)(2)(5) = 40 ways

Tree Diagram for Car Selections

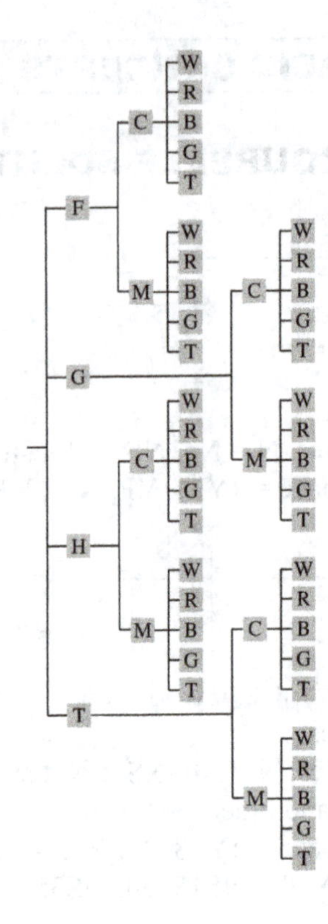

4. (1) Each letter is an event (26 choices for each).
$26 \cdot 26 \cdot 26 \cdot 26 \cdot 26 \cdot 26 = 308{,}915{,}776$

(2) Each letter is an event (26, 25, 24, 23, 22, and 21 choices).
$26 \cdot 25 \cdot 24 \cdot 23 \cdot 22 \cdot 21 = 165{,}765{,}600$

(3) Each letter is an event (22, 26, 26, 26, 26, and 26 choices).
$22 \cdot 26 \cdot 26 \cdot 26 \cdot 26 \cdot 26 = 261{,}390{,}272$

(4) Each digit is an event and each letter is an event (9, 26, 26, 26, 26, and 26 choices).
$9 \cdot 26 \cdot 26 \cdot 26 \cdot 26 \cdot 26 = 106{,}932{,}384$

5. (1) $P(\text{9 of clubs}) = \dfrac{1}{52} \approx 0.019$

(2) $P(\text{heart}) = \dfrac{13}{52} = 0.25$

(3) $P(\text{diamond, heart, club, or spade}) = \dfrac{52}{52} = 1$

6. The event is "read only digital books." The frequency is 91.
Total Frequency = 1490

$P(\text{read only digital books}) = \dfrac{91}{1490} \approx 0.061$

7. Frequency = 254
Total of the Frequencies = 975

$P(\text{age 36 to 49}) = \dfrac{254}{975} \approx 0.261$

8. The event is "salmon successfully passing through a dam on the Columbia River."
The probability is estimated from the results of an experiment.
Empirical probability

9. $P(\text{age 18 to 22}) = \dfrac{156}{975} = 0.16$

$P(\text{age is not 18 to 22}) = 1 - \dfrac{156}{975} = \dfrac{819}{975} = 0.84$

$\dfrac{819}{975} = \dfrac{21}{25}$ or 0.84

10. There are 5 outcomes in the event: {T1, T2, T3, T4, T5}.

$P(\text{tail and less than 6}) = \dfrac{5}{16} \approx 0.313$

11. $10 \cdot 10 \cdot 10 \cdot 10 \cdot 10 \cdot 10 \cdot 10 = 10{,}000{,}000$

$\dfrac{1}{10{,}000{,}000}$

3.1 EXERCISE SOLUTIONS

1. An outcome is the result of a single trial in a probability experiment, whereas an event is a set of one or more outcomes.

3. It is impossible to have more than a 100% chance of rain.

5. The law of large numbers states that as an experiment is repeated over and over, the probabilities found in the experiment will approach the actual probabilities of the event. Examples will vary.

7. False. The event "choosing false on a true or false question and choosing A or B on a multiple choice question" is not simple because it consists of two possible outcomes and can be represented as A = {FA, FB}.

9. False. A probability of less than $\dfrac{1}{20} = 0.05$ indicates an unusual event.

11. d **13.** b **15.** a

17. $P(E')=1-P(E)=1-\frac{19}{23}=\frac{23}{23}-\frac{19}{23}=\frac{4}{23}$

19. $P(E')=1-P(E)=1-0.03=0.97$

21. $P(E)=1-P(E')=1-0.95=0.05$

23. $P(E)=1-P(E')=1-\frac{3}{4}=\frac{4}{4}-\frac{3}{4}=\frac{1}{4}$

25. {A, B, C, D, E, F, G, H, I, J, K, L, M, N, O, P, Q, R, S, T, U, V, W, X, Y, Z}; 26

27. {A♥, K♥, Q♥, J♥, 10♥, 9♥, 8♥, 7♥, 6♥, 5♥, 4♥, 3♥, 2♥,
A♦, K♦, Q♦, J♦, 10♦, 9♦, 8♦, 7♦, 6♦, 5♦, 4♦, 3♦, 2♦,
A♠, K♠, Q♠, J♠, 10♠, 9♠, 8♠, 7♠, 6♠, 5♠, 4♠, 3♠, 2♠,
A♣, K♣, Q♣, J♣, 10♣, 9♣, 8♣, 7♣, 6♣, 5♣, 4♣, 3♣, 2♣}; 52

29.

H — H, T
T — H, T

{HH, HT, TH, TT }; 4

31. {(1, 1), (1, 2), (1, 3), (1, 4), (1, 5), (1, 6), (2, 1), (2, 2,), (2, 3), (2, 4), (2, 5), (2, 6), (3, 1), (3, 2), (3, 3), (3, 4), (3, 5), (3, 6), (4, 1), (4, 2), (4, 3), (4, 4), (4, 5), (4, 6), (5, 1), (5, 2), (5, 3), (5, 4), (5, 5), (5, 6), (6, 1), (6, 2), (6, 3), (6, 4), (6, 5), (6, 6)}; 36

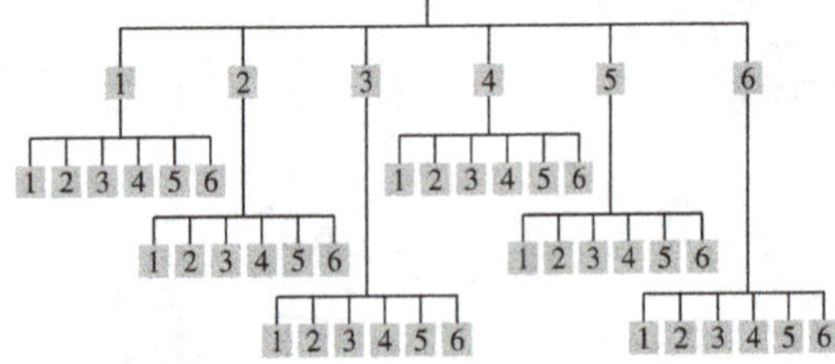

33. 1 outcome; simple event because it is an event that consists of a single outcome.

35. There are 13 diamonds or 13 outcomes. Not a simple event because it is an event that consists of more than a single outcome.

37. $(6)(12)(8)=576$

39. $(9)(10)(10)(5)=4500$

41. $P(A)=\frac{1}{12}\approx 0.083$

43. $P(C)=\frac{8}{12}=\frac{2}{3}\approx 0.667$

45. $P(E) = \dfrac{4}{12} \approx 0.333$

47. $P(\text{does not have a tattoo}) = \dfrac{1584}{2225} \approx 0.712$

49. $P(18 \text{ to } 29) = \dfrac{48.9}{226.9} \approx 0.216$

51. $P(45 \text{ to } 64) = \dfrac{78.1}{226.9} \approx 0.344$

53. Empirical probability because company records were used to calculate the frequency of a washing machine breaking down.

55. Subjective probability because it is most likely based on an educated guess.

57. Classical probability because each outcome in the sample space is equally likely to occur.

59. $P(A) = 1 - P(A') = 1 - \dfrac{123}{777} \approx 1 - 0.158 \approx 0.842$

61. $P(C) = 1 - P(C') = 1 - \dfrac{173}{777} \approx 1 - 0.223 \approx 0.777$

63-65.

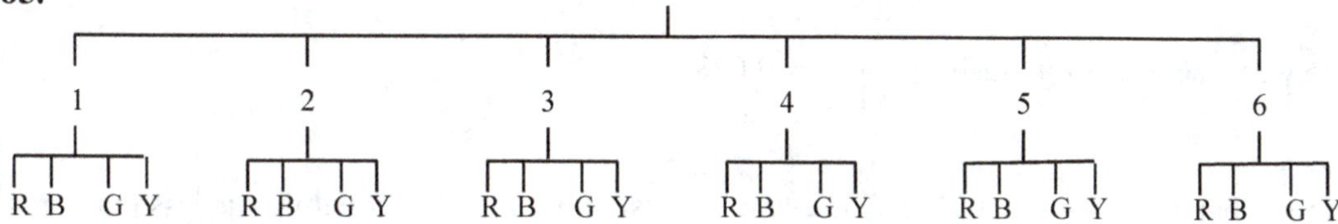

63. $P(A) = \dfrac{1}{24} \approx 0.042$; unusual

65. $P(C) = \dfrac{5}{24} \approx 0.208$; not unusual

67. (a) $\dfrac{1}{1000} = 0.001$ (b) $1 - \dfrac{1}{1000} = \dfrac{999}{1000} = 0.999$

69. There are 8 outcomes in the sample space: {SSS, SSR, SRS, SRR, RSS, RSR, RRS, RRR}

$P(\text{SSS}) = \dfrac{1}{8} = 0.125$

71. There are 8 outcomes in the sample space: {SSS, SSR, SRS, SRR, RSS, RSR, RRS, RRR}

$P(\text{SSR, SRS, RSS}) = \dfrac{3}{8} = 0.375$

73. $P(\text{voted Republican}) = \dfrac{1.8}{1.8+2.2} = 0.450$

75. ($P(\text{doctorate}) = \dfrac{3}{91} \approx 0.033$

77. $P(\text{master's degree}) = \dfrac{25}{91} \approx 0.275$

79. No; None of the events have a probability of 0.05 or less.

81. (a) $P(\text{pink}) = \dfrac{2}{4} = 0.5$ (b) $P(\text{red}) = \dfrac{1}{4} = 0.25$ (c) $P(\text{white}) = \dfrac{1}{4} = 0.25$

83. Total workers (in thousands) = $120{,}221 + 2422 + 15{,}338 + 10{,}852 = 148{,}833$

$P(\text{services industry}) = \dfrac{120{,}221}{148{,}883} \approx 0.808$

85. Total workers (in thousands) = $120{,}221 + 2422 + 15{,}338 + 10{,}852 = 148{,}833$

$P(\text{manufacturing industry}) = \dfrac{15{,}338}{148{,}833} \approx 0.103$

87. (a) $P(\text{at least } 51) = \dfrac{40}{128} \approx 0.313$

(b) $P(\text{between 20 and 30 inclusive}) = \dfrac{10}{128} = 0.078$

(c) $P(\text{more than } 63) = \dfrac{4}{128} \approx 0.031$; This event is unusual because its probability is less than or equal to 0.05.

89. The probability of randomly choosing a tea drinker who does not have a college degree.

91. No, the odds of winning a prize are 1 : 6 (one winning cap and six losing caps). So, the statement should read, "one in seven game pieces win a prize."

93. (a) $P(\text{event will occur}) = \dfrac{4}{9} \approx 0.444$

(b) $P(\text{event will not occur}) = \dfrac{5}{9} \approx 0.556$

95. $39:13 = 3:1$

97. (a)

Sum	Outcomes	P(sum)	Probability
2	(1, 1)	1/36	0.028
3	(1, 2), (2, 1)	2/36	0.056
4	(1, 3), (2, 2,), (3, 1)	3/36	0.083
5	(1, 4), (2, 3), (3, 2), (4, 1)	4/36	0.111
6	(1, 5), (2, 4), (3, 3), (4, 2), (5, 1)	5/36	0.139
7	(1, 6), (2, 5), (3, 4), (4, 3), (5, 2), (6, 1)	6/36	0.167
8	(2, 6), (3, 5), (4, 4), (5, 3), (6, 2)	5/36	0.139
9	(3, 6), (4, 5), (5, 4), (6, 3)	4/36	0.111
10	(4, 6), (5, 5), (6, 4)	3/36	0.083
11	(5, 6), (6, 5)	2/36	0.056
12	(6, 6)	1/36	0.028

(b) Answers will vary.

(c) Answers will vary.

3.2 CONDITIONAL PROBABILITY AND THE MULTIPLICATION RULE

3.2 TRY IT YOURSELF SOLUTIONS

1. $P(\text{female} \mid \text{not offended by something on social media}) = \dfrac{549}{1125} = 0.488$

2. (1) Dependent (2) Independent

3. (1) Let A = {first salmon swims successfully through the dam}
B = {second salmon swims successfully through the dam}
$P(A \text{ and } B) = P(A) \cdot P(B) = (0.85) \cdot (0.85) \approx 0.723$

(2) Let A = {selecting a heart}
B = {selecting a second heart}

$$P(A \text{ and } B) = P(A) \cdot P(B|A) = \left(\frac{13}{52}\right) \cdot \left(\frac{12}{51}\right) \approx 0.059$$

4. (1) $P(3 \text{ surgeries are successful}) = (0.90) \cdot (0.90) \cdot (0.90) = 0.729$

(2) $P(\text{none are successful}) = (0.10) \cdot (0.10) \cdot (0.10) = 0.001$

(3) $P(\text{at least one rotator cuff surgery is successful}) = 1 - P(\text{none are successful})$
$= 1 - 0.001 = 0.999$

5. A = {is female}; B = {works in health field}

(1) $P(A \text{ and } B) = P(A) \cdot P(B|A) = (0.65) \cdot (0.25) = 0.163$

(2) $P(A \text{ and } B') = P(A) \cdot P(B'|A) = P(A) \cdot (1 - P(B|A)) = (0.65) \cdot (0.75) = 0.488$

The events are not unusual because their probabilities are not less than or equal to 0.05.

3.2 EXERCISE SOLUTIONS

1. Two events are independent if the occurrence of one of the events does not affect the probability of the occurrence of the other event, whereas two events are dependent if the occurrence of one of the events does affect the probability of the occurrence of the other event.

3. The notation $P(B|A)$ means the probability of event B occurring, given that event A has occurred.

5. False. If two events are independent, $P(A|B) = P(A)$.

7. M = {student is male}; B = {student received business degree}; F = {student is female}

(a) $P(M \mid B) = \dfrac{191,310}{363,799} \approx 0.526$ (b) $P(B \mid F) = \dfrac{172,489}{1,082,265} \approx 0.159$

9. Independent. The outcome of the first card drawn does not affect the outcome of the second card drawn.

11. Dependent. The outcome of returning a movie after its due date affects the outcome of receiving a late fee.

13. Dependent. The sum of the rolls depends on which numbers came up on the first and second rolls.

15. Events: obstructive sleep apnea, heart disease; Dependent. People with obstructive sleep apnea are more likely to have heart disease

17. Events: playing violent video games, aggressive or bullying behavior; Independent. Playing violent video games does not cause aggressive or bullying behavior in teens.

19. Let A = {card is a heart} and B = {card is a club | a heart card has been removed from the deck}. Since the card was not replaced, there are only 51 cards remaining for the second selection.

$P(A) = \dfrac{13}{52} = \dfrac{1}{4}$ and $P(B) = \dfrac{13}{51}$. Thus, $P(A \text{ and } B) = \left(\dfrac{1}{4}\right) \cdot \left(\dfrac{13}{51}\right) \approx 0.064$

21. Let A = {woman carries mutation of BRCA1 gene} and B = {woman develops breast cancer}.

$P(A) = \dfrac{1}{600}$ and $P(B \mid A) = \dfrac{6}{10} = \dfrac{3}{5}$. Thus, $P(A \text{ and } B) = P(A) \cdot P(B \mid A) = \left(\dfrac{1}{600}\right) \cdot \left(\dfrac{3}{5}\right) \approx 0.001$

23. Let A = {first adult thinks that most Hollywood celebrities are good role models} and B = {second adult thinks that most Hollywood celebrities are good role models}.

$P(A)=\dfrac{200}{1000}$ and $P(B|A)=\dfrac{199}{999}$

(a) $P(A \text{ and } B)=P(A)\cdot P(B|A)=\left(\dfrac{200}{1000}\right)\cdot\left(\dfrac{199}{999}\right)\approx 0.040$

(b) $P(A' \text{ and } B')=P(A')\cdot P(B'|A')=\left(\dfrac{800}{1000}\right)\cdot\left(\dfrac{799}{999}\right)\approx 0.640$

(c) at least one of the two adults thinks that most Hollywood celebrities are good role models)
$=1-P(A' \text{ and } B')\approx 1-0.640\approx 0.360$

25. Let A = {first adult said John Kennedy was the best president} and B = {second adult said John Kennedy was the best president}.
$P(A)=\dfrac{217}{1446}$ and $P(B|A)=\dfrac{216}{1445}$.

(a) $P(A \text{ and } B)=P(A)\cdot P(B|A)=\left(\dfrac{217}{1446}\right)\cdot\left(\dfrac{216}{1445}\right)\approx 0.022$

(b) $P(A' \text{ and } B')=P(A')\cdot P(B'|A')=\left(\dfrac{1229}{1446}\right)\cdot\left(\dfrac{1228}{1445}\right)\approx 0.722$

(c) $P(\text{at least 1 of 2 adults says John Kennedy was the best president})$
$=1-P(A' \text{ and } B')\approx 1-0.722\approx 0.278$

(d) The event in part (a) is unusual because its probability is less than or equal to 0.05.

27. (a) $P(\text{all six have O+})=(0.47)\cdot(0.47)\cdot(0.47)\cdot(0.47)\cdot(0.47)\cdot(0.47)\approx 0.011$

(b) $P(\text{none have O+})=(0.53)\cdot(0.53)\cdot(0.53)\cdot(0.53)\cdot(0.53)\cdot(0.53)\approx 0.022$

(c) $P(\text{at least one has O+})=1-P(\text{none have O+})\approx 1-0.022=0.978$

(d) The events in parts (a) and (b) are unusual because their probabilities are less than or equal to 0.05.

29. Let A = {IVF procedure resulted in pregnancy} and B = {IVF pregnancy resulted in multiple birth}.
$P(A)=0.016$ and $P(B|A)=0.411$

(a) $P(A \text{ and } B)=P(A)\cdot P(B|A)=(0.016)\cdot(0.411)\approx 0.007$

(b) $P(B'|A)=1-P(B|A)=1-0.411=0.589$

(d) Yes, this is usual because the probability is less than or equal to 0.05.

31. Let A = {use digital content} and B = {use as part of curriculum}.

$P(A)=0.80$ and $P(B \mid A)=\frac{4}{10}=0.4$

$P(A \text{ and } B)=P(A)\cdot P(B \mid A)=(0.8)\cdot(0.4)=0.32$

33. $$P(A|B)=\frac{P(A)\cdot P(B|A)}{P(A)\cdot P(B|A)+P(A')\cdot P(B|A')}$$
$$=\frac{\left(\frac{2}{3}\right)\cdot\left(\frac{1}{5}\right)}{\left(\frac{2}{3}\right)\cdot\left(\frac{1}{5}\right)+\left(\frac{1}{3}\right)\cdot\left(\frac{1}{2}\right)}=\frac{\frac{2}{15}}{\frac{3}{10}}=\frac{4}{9}\approx 0.444$$

35. $$P(A|B)=\frac{P(A)\cdot P(B|A)}{P(A)\cdot P(B|A)+P(A')\cdot P(B|A')}$$
$$=\frac{(0.25)\cdot(0.3)}{(0.25)\cdot(0.3)+(0.75)\cdot(0.5)}=\frac{0.075}{0.45}\approx 0.167$$

37. $$P(A|B)=\frac{P(A)\cdot P(B|A)}{P(A)\cdot P(B|A)+P(A')\cdot P(B|A')}$$
$$=\frac{(0.73)\cdot(0.46)}{(0.73)\cdot(0.46)+(0.17)\cdot(0.52)}=\frac{0.3358}{0.4242}\approx 0.792$$

39. $P(A)=\frac{1}{200}=0.005$, $P(B|A)=0.8$, and $P(B|A')=0.05$

(a) $$P(A|B)=\frac{P(A)\cdot P(B|A)}{P(A)\cdot P(B|A)+P(A')\cdot P(B|A')}$$
$$=\frac{(0.005)\cdot(0.8)}{(0.005)\cdot(0.8)+(0.995)\cdot(0.05)}=\frac{0.004}{0.05375}\approx 0.074$$

(b) $$P(A'|B')=\frac{P(A')\cdot P(B'|A')}{P(A')\cdot P(B'|A')+P(A)\cdot P(B'|A)}$$
$$=\frac{(0.995)\cdot(0.95)}{(0.995)\cdot(0.95)+(0.005)\cdot(0.2)}=\frac{0.94525}{0.94625}\approx 0.999$$

41. Let A = {flight departs on time} and B = {flight arrives on time}.

$$P(A|B)=\frac{P(A \text{ and } B)}{P(B)}=\frac{0.83}{0.87}\approx 0.954$$

3.3 THE ADDITION RULE

3.3 TRY IT YOURSELF SOLUTIONS

1 (1) Not mutually exclusive; The events can occur at the same time.

(2) Mutually exclusive; The events cannot occur at the same time.

2. (1) Mutually exclusive
Let $A = \{6\}$ and $B = \{\text{odd}\}$.

$$P(A) = \frac{1}{6} \text{ and } P(B) = \frac{3}{6} = \frac{1}{2}$$

$$P(A \text{ or } B) = P(A) + P(B) = \frac{1}{6} + \frac{1}{2} \approx 0.667$$

(2) Not mutually exclusive
Let $A = \{\text{face card}\}$ and $B = \{\text{heart}\}$.

$$P(A) = \frac{12}{52},\ P(B) = \frac{13}{52}, \text{ and } P(A \text{ and } B) = \frac{3}{52}$$

$$P(A \text{ or } B) = P(A) + P(B) - P(A \text{ and } B) = \frac{12}{52} + \frac{13}{52} - \frac{3}{52} \approx 0.423$$

3. Let $A = \{\text{sales between \$0 and \$24,999}\}$
Let $B = \{\text{sales between \$25,000 and \$49,999}\}$.
A and B cannot occur at the same time. So A and B are mutually exclusive.

$$P(A) = \frac{3}{36} \text{ and } P(B) = \frac{5}{36}$$

$$P(A \text{ or } B) = P(A) + P(B) = \frac{3}{36} + \frac{5}{36} \approx 0.222$$

4. (1) Let $A = \{\text{type B}\}$ and $B = \{\text{type AB}\}$.
A and B cannot occur at the same time. So, A and B are mutually exclusive.

$$P(A) = \frac{45}{409} \text{ and } P(B) = \frac{16}{409}$$

$$P(A \text{ or } B) = P(A) + P(B) = \frac{45}{409} + \frac{16}{409} \approx 0.149$$

(2) If a donor does not have type O or type A blood, then the donor must have type B or type AB blood. See part (1). The probability is about 0.149.

(3) Let $A = \{\text{type O}\}$ and $B = \{\text{Rh-positive}\}$.
A and B can occur at the same time. So, A and B are not mutually exclusive.

$$P(A) = \frac{184}{409},\ P(B) = \frac{344}{409}, \text{ and } P(A \text{ and } B) = \frac{156}{409}$$

$$P(A \text{ or } B) = P(A) + P(B) - P(A \text{ and } B) = \frac{184}{409} + \frac{344}{409} - \frac{156}{409} \approx 0.910$$

(4) Let A = {type A} and B = {Rh-negative}
A and B can occur at the same time. So, A and B are not mutually exclusive.
$P(A)=\frac{164}{409}$, $P(B)=\frac{65}{409}$, and $P(A \text{ and } B)=\frac{25}{409}$
$P(A \text{ or } B)=P(A)+P(B)-P(A \text{ and } B)=\frac{164}{409}+\frac{65}{409}-\frac{25}{409}\approx 0.499$

5. Let A = {linebacker} and B = {quarterback}.
$P(A \text{ or } B)=P(A)+P(B)=\frac{34}{253}+\frac{15}{253}\approx 0.194$
P(not a linebacker or quarterback) = 1 – $P(A \text{ or } B) \approx 1-0.194=0.806$

3.3 EXERCISE SOLUTIONS

1. $P(A \text{ and } B) = 0$ because A and B cannot occur at the same time.

3. True

5. False. The probability that event A or event B will occur is $P(A \text{ or } B)=P(A)+P(B)-P(A \text{ and } B)$.

7. Not mutually exclusive. A presidential candidate can lose the popular vote and win the election.

9. Not mutually exclusive. A male psychology major can be 20 years old.

11. Mutually exclusive. A person cannot be both a Republican and a Democrat.

13. Let A = {physics major} and B = {female}.
$P(A)=\frac{12}{40}$, $P(B)=\frac{16}{40}$, and $P(B \mid A)=\frac{3}{12}$.
Thus, $P(A \text{ and } B)=P(A)\cdot P(B \mid A)=\frac{12}{40}\cdot\frac{3}{12}=\frac{3}{40}$.
$P(A \text{ or } B)=P(A)+P(B)-P(A \text{ and } B)=\frac{12}{40}+\frac{16}{40}-\frac{3}{40}=\frac{25}{40}=0.625$

15. Let A = {puncture} and B = {smashed corner}.
$P(A)=0.05$, $P(B)=0.08$, and $P(A \text{ and } B)=0.004$.
$P(A \text{ or } B)=P(A)+P(B)-P(A \text{ and } B)=0.05+0.08-0.004=0.126$

17. (a) $P(\text{club or } 3)=P(\text{club})+P(3)-P(\text{club and } 3)=\frac{13}{52}+\frac{4}{52}-\frac{1}{52}\approx 0.308$

(b) $P(\text{red or king})=P(\text{red})+P(\text{king})-P(\text{red and king})=\frac{26}{52}+\frac{4}{52}-\frac{2}{52}\approx 0.538$

(c) $P(9 \text{ or face card})=P(9)+P(\text{face card})-P(9 \text{ and face card})=\frac{4}{52}+\frac{12}{52}-0\approx 0.308$

19. (a) $P(\text{under } 5) = 0.06$

(b) $P(45+) = P(45-64) + P(65-74) + P(75+) = 0.236 + 0.107 + 0.083 = 0.426$

(c) $P(\text{not } 65+) = 1 - P(65+) = 1 - P(65-74) - P(75+) = 1 - 0.107 - 0.083 = 0.81$

(d) $P(\text{between 20 and 34}) = P(20-24) + P(25-34) = 0.064 + 0.137 = 0.201$

21. (a) $P(\text{not A or B}) = 1 - P(\text{A or B}) = 1 - \frac{276}{1254} = \frac{978}{1254} \approx 0.780$

(b) $P(\text{better than D}) = P(\text{A or B}) + P(\text{C}) = \frac{276}{1254} + \frac{238}{1254} = \frac{514}{1254} \approx 0.410$

(c) $P(\text{D or F}) = P(\text{D}) + P(\text{F}) = \frac{263}{1254} + \frac{477}{1254} = \frac{740}{1254} \approx 0.590$

(d) $P(\text{C or D}) = P(\text{C}) + P(\text{D}) = \frac{238}{1254} + \frac{263}{1254} = \frac{501}{1254} \approx 0.400$

23. Let $A = \{\text{male}\}$; $B = \{\text{business major}\}$

(a) $P(A \text{ or } B) = P(A) + P(B) - P(A \text{ and } B) = \frac{812,669}{1,894,934} + \frac{363,799}{1,894,934} - \frac{191,310}{1,894,934} \approx 0.520$

(b) $P(A' \text{ or } B') = P(A') + P(B') - P(A' \text{ and } B') = \frac{1,082,265}{1,894,934} + \frac{1,531,135}{1,894,934} - \frac{909,776}{1,894,934} \approx 0.899$

(c) $P(A \text{ or } B') = P(A) + P(B') - P(A \text{ and } B') = \frac{812,669}{1,894,934} + \frac{1,531,135}{1,894,934} - \frac{621,359}{1,894,934} \approx 0.909$

25. $A = \{\text{frequently}\}$; $B = \{\text{occasionally}\}$; $C = \{\text{not at all}\}$; $D = \{\text{male}\}$

(a) $P(D \text{ or } A) = P(D) + P(A) - P(D \text{ and } A) = \frac{1472}{2850} + \frac{428}{2850} - \frac{221}{2850} \approx 0.589$

(b) $P(D' \text{ or } C) = P(D') + P(C) - P(D' and C) = \frac{1378}{2850} + \frac{1536}{2850} - \frac{741}{2850} \approx 0.762$

(c) $P(A \text{ or } B) = \frac{428}{2850} + \frac{886}{2850} \approx 0.461$

(d) $P(D' \text{ or } A') = P(D') + P(A') - P(D' \text{ and } A') = \frac{1378}{2850} + \frac{886 + 1536}{2850} - \frac{430 + 741}{2850} \approx 0.922$

27. $P(A \text{ or } B \text{ or } C) = P(A) + P(B) + P(C) - P(A \text{ and } B) - P(A \text{ and } C)$

$$- P(B \text{ and } C) + P(A \text{ and } B \text{ and } C)$$

$$= 0.40 + 0.10 + 0.50 - 0.05 - 0.25 - 0.10 + 0.03 = 0.63$$

29. If events A, B and C are not mutually exclusive, $P(A \text{ and } B \text{ and } C)$ must be added because $P(A) + P(B) + P(C)$ counts the intersection of all three events three times and $-P(A \text{ and } B) - P(A \text{ and } C) - P(B \text{ and } C)$ subtracts the intersection of all three events three times. So, if $P(A \text{ and } B \text{ and } C)$ is not added at the end, it will not be counted.

3.4 ADDITIONAL TOPICS IN PROBABILITY AND COUNTING

3.4 TRY IT YOURSELF SOLUTIONS

1. $n = 10$ schools
The number of permutations is $10! = 3{,}628{,}800$.

2. $n = 8, r = 3$

$$\frac{8!}{(8-3)!} = \frac{8!}{5!} = \frac{8 \cdot 7 \cdot 6 \cdot 5 \cdot 4 \cdot 3 \cdot 2 \cdot 1}{5 \cdot 4 \cdot 3 \cdot 2 \cdot 1} = 8 \cdot 7 \cdot 6 = 336$$

There are 336 possible ways that the subject can pick a first, second, and third activity.

3. $n = 12, r = 4$

$${}_{12}P_4 = \frac{12!}{(12-4)!} = \frac{12!}{8!} = 12 \cdot 11 \cdot 10 \cdot 9 = 11{,}880$$

4. $$\frac{n!}{n_1! \cdot n_2! \cdot n_3!} = \frac{20!}{6! \cdot 9! \cdot 5!} = 77{,}597{,}520$$

5. $${}_{20}C_3 = \frac{20!}{(20-3)!3!} = \frac{20!}{17!3!} = \frac{20 \cdot 19 \cdot 18 \cdot 17!}{17! \cdot 3!} = 1140$$

6. $${}_{20}P_2 = \frac{20!}{(20-2)!} = \frac{20!}{18!} = \frac{20 \cdot 19 \cdot 18!}{18!} = 20 \cdot 19 = 380$$

$$P(\text{selecting the two members}) = \frac{1}{380} \approx 0.003$$

7. $${}_{15}C_5 = \frac{15!}{5!(15-5)!} = 3003$$

$${}_{54}C_5 = \frac{54!}{5!(54-5)!} = 3{,}162{,}510$$

$$P(5 \text{ diamonds from deck with 2 jokers}) = \frac{{}_{15}C_5}{{}_{54}C_5} == \frac{3003}{3{,}162{,}510} \approx 0.0009$$

8. $_5C_3 \cdot {_7C_0} = \dfrac{5!}{3!2!} \cdot \dfrac{7!}{0!7!} = 10 \cdot 1 = 10$

$_{12}C_3 = \dfrac{12!}{3!9!} = 220$

$P(\text{all three are men}) = \dfrac{10}{220} \approx 0.045$

3.4 EXERCISE SOLUTIONS

1. The number of ordered arrangements of n objects taken r at a time.
Sample answer: An example of a permutation is the number of seating arrangements of you and three friends.

3. False. A permutation is an ordered arrangement of objects.

5. True

7. $_9P_5 = \dfrac{9!}{(9-5)!} = \dfrac{9!}{4!} = 9 \cdot 8 \cdot 7 \cdot 6 \cdot 5 = 15{,}120$

9. $_8C_3 = \dfrac{8!}{(8-3)!3!} = \dfrac{8!}{5!3!} = \dfrac{8 \cdot 7 \cdot 6 \cdot 5!}{5! \cdot 3!} = 56$

11. $\dfrac{_8C_4}{_{12}C_6} = \dfrac{\dfrac{8!}{(8-4)!4!}}{\dfrac{12!}{(12-6)!6!}} = \dfrac{\dfrac{8!}{4!4!}}{\dfrac{12!}{6!6!}} = \dfrac{70}{924} \approx 0.076$

13. $\dfrac{_3P_2}{_{13}P_1} = \dfrac{\dfrac{3!}{(3-2)!}}{\dfrac{13!}{(13-1)!}} = \dfrac{\dfrac{3!}{1!}}{\dfrac{13!}{12!}} = \dfrac{3 \cdot 2}{13} \approx 0.462$

15. Permutation. The order of the 16 floats in line matters.

17. Combination. The order does not matter because the position of one captain is the same as the other.

19. $7! = 5040$

21. $6! = 720$

23. $_{50}P_3 = \dfrac{50!}{(50-3)!} = \dfrac{50!}{47!} = 117{,}600$

25. $_{24}P_6 = \dfrac{24!}{(24-6)!} = \dfrac{24!}{18!} = 96{,}909{,}120$

27. $\dfrac{17!}{8!6!3!} = 2,042,040$

29. 3 *S*'s, 3 *T*'s, 1 *A*, 2 *I*'s, 1 *C*

$$\frac{10!}{3!3!1!2!1!} = 50,400$$

31. ${}_{20}C_4 = 4845$

33. ${}_{40}C_3 = \dfrac{40!}{(40-3)!3!} = 9880$

35. $10 \cdot 8 \cdot {}_{13}C_2 = 10 \cdot 8 \cdot \dfrac{13!}{(13-2)!2!} = 6240$

37. ${}_5C_1 \cdot {}_{75}C_5 = \dfrac{5!}{1!4!} \cdot \dfrac{75!}{70!5!} = 86,296,950$

39. $\dfrac{1}{{}_{15}P_2} = \dfrac{1}{210} \approx 0.005$

41. $\dfrac{1}{{}_{12}C_3} = \dfrac{1}{220} \approx 0.005$

43. (a) $\dfrac{{}_{15}C_3}{{}_{56}C_3} = \dfrac{455}{27,720} \approx 0.016$

(b) $\dfrac{{}_{41}C_3}{{}_{56}C_3} = \dfrac{10,660}{27,720} \approx 0.385$

45. $(3\%)(1500) = (0.03)(1500) = 45$ of the 1500 say most have food allergies.

$P(\text{2 say most}) = \dfrac{{}_{45}C_2}{{}_{1500}C_2} = \dfrac{990}{1,124,250} \approx 0.0009$

47. $(21\%)(1500) = (0.21)(1500) = 315$ of the 1500 say some have food allergies. $\Rightarrow 1500 - 315 = 1185$ do not say some have food allergies.

$P(\text{none of the 6 say some}) = \dfrac{{}_{1185}C_6}{{}_{1500}C_6} \approx 0.242$

49. ${}_{40}C_5 = 658,008$, $P(\text{win}) = \dfrac{1}{658,008} \approx 0.0000015$

51. $\dfrac{({}_{24}C_5)({}_{30}C_3)}{{}_{54}C_8} = \dfrac{(42,504)(4060)}{1,040,465,790} \approx 0.166$

53. $\dfrac{({}_{13}C_4)({}_{41}C_4)}{{}_{54}C_8} = \dfrac{(715)(101,270)}{1,040,465,790} \approx 0.070$

55. $\dfrac{{}_8C_3}{{}_{10}C_3} + \dfrac{({}_8C_2)({}_2C_1)}{{}_{10}C_3} = \dfrac{56}{120} + \dfrac{(28)(2)}{120} = \dfrac{112}{120} \approx 0.933$

57. ${}_4C_2\left(\dfrac{({}_2C_2)({}_2C_2)({}_2C_0)({}_2C_0)}{{}_8C_4}\right) = 6 \cdot \dfrac{(1)(1)(1)(1)}{70} \approx 0.086$

59. $\dfrac{\left({}_{13}C_2\right)\left({}_{13}C_1\right)\left({}_{13}C_1\right)\left({}_{13}C_1\right)}{{}_{52}C_5} = \dfrac{(78)(13)(13)(13)}{2{,}598{,}960} \approx 0.066$

61. $\left({}_{13}C_2\right)\left(\dfrac{\left({}_{4}C_3\right)\left({}_{4}C_2\right)}{{}_{52}C_5}\right) = 78\left(\dfrac{(4)(6)}{2{,}598{,}960}\right) \approx 0.001$

CHAPTER 3 REVIEW EXERCISE SOLUTIONS

1. Sample space:
{HHHH, HHHT, HHTH, HHTT, HTHH, HTHT, HTTH, HTTT, THHH, THHT, THTH, HTT, TTHH, TTHT, TTTH, TTTT}

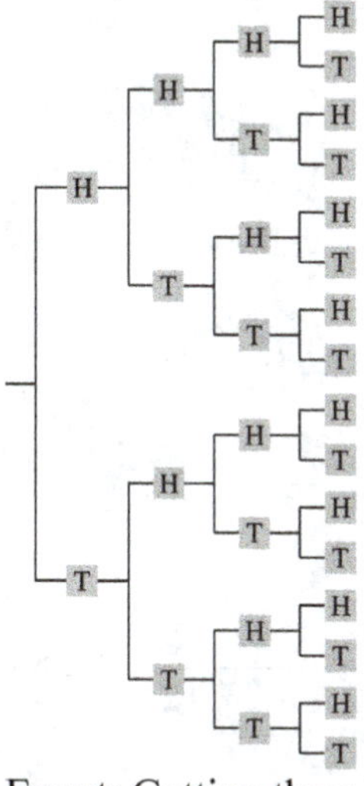

Event: Getting three heads
{HHHT, HHTH, HTHH, THHH}
There are 4 outcomes.

3. Sample space: {January, February, March, April, May, June, July, August, September, October, November, December}
Event: {January, June, July}
There are 3 outcomes.

5. $(7)(4)(3) = 84$

7. Empirical probability because prior counts were used to calculate the frequency of a part being defective.

9. Subjective probability because it is based on opinion.

11. Classical probability because all the outcomes in the event and the sample space can be counted.

13. $P(\text{business or psychology}) = P(\text{business}) + P(\text{psychology}) = \dfrac{361}{1840} + \dfrac{114}{1840} \approx 0.258$

15. $\dfrac{1}{(8)(10)(10)(10)(10)(10)(10)} = 1.25 \times 10^{-7}$

17. $P(\text{first}|\text{failed}) = \dfrac{13,194}{23,775} \approx 0.555$

19. Independent. The outcomes of the first four coin tosses do not affect the outcome of the fifth coin toss.

21. Dependent. The outcome of taking a driver's education course affects the outcome of passing the driver's license exam.

23. $P(\text{correct toothpaste and correct dental rinse}) = P(\text{correct toothpaste}) \cdot P(\text{correct dental rinse})$

$$= P(\text{correct toothpaste})P(\text{correct dental rinse}) = \frac{1}{8} \cdot \frac{1}{5} = \frac{1}{40} = 0.025$$

The event is unusual because its probability is less than or equal to 0.05.

25. Mutually exclusive. A jelly bean cannot be both completely red and completely yellow.

27. $P(\text{home or work}) = P(\text{work}) + P(\text{home}) - P(\text{home and work}) = 0.74 + 0.88 - 0.72 = 0.9$

29. $P(4-8 \text{ or club}) = P(4-8) + P(\text{club}) - P(4-8 \text{ and club}) = \dfrac{20}{52} + \dfrac{13}{52} - \dfrac{5}{52} \approx 0.538$

31. $P(\text{odd or less than } 4) = P(\text{odd}) + P(\text{less than } 4) - P(\text{odd and less than } 4) = \dfrac{6}{12} + \dfrac{3}{12} - \dfrac{2}{12} \approx 0.583$

33. $P(\text{fewer than } 500) = P(300-499) + P(\text{fewer than } 300) = 0.274 + 0.302 = 0.576$

35. $P(\text{Middle class or Upper-middle class}) = \dfrac{1329}{3078} + \dfrac{468}{3078} \approx 0.584$

37. $P(\text{not Middle class}) = 1 - P(\text{Middle class}) = 1 - \dfrac{1329}{3078} \approx 0.568$

39. No. You do not know whether events *A* and *B* are mutually exclusive.

41. ${}_{11}P_2 = \dfrac{11!}{(11-2)!} = \dfrac{11!}{9!} = 11 \cdot 10 = 110$

43. ${}_7C_4 = \dfrac{7!}{(7-4)!4!} = \dfrac{7!}{3!4!} = 35$

45. Order is important: ${}_{15}P_3 = \dfrac{15!}{(15-3)!} = \dfrac{15!}{12!} = 2730$

47. Order is not important: $_{17}C_4 = \dfrac{17!}{(17-4)!4!} = 2380$

49. $P(\text{3 kings and 2 queens}) = \dfrac{({}_4C_3)\cdot({}_4C_2)}{{}_{52}C_5} = \dfrac{4\cdot 6}{2{,}598{,}960} \approx 0.000009$

51. (a) $P(\text{no defective}) = \dfrac{{}_{197}C_3}{{}_{200}C_3} = \dfrac{1{,}254{,}890}{1{,}313{,}400} \approx 0.955$

(b) $P(\text{all defective}) = \dfrac{{}_3C_3}{{}_{200}C_3} = \dfrac{1}{1{,}313{,}400} \approx 0.0000008$

(c) $P(\text{at least one defective}) = 1 - P(\text{no defective}) \approx 1 - 0.955 = 0.045$

(d) $P(\text{at least one non-defective}) = 1 - P(\text{all defective}) \approx 1 - 0.0000008 \approx 0.9999992$

53. (a) $P(\text{4 men}) = \dfrac{{}_6C_4}{{}_{10}C_4} = \dfrac{15}{210} \approx 0.071$

(b) $P(\text{4 women}) = \dfrac{{}_4C_4}{{}_{10}C_4} = \dfrac{1}{210} \approx 0.005$

(c) $P(\text{2 men and 2 women}) = \dfrac{({}_6C_2)\cdot({}_4C_2)}{{}_{10}C_4} = \dfrac{15\cdot 6}{210} \approx 0.429$

(d) $P(\text{1 man and 3 women}) = \dfrac{({}_6C_1)\cdot({}_4C_3)}{{}_{10}C_4} = \dfrac{6\cdot 4}{210} \approx 0.114$

CHAPTER 3 QUIZ SOLUTIONS

1. $(9)(10)(10)(10)(10)(5) = 450{,}000$

2. (a) $P(\text{bachelor's degree}) = \dfrac{319.2}{447.4} \approx 0.713$

(b) $P(\text{bachelor's degree} \mid \text{computer science/engineering}) = \dfrac{164.3}{248.3} \approx 0.662$

(c) $P(\text{bachelor's degree} \mid \text{not computer science/engineering}) = \dfrac{154.9}{199.1} \approx 0.778$

(d) $P(\text{bachelor's degree or master's degree}) = P(\text{bachelor's degree}) + P(\text{master's degree})$

$$= \frac{319.2}{447.4} + \frac{100.1}{447.4} \approx 0.937$$

(e) $P(\text{doctorate degree} \mid \text{computer science/engineering}) = \frac{12.1}{248.3} \approx 0.049$

(f) $P(\text{master's or degree in natural sciences/mathematics})$

$= P(\text{master's}) + P(\text{degree in natural sciences/mathematics})$

$- P(\text{master's and degree in natural sciences/mathematics})$

$$= \frac{100.1}{447.4} + \frac{199.1}{447.4} - \frac{28.2}{447.4} = \frac{271}{447.4} \approx 0.606$$

(g) $P(\text{bachelor's and degree in natural sciences/mathematics}) = \frac{154.9}{447.4} \approx 0.346$

(h) $P(\text{degree in computer science/engineering} \mid \text{bachelor's degree}) = \frac{164.3}{319.2} \approx 0.515$

3. The event in part (e) is unusual because its probability is less than or equal to 0.05.

4. Not mutually exclusive. A bowler can have the highest game in a 40-game tournament and still lose the tournament.
Dependent. One event can affect the occurrence of the second event.

5. $_{30}P_4 = \frac{30!}{(30-4)!} = 657{,}720$

6. (a) $_{247}C_3 = \frac{247!}{(247-3)!3!} = 2{,}481{,}115$

(b) $_3C_3 = \frac{3!}{(3-3)!3!} = 1$

(c) $(_{250}C_3) - (_3C_3) = \frac{250!}{(250-3)!3!} - 1 = 2{,}573{,}000 - 1 = 2{,}572{,}999$

7. (a) $\frac{_{247}C_3}{_{250}C_3} = \frac{2{,}481{,}115}{2{,}573{,}000} \approx 0.964$

(b) $\frac{_3C_3}{_{250}C_3} = \frac{1}{2{,}573{,}000} \approx .0000004$

(c) $\frac{_{250}C_3 - {_3C_3}}{_{250}C_3} = \frac{2{,}572{,}999}{2{,}573{,}000} \approx 0.9999996$

Discrete Probability Distributions

4.1 PROBABILITY DISTRIBUTIONS

4.1 TRY IT YOURSELF SOLUTIONS

1. (1) The random variable is continuous because x can be any speed up to the maximum speed of a rocket.

(2) The random variable is discrete because the number of calves born on a farm in one year is countable.

(3) The random variable is discrete because the number of days of rain for the next three days is countable.

2.

x	f	$P(x)$
0	16	0.16
1	19	0.19
2	15	0.15
3	21	0.21
4	9	0.09
5	10	0.10
6	8	0.08
7	2	0.02
	$n = 100$	$\sum P(x) = 1$

3. Each $P(x)$ is between 0 and 1.
$\sum P(x) = 1$
Because both conditions are met, the distribution is a probability distribution.

4. (1) Probability distribution; The probability of each outcome is between 0 and 1, and the sum of all the probabilities is 1.

(2) Not a probability distribution; The sum of all the probabilities is not 1.

5

x	$P(x)$	$xP(x)$
0	0.16	$(0)(0.16) = 0.00$
1	0.19	$(1)(0.19) = 0.19$
2	0.15	$(2)(0.15) = 0.30$
3	0.21	$(3)(0.21) = 0.63$
4	0.09	$(4)(0.09) = 0.36$
5	0.10	$(5)(0.10) = 0.50$
6	0.08	$(6)(0.08) = 0.48$
7	0.02	$(7)(0.02) = 0.14$
	$\sum P(x) = 1$	$\sum xP(x) = 2.60$

$\mu = \sum xP(x) = 2.6$

On average, a new employee makes 2.6 sales per day.

6. From 5, $\mu = 2.6$,

x	$P(x)$	$x - \mu$	$(x - \mu)^2$	$P(x)(x - \mu)^2$
0	0.16	−2.6	6.76	(0.16)(6.76) = 1.0816
1	0.19	−1.6	2.56	(0.19)(2.56) = 0.4864
2	0.15	−0.6	0.36	(0.15)(0.36) = 0.0540
3	0.21	0.4	0.16	(0.21)(0.16) = 0.0336
4	0.09	1.4	1.96	(0.09)(1.96) = 0.1764
5	0.10	2.4	5.76	(0.10)(5.76) = 0.5760
6	0.08	3.4	11.56	(0.08)(11.56) = 0.9248
7	0.02	4.4	19.36	(0.02)(19.36) = 0.3872
	$\sum P(x) = 1$			$\sum P(x)(x - \mu)^2 = 3.72$

$\sigma^2 \approx 3.7$

$\sigma = \sqrt{\sigma^2} \approx \sqrt{3.7} \approx 1.9$

Most of the data values differ from the mean by no more than 1.9 sales per day.

7.

Gain, x	$P(x)$	$xP(x)$
\$1995	$\frac{1}{2000}$	$\frac{1995}{2000}$
\$ 995	$\frac{1}{2000}$	$\frac{995}{2000}$
\$ 495	$\frac{1}{2000}$	$\frac{495}{2000}$
\$ 245	$\frac{1}{2000}$	$\frac{245}{2000}$
\$ 95	$\frac{1}{2000}$	$\frac{95}{2000}$
\$ −5	$\frac{1995}{2000}$	$-\frac{9975}{2000}$
	$\sum P(x) = 1$	$\sum xP(x) \approx -3.08$

$E(x) = \sum xP(x) \approx -\$3.08$

Because the expected value is negative, you can expect to lose an average of \$3.08 for each ticket you buy.

4.1 EXERCISE SOLUTIONS

1. A random variable represents a numerical value associated with each outcome of a probability experiment. Examples: Answers will vary.

3. No; The expected value may not be a possible value of x for one trial, but it represents the average value of x over a large number of trials.

5. False. In most applications, discrete random variables represent counted data, while continuous random variables represent measured data.

7. False. The mean of the random variable of a probability distribution describes a typical outcome. The variance and standard deviation of the random variable of a probability distribution describe how the outcomes vary.

9. Discrete; Attendance is a random variable that is countable.

11. Continuous; The distance a baseball travels after being hit is a random variable that must be measured.

13. Discrete; The number of cars in a university parking lot is a random variable that is countable.

15. Continuous; The volume of blood drawn for a blood test is a random variable that must be measured.

17. Discrete; The number of texts a student sends in one day is a random variable that is countable.

19. (a)

x	f	$P(x)$
0	26	0.01
1	442	0.17
2	728	0.28
3	1404	0.54
	$n = 2600$	$\sum P(x) = 1$

(b)

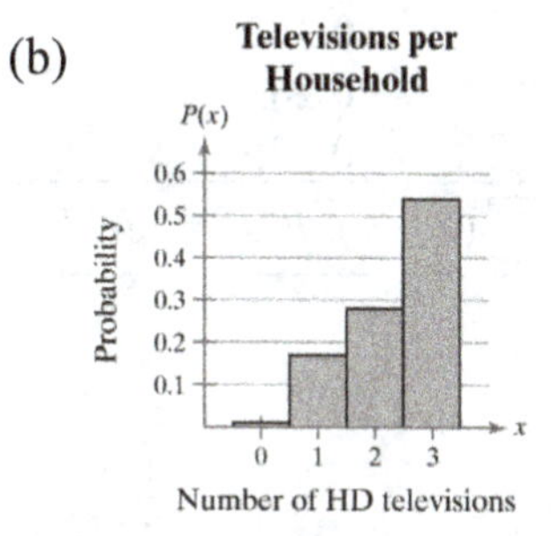

Skewed left

21. (a) $P(1 \text{ or } 2) = 0.17 + 0.28 = 0.45$

(b) $P(2 \text{ or more}) = 0.28 + 0.54 = 0.82$

(c) $P(1 \le x \le 3) = 0.17 + 0.28 + 0.54 = 0.99$

(d) $P(x \le 2) = 1 - P(x = 3) = 1 - 0.54 = 0.46$

23. $P(0) = 0.01$. Yes, because the probability is less than 0.05.

25. $\sum P(x) = 1 \rightarrow P(3) = 1 - (0.06 + 0.12 + 0.18 + 0.30) = 0.34$

27. Yes, because each $P(x)$ is between 0 and 1, and $\sum P(x) = 1$, the distribution is a probability distribution.

29. (a)

x	$P(x)$	$xP(x)$	$(x-\mu)$	$(x-\mu)^2$	$(x-\mu)^2 P(x)$
0	0.686	0	−0.497	0.247	0.169
1	0.195	0.195	0.503	0.253	0.049
2	0.077	0.154	1.503	2.259	0.174
3	0.022	0.066	2.503	6.265	0.138
4	0.013	0.052	3.503	12.271	0.160
5	0.006	0.030	4.503	20.277	0.122
	$\sum P(x) \approx 1$	$\sum xP(x) = 0.497$			$\sum (x-\mu)^2 P(x) = 0.812$

$\mu = \sum xP(x) = 0.497 \approx 0.5$

$\sigma^2 = \sum (x-\mu)^2 P(x) = 0.812 \approx 0.8$

$\sigma = \sqrt{\sigma^2} = \sqrt{0.812} \approx 0.9$

(b) The mean is 0.5, so the average number of dogs per household is about 0 or 1 dog. The standard deviation is 0.9, so most of the households differ from the mean by no more than 1 dog.

31. (a)

x	$P(x)$	$xP(x)$	$(x-\mu)$	$(x-\mu)^2$	$(x-\mu)^2 P(x)$
0	0.263	0	−1.467	2.152	0.566
1	0.285	0.285	−0.467	0.218	0.062
2	0.243	0.486	0.533	0.284	0.069
3	0.154	0.462	1.533	2.350	0.362
4	0.041	0.164	2.533	6.416	0.263
5	0.014	0.070	3.533	12.482	0.175
		$\sum xP(x) = 1.467$			$\sum (x-\mu)^2 P(x) = 1.497$

$\mu = \sum xP(x) \approx 1.5$

$\sigma^2 = \sum (x-\mu)^2 P(x) = 1.497 \approx 1.5$

$\sigma = \sqrt{\sigma^2} = \sqrt{1.497} \approx 1.2$

(b) The mean is 1.5, so the average batch of 1000 machine parts has about 1 or 2 defects. The standard deviation is 1.2, so most of the batches of 1000 differ from the mean by no more than about 1 defect.

33. (a)

x	$P(x)$	$xP(x)$	$(x-\mu)$	$(x-\mu)^2$	$(x-\mu)^2 P(x)$
1	0.411	0.411	−0.996	0.992	0.408
2	0.279	0.558	0.004	0.000	0.000
3	0.223	0.669	1.004	1.008	0.225
4	0.077	0.308	2.004	4.016	0.309
5	0.010	0.050	3.004	9.024	0.090
		$\sum xP(x) = 1.996$			$\sum (x-\mu)^2 P(x) = 1.032$

$$\mu = \sum xP(x) = 1.996 \approx 2.0$$
$$\sigma^2 = \sum (x-\mu)^2 P(x) = 1.032 \approx 1.0$$
$$\sigma = \sqrt{\sigma^2} = \sqrt{1.032} \approx 1.0$$

(b) The mean is 2.0, so an average hurricane that hits the U.S. mainland is expected to be a category 2 hurricane. The standard deviation is 1.0, so most of the hurricanes differ from the mean by no more than 1 category level.

35. An expected value of 0 means that the money gained is equal to the money spent, representing the breakeven point.

37. $E(x) = \mu = \sum xP(x) = (-\$1)\cdot\left(\frac{37}{38}\right) + (\$35)\cdot\left(\frac{1}{38}\right) \approx -\$0.05$

39. $\mu_y = a + b\mu_x = 600 + 1.03(46{,}000) = \$47{,}980$

41. $\mu_{x+y} = \mu_x + \mu_y = 524 + 494 = 1018$

$\mu_{x-y} = \mu_x - \mu_y = 524 - 494 = 30$

4.2 BINOMIAL DISTRIBUTIONS

4.2 TRY IT YOURSELF SOLUTIONS

1. Trial: answering a question
Success: the question answered correctly
Yes, the experiment satisfies the four conditions of a binomial experiment.
It is a binomial experiment.
$n = 10,\ p = 0.25,\ q = 0.75,\ x = 0,1,2,3,4,5,6,7,8,9,10$

2. Trial: drawing a card with replacement
Success: card drawn is a club
Failure: card drawn is not a club
$n = 5,\ p = 0.25,\ q = 0.75,\ x = 3$

$$P(3) = \frac{5!}{(5-3)!3!}(0.25)^3(0.75)^2 \approx 0.088$$

3. Trial: Selecting an adult and asking a question
Success: selecting an adult who uses the social media platform Instagram
Failure: selecting an adult who does not use the social media platform Instagram
$n = 5,\ p = 0.28,\ q = 0.72,\ x = 0,1,2,3,4,5$

$$P(0) = {}_5C_0(0.28)^0(0.72)^5 \approx 0.193$$
$$P(1) = {}_5C_1(0.28)^1(0.72)^4 \approx 0.376$$

$P(2) = {}_5C_2(0.28)^2(0.72)^3 \approx 0.293$
$P(3) = {}_5C_3(0.28)^3(0.72)^2 \approx 0.114$
$P(4) = {}_5C_4(0.28)^4(0.72)^1 \approx 0.022$
$P(5) = {}_5C_5(0.28)^5(0.72)^0 \approx 0.002$

x	$P(x)$
0	0.193
1	0.376
2	0.293
3	0.114
4	0.022
5	0.002
	$\sum xP(x) = 1$

4. $n = 150,\ p = 0.52,\ x = 65$
$P(65) \approx 0.007$

5. (1) $P(2) = {}_5C_2(0.27)^2(0.73)^3 \approx 0.284$

(2) $P(2) = {}_5C_2(0.27)^2(0.73)^3 \approx 0.284$; $P(3) = {}_5C_3(0.27)^3(0.73)^2 \approx 0.105$
$P(4) = {}_5C_4(0.27)^4(0.73)^1 \approx 0.019$; $P(5) = {}_5C_5(0.27)^5(0.73)^0 \approx 0.001$

$P(x \geq 2) = P(2) + P(3) + P(4) + P(5) = 0.284 + 0.105 + 0.019 + 0.001 = 0.409$

(3) $P(0) = {}_5C_0(0.27)^0(0.73)^5 \approx 0.2073$; $P(1) = {}_5C_1(0.27)^1(0.73)^4 \approx 0.3834$;
$P(x < 2) = P(0) + P(1) \approx 0.2073 + 0.3834 \approx 0.591$

6. $n = 6,\ p = 0.05,\ x = 2$
From Table 2 in Appendix B, the probability is 0.031.

7. $P(0) = {}_4C_0(0.28)^0(0.72)^4 \approx 0.269$
$P(1) = {}_4C_1(0.28)^1(0.72)^3 \approx 0.418$
$P(2) = {}_4C_2(0.28)^2(0.72)^2 \approx 0.244$
$P(3) = {}_4C_3(0.28)^3(0.72)^1 \approx 0.063$
$P(4) = {}_4C_4(0.28)^4(0.72)^0 \approx 0.006$

x	$P(x)$
0	0.269
1	0.418
2	0.244
3	0.063
4	0.006

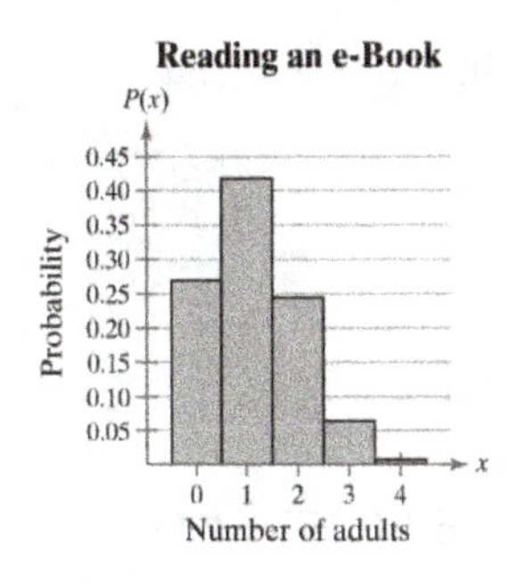

8. $n = 31, p = 0.44, q = 0.56$
$\mu = np = (31)(0.44) \approx 13.6$
$\sigma^2 = npq = (31)(0.44)(0.56) \approx 7.6$
$\sigma = \sqrt{npq} = \sqrt{(31)(0.44)(0.56)} \approx 2.8$
On average, there are about 14 clear days during the month of May. The standard deviation is about 3 days.
Values that are more than 2 standard deviations from the mean are considered unusual. Because 13.6 − 2(2.8) = 8 and 13.6 + 2(2.8) = 19.2, a May with fewer than 8 clear days, or more than 19 clear days would be unusual.

4.2 EXERCISE SOLUTIONS

1. Each trial is independent of the other trials when the outcome of one trial does not affect the outcome of any of the other trials.

3. c; Because the probability is greater than 0.5, the distribution is skewed left.

5. a; Because the probability is less than 0.5, the distribution is skewed right.

7. a; The histogram shows probabilities for 4 trials.

9. Unusual values have probabilities that are less than 0.05.
(3) $x = 0, 1$ (4) $x = 0, 5$ (5) $x = 4, 5$

11. $\mu = np = (50)(0.4) = 20$
$\sigma^2 = npq = (50)(0.4)(0.6) = 12$
$\sigma = \sqrt{npq} = \sqrt{(50)(0.4)(0.6)} \approx 3.5$

13. $\mu = np = (124)(0.26) \approx 32.2$
$\sigma^2 = npq = (124)(0.26)(0.74) \approx 23.9$
$\sigma = \sqrt{npq} = \sqrt{(124)(0.26)(0.74)} \approx 4.9$.

15. It is a binomial experiment.
Success: frequent gamer who plays video games on smartphone
$n = 10$, $p = 0.36$, $q = 0.64$, $x = 0, 1, 2, 3, 4, 5, 6, 7, 8, 9, 10$

17. Not a binomial experiment because the probability of a success is not the same for each trial.

19. $n = 8,\ p = 0.34$

(a) $P(6) \approx 0.019$

(b) $P(x \geq 4) = P(4) + P(5) + P(6) + P(7) + P(8) \approx 0.177 + 0.073 + 0.019 + 0.003 + 0.000 \approx 0.272$

(c) $P(x < 5) = P(0) + P(1) + P(2) + P(3) + P(4) \approx 0.036 + 0.148 + 0.268 + 0.276 + 0.177 \approx 0.905$

21. $n = 10,\ p = 0.56$

(a) $P(4) \approx 0.150$

(b) $P(x \geq 5) = P(5) + P(6) + P(7) + P(8) + P(9) + P(10)$
$\approx 0.2289 + 0.2427 + 0.1765 + 0.0843 + 0.0238 + 0.0030 \approx 0.759$

(c) $P(x < 7) = 1 - P(x \geq 7) = 1 - [P(7) + P(8) + P(9) + P(10)]$
$\approx 1 - (0.1765 + 0.0843 + 0.0238 + 0.0030) \approx 0.712$

23. $n = 11,\ p = 0.40$

(a) $P(5) \approx 0.221$

(b) $P(x > 5) = P(6) + P(7) + P(8) + P(9) + P(10) + P(11)$
$\approx 0.1471 + 0.0701 + 0.0234 + 0.0052 + 0.0007 + 0.0000 \approx 0.247$

(c) $P(x \leq 5) = 1 - P(x > 5) = 1 - 0.247 \approx 0.753$

25 $n = 14,\ p = 0.04$

(a) $P(2) \approx 0.089$

(b) $P(x > 2) = 1 - P(x \leq 2) = 1 - [P(0) + P(1) + P(2)]$
$\approx 1 - (0.5647 + 0.3294 + 0.0892) \approx 0.017$

(c) $P(2 \leq x \leq 5) = P(2) + P(3) + P(4) + P(5) \approx 0.0892 + 0.0149 + 0.0017 + 0.0001 \approx 0.106$

27. (a)

x	$P(x)$
0	0.008974
1	0.060355
2	0.173965
3	0.278572
4	0.267647
5	0.154291
6	0.049413
7	0.006782

(b)

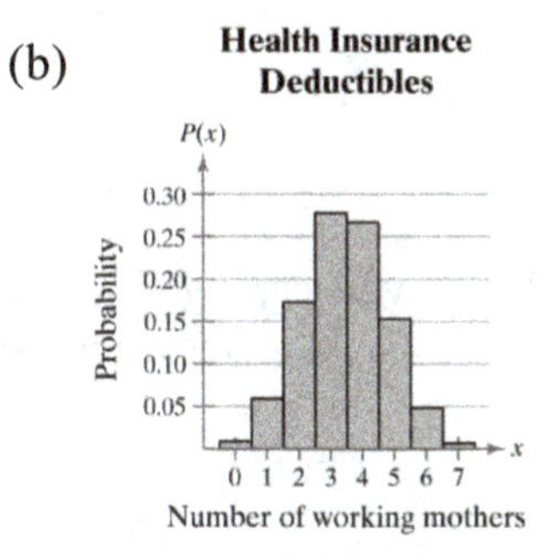

Approximately symmetric

(c) The values 0, 6, and 7 are unusual because their probabilities are less than 0.05.

29. (a)

x	$P(x)$
0	0.00064
1	0.01077
2	0.07214
3	0.24151
4	0.40426
5	0.27068

(b)

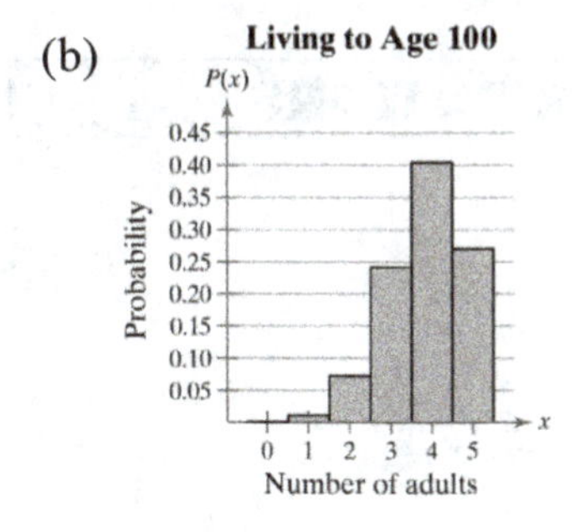

Skewed Left

(c) The values 0 and 1 are unusual because their probabilities are less than 0.05.

31. $\mu = np = (7)(0.71) \approx 5.0$

$\sigma^2 = npq = (7)(0.71)(0.29) \approx 1.4$

$\sigma = \sqrt{npq} = \sqrt{(7)(0.71)(0.29)} \approx 1.2$

On average, 5 out of every 7 U.S. adults think that political correctness is a problem in America today. The standard deviation is 1.2, so most samples of 7 U.S. adults would differ from the mean by at most 1.2 U.S. adults.

33. $\mu = np = (8)(0.79) \approx 6.3$

$\sigma^2 = npq = (8)(0.79)(0.21) \approx 1.3$

$\sigma = \sqrt{npq} = \sqrt{(8)(0.79)(0.21)} \approx 1.2$

On average, 6.3 out of every 8 adults believe that life on other planets is possible. The standard deviation is 1.2, so most samples of 8 adults would differ from the mean by at most 1.2 adults.

35. $\mu = np = (6)(0.32) \approx 1.9$

$\sigma^2 = npq = (6)(0.32)(0.68) \approx 1.3$

$\sigma = \sqrt{npq} = \sqrt{(6)(0.32)(0.68)} \approx 1.1$

On average, 1.9 out of every 6 U.S. employees who are late for work blame oversleeping. The standard deviation is 1.1, so most samples of 6 U.S. employees who are late would differ from the mean by at most 1.1 U.S. employees.

37. $P(5,2,2,1) = \frac{10!}{5!2!2!1!}\left(\frac{9}{16}\right)^5\left(\frac{3}{16}\right)^2\left(\frac{3}{16}\right)^2\left(\frac{1}{16}\right)^1 = 0.033$

39. (a) P(0 defective) = P(1st not defective) + P(2nd not defective given 1st not defective) + P(3rd not defective given first 2 not defective) +… + P(10th not defective given the first 9 not defective)

$$P(0) = \frac{8000}{10{,}000} \times \frac{7999}{9999} \times \frac{7998}{9998} \times \ldots \times \frac{7991}{9991} \approx 0.107$$

(b) $n = 10,\ p = 0.20$

$P(0) \approx 0.107$

(c) The results are the same.

4.3 MORE DISCRETE PROBABILITY DISTRIBUTIONS

4.3 TRY IT YOURSELF SOLUTIONS

1. $p = 0.14, q = 0.86, x = 6$

$P(6) = (0.14)(0.86)^{6-1} \approx 0.066$

2. $P(0) \approx \frac{3^0(2.71828)^{-3}}{0!} \approx 0.050$ $\quad P(1) \approx \frac{3^1(2.71828)^{-3}}{1!} \approx 0.149$

$P(2) \approx \frac{3^2(2.71828)^{-3}}{2!} \approx 0.224$ $\quad P(3) \approx \frac{3^3(2.71828)^{-3}}{3!} \approx 0.224$

$P(4) \approx \frac{3^4(2.71828)^{-3}}{4!} \approx 0.168$

$P(0) + P(1) + P(2) + P(3) + P(4) \approx 0.050 + 0.149 + 0.224 + 0.224 + 0.168 = 0.815$

$P(x > 4) = 1 - P(x \le 4) = 1 - 0.815 = 0.185$

3. $\mu = \frac{2000}{20{,}000} = 0.10, x = 3$

$P(3) = 0.0002$

The probability of finding three brown trout in any given cubic meter of the lake is 0.0002. Because 0.0002 is less than 0.05, this can be considered an unusual event.

4.3 EXERCISE SOLUTIONS

1. $P(3) = (0.65)(0.35)^{3-1} \approx 0.080$

3. $P(5) = (0.09)(0.91)^{5-1} \approx 0.062$

5. $P(4) \approx \dfrac{(5)^4 (2.71828)^{-5}}{4!} \approx 0.175$

7. $P(2) \approx \dfrac{(1.5)^2 (2.71828)^{-1.5}}{2!} \approx 0.251$

9. In a binomial distribution, the value of x represents the number of successes in n trials. In a geometric distribution the value of x represents the first trial that results in a success.

11. $p = 0.19$

(a) $P(5) = (0.19)(0.81)^{5-1} \approx 0.082$

(b) $P(\text{sale on } 1^{\text{st}},\ 2^{\text{nd}}, \text{or } 3^{\text{rd}} \text{ call}) = P(1) + P(2) + P(3)$
$= (0.19)(0.81)^{1-1} + (0.19)(0.81)^{2-1} + (0.19)(0.81)^{3-1} \approx 0.469$

(c) $P(\text{no sale on first 3 calls}) = 1 - P(\text{sale on } 1^{\text{st}},\ 2^{\text{nd}}, \text{or } 3^{\text{rd}} \text{ call}) \approx 1 - 0.469 \approx 0.531$

13. $\mu = 2$

(a) $P(5) \approx \dfrac{(2)^5 (2.71828)^{-2}}{5!} \approx 0.036$

This event is unusual because its probability is less than 0.05.

(b) $P(x \geq 5) = 1 - \big(P(0) + P(1) + P(2) + P(3) + P(4)\big)$
$\approx 1 - \big(0.135 + 0.271 + 0.271 + 0.180 + 0.090\big) \approx 0.053$

This event is unusual because its probability is less than 0.05.

(c) $P(x > 5) = 1 - P(x \leq 5) = 1 - \big(P(0) + P(1) + P(2) + P(3) + P(4) + P(5)\big)$
$\approx 1 - \big(0.135 + 0.271 + 0.271 + 0.180 + 0.090 + 0.036\big) \approx 0.017$

This event is unusual because its probability is less than 0.05

15. $p = 0.641$

(a) $P(2) = (0.641)(0.359)^{2-1} \approx 0.230$.

(b) $P(\text{completes } 1^{\text{st}} \text{ or } 2^{\text{nd}} \text{ pass}) = P(1) + P(2)$

$$= (0.641)(0.359)^{1-1} + (0.641)(0.359)^{2-1} \approx 0.641 + 0.230 \approx 0.871$$

(c) $P(\text{does not complete his } 1^{\text{st}} \text{ two passes}) = 1 - P(\text{completes } 1^{\text{st}} \text{ or } 2^{\text{nd}} \text{ pass}) \approx 1 - 0.871 = 0.129$

17. $p = \dfrac{1}{500} = 0.002$

(a) $P(10) = (0.002)(0.998)^{10-1} \approx 0.002$

This event is unusual because its probability is less than 0.05.

(b) $P\left(1^{\text{st}}, 2^{\text{nd}}, \text{ or } 3^{\text{rd}} \text{ part is warped}\right) = P(1) + P(2) + P(3)$

$$= (0.002)(0.998)^{1-1} + (0.002)(0.998)^{2-1} + (0.002)(0.998)^{3-1} \approx 0.002 + 0.002 + 0.002 \approx 0.006$$

This event is unusual because its probability is less than 0.05.

(c) $P(x > 10) = 1 - P(x \leq 10) = 1 - \left(P(1) + P(2) + \cdots + P(10)\right)$

$$\approx 1 - \left(0.002 + 0.002 + 0.002 + 0.002 + 0.002 + 0.002 + 0.002 + 0.002 + 0.002 + 0.002\right)$$

$$\approx 1 - 0.020 \approx 0.980$$

19. $\mu = 1.7$

(a) $P(1) \approx \dfrac{(1.7)^1 (2.71828)^{-1.7}}{1!} \approx 0.311$

(b) $P(x \leq 1) = P(0) + P(1) \approx 0.1827 + 0.3106 \approx 0.493$

(c) $P(x > 1) = 1 - P(x \leq 1) \approx 1 - 0.493 \approx 0.507$

21. $n = 8,\ p = 0.51$

(a) $P(4) = \dfrac{8!}{4!4!}(0.51)^4 (0.49)^4 \approx 0.273$

(b) $P(x < 5) = P(0) + P(1) + P(2) + P(3) + P(4)$

$$= \frac{8!}{8!0!}(0.51)^0 (0.49)^8 + \frac{8!}{7!1!}(0.51)^1 (0.49)^7 + \frac{8!}{6!2!}(0.51)^2 (0.49)^6$$

$$+ \frac{8!}{5!3!}(0.51)^3 (0.49)^5 + \frac{8!}{4!4!}(0.51)^4 (0.49)^4$$

$$\approx 0.0033 + 0.0277 + 0.1008 + 0.2098 + 0.2730 \approx 0.615$$

(c) $P(x \geq 3) = 1 - \left(P(0) + P(1) + P(2)\right) \approx 1 - \left(0.0033 + 0.0277 + 0.1008\right) \approx 1 - 0.132 \approx 0.868$

23. $n=5$, $p=0.68$

(a) $P(3)=\frac{5!}{2!3!}(0.68)^3(0.32)^2 \approx 0.322$

(b) $P(x<4)=P(x\leq 3)=P(0)+P(1)+P(2)+P(3)\approx 0.0034+0.0357+0.1515+0.3220\approx 0.513$

(c) $P(x\geq 3)=P(3)+P(4)+P(5)\approx 0.322+0.342+0.145\approx 0.809$

25. $\mu=14$

(a) $P(17)\approx\frac{(14)^{17}(2.71828)^{-14}}{17!}\approx 0.071$

(b) $P(x\leq 17)=P(0)+P(1)+\cdots+P(17)\approx 0.827$

(c) $P(x>17)=1-P(x\leq 17)\approx 1-0.827\approx 0.173$

27. (a) $n=6000$, $p=\frac{1}{2500}=0.0004$

$P(4)=\frac{6000!}{5996!4!}(0.0004)^4(0.9996)^{5996}\approx 0.12542$

(b) $\mu=\frac{6000}{2500}=2.4$ cars with defects per 6000.

$P(4)=\frac{(2.4)^4(2.71828)^{-2.4}}{4!}\approx 0.12541$

The results are approximately the same.

29. $p=\frac{1}{1000}=0.001$

(a) $\mu=\frac{1}{p}=\frac{1}{0.001}=1000$

$\sigma^2=\frac{q}{p^2}=\frac{0.999}{(0.001)^2}=999{,}000$

$\sigma=\sqrt{\sigma^2}\approx 999.5$

(b) 1000 times, because it is the mean.

(c) You would expect to lose money, because, on average, you would win \$500 every 1000 times you play the lottery and pay \$1000 to play it. So, the net gain would be –\$500.

31. $\mu=4.1$

(a) $\sigma^2=\mu=4.1$

$\sigma=\sqrt{\sigma^2}\approx 2.0$

The standard deviation is 2.0 strokes, so most of Steven's scores per hole differ from the mean by no more than 2 strokes.

(b) For 18 holes, Steven's average would be $(18)(4.1) = 73.8$ strokes.
So, $\mu = 73.8$.
$P(x > 72) = 1 - P(x \le 72) \approx 1 - 0.447 \approx 0.553$

CHAPTER 4 REVIEW EXERCISE SOLUTIONS

1. Discrete; The number of pumps in use at a gas station is a random variable that is countable.

3. (a)

x	f	$P(x)$
0	29	0.207
1	62	0.443
2	33	0.236
3	12	0.086
4	3	0.021
5	1	0.007
	$n = 140$	$\sum P(x) = 1$

(b)

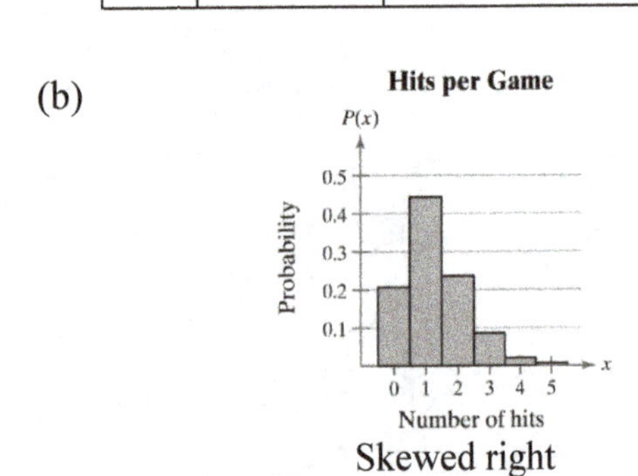

Skewed right

5. Yes.

7. (a)

x	$P(x)$	$xP(x)$	$(x-\mu)$	$(x-\mu)^2$	$(x-\mu)^2P(x)$
0	0.020	0	−2.804	7.862	0.157
1	0.140	0.140	−1.804	3.254	0.456
2	0.272	0.544	−0.804	0.646	0.176
3	0.292	0.876	0.196	0.038	0.011
4	0.168	0.672	1.196	1.430	0.240
5	0.076	0.380	2.196	4.822	0.367
6	0.032	0.192	3.196	10.214	0.327
		$\sum xP(x) = 2.804$			$\sum (x-\mu)^2 P(x) = 1.734$

$\mu = \sum xP(x) = 2.804 \approx 2.8$

$\sigma^2 = \sum (x-\mu)^2 P(x) = 1.734 \approx 1.7$; $\sigma = \sqrt{\sigma^2} = \sqrt{1.734} \approx 1.3$

(b) The mean is 2.8, so the average number of cellular phones per household is about 3. The standard deviation is 1.3, so most of the households differ from the mean by no more than about 1 cellular phone.

9. (a)

x	$P(x)$	$xP(x)$
100	0.125	12.500
0	0.250	0.000
–25	0.625	–15.625
		$\sum xP(x) = -3.125$

$\mu = \sum xP(x) = -3.125 \approx -\$3.13$

11. It is a binomial experiment.
Success: a green candy is selected
$n = 12,\ p = 0.16,\ q = 0.84,\ x = 0,1,2,3,4,5,6,7,8,9,10,11,12$

13. $n = 8,\ p = 0.53$

(a) $P(3) \approx 0.191$

(b) $P(x \geq 3) = 1 - P(x < 3) = 1 - (P(0) + P(1) + P(2))$
$\approx 1 - (0.0024 + 0.0215 + 0.0848) \approx 1 - 0.109 \approx 0.891$

(c) $P(x > 3) = 1 - P(x \leq 3) = 1 - (P(0) + P(1) + P(2) + P(3))$
$\approx 1 - (0.0024 + 0.0215 + 0.0848 + 0.1912) \approx 1 - 0.300 \approx 0.700$

15. $n = 9,\ p = 0.88$

(a) $P(6) \approx 0.067$

(b) $P(x \geq 6) = P(6) + P(7) + P(8) + P(9)$
$\approx 0.0674 + 0.2119 + 0.3884 + 0.3165 \approx 0.984$

(c) $P(x > 6) = P(7) + P(8) + P(9)$
$\approx 0.2119 + 0.3884 + 0.3165 \approx 0.917$

17. (a)

x	$P(x)$
0	0.0008
1	0.0126
2	0.0798
3	0.2529
4	0.4003
5	0.2536

(b)

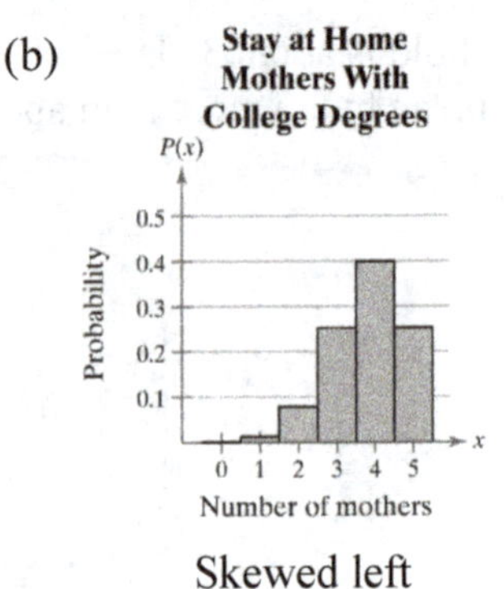

Skewed left

(c) The values 0 and 1 are unusual because their probabilities are less than 0.05.

19. $n=8,\ p=0.13$

$\mu = np = (8)(0.13) \approx 1.0$

$\sigma^2 = npq = (8)(0.13)(0.87) \approx 0.9$

$\sigma = \sqrt{npq} = \sqrt{(8)(0.13)(0.87)} \approx 1.0$

On average, 1 out of every 8 drivers are uninsured. The standard deviation is 1.0, so most samples of 8 drivers would differ from the mean by at most 1 driver.

21. $p=0.82$

(a) $P(2) = (0.82)(0.18)^1 \approx 0.148$

(b) $P(4^{\text{th}} \text{ or } 5^{\text{th}}) = P(4) + P(5) = (0.82)(0.18)^3 + (0.82)(0.18)^4 \approx 0.006$

This event is unusual because its probability is less than 0.05.

(c) $P(\text{not one of the } 2^{\text{nd}} \text{through } 7^{\text{th}}) = P(1^{\text{st}} \text{ or } 8^{\text{th}} \text{or } 9^{\text{th}} \text{or } 10^{\text{th}})$

$\approx 0.820 + 0.000 + 0.000 + 0.000 \approx 0.820$

23. $n=7,\ p=0.36$

(a) $P(4) \approx 0.154$

(b) $P(x<2) = P(0) + P(1) \approx 0.044 + 0.173 \approx 0.217$

(c) $P(x \geq 6) = P(6) + P(7) \approx 0.010 + 0.001 \approx 0.011$

This event would be unusual because its probability is less than 0.05.

25. $\mu = 6.1$

(a) $P(3) \approx \dfrac{(6.1)^3(2.71828)^{-6.1}}{3!} \approx 0.085$

(b) $P(x>6) = 1 - P(x \leq 6) = 1 - \left(P(0) + P(1) + P(2) + P(3) + P(4) + P(5) + P(6)\right)$

$\approx 1 - \left(0.002 + 0.014 + 0.042 + 0.085 + 0.129 + 0.158 + 0.160\right) \approx 0.410$

(c) $P(x \le 5) = P(0) + P(1) + P(2) + P(3) + P(4) + P(5)$
$\approx 0.002 + 0.014 + 0.042 + 0.085 + 0.129 + 0.158 \approx 0.430$

CHAPTER 4 QUIZ SOLUTIONS

1. (a) Discrete; The number of lightning strikes that occur in Wyoming during the month of June is a random variable that is countable.

(b) Continuous; The fuel (in gallons) used by a jet during takeoff is a random variable that has an infinite number of possible outcomes and cannot be counted.

(c) Discrete; The number of die rolls required for an individual to roll a five is a random variable that is countable.

2. (a)

x	f	$P(x)$
0	277	0.238
1	471	0.405
2	243	0.209
3	105	0.090
4	46	0.040
5	22	0.019
	$n = 1164$	$\sum P(x) \approx 1$

(b)

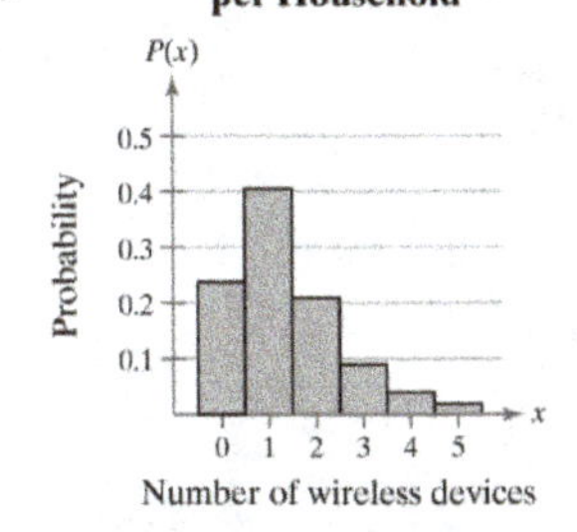

Skewed right

(c)

$xP(x)$	$(x-\mu)$	$(x-\mu)^2$	$(x-\mu)^2 P(x)$
0.000	−1.345	1.809	0.430
0.405	−0.345	0.119	0.048
0.418	0.655	0.429	0.090
0.271	1.655	2.739	0.247
0.158	2.655	7.049	0.279
0.095	3.655	13.359	0.252
$\sum xP(x) = 1.345$			$\sum (x-\mu)^2 P(x) = 1.346$

$\mu = \sum xP(x) = 1.345 \approx 1.3$

$\sigma^2 = \sum (x-\mu)^2 P(x) = 1.346 \approx 1.3$

$\sigma = \sqrt{\sigma^2} = \sqrt{1.346} \approx 1.2$

The mean is 1.3, so the average number of wireless devices per household is 1.3. The standard deviation is 1.2, so most households will differ from the mean by no more than about 1.2 wireless devices.

(d) $P(x \geq 4) = P(4) + P(5) = \frac{46}{1164} + \frac{22}{1164} \approx 0.058$

3. $n = 9,\ p = 0.36$

(a) $P(3) \approx 0.269$

(b) $P(x \leq 4) = P(0) + P(1) + P(2) + P(3) + P(4) \approx 0.0180 + 0.0912 + 0.2052 + 0.2693 + 0.2272 \approx 0.811$

(c) $P(x > 5) = P(6) + P(7) + P(8) + P(9) \approx 0.0479 + 0.0116 + 0.0016 + 0.0001 \approx 0.061$

4. $n = 6,\ p = 0.86$

(a)

x	$P(x)$
0	0.000008
1	0.000278
2	0.004262
3	0.034907
4	0.160820
5	0.395159
6	0.404567

(b)

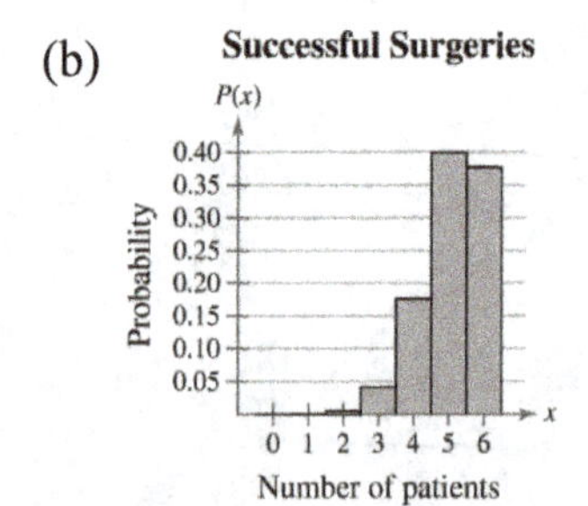

Skewed left

(c) $\mu = np = (6)(0.86) = 5.2$

$\sigma^2 = npq = (6)(0.86)(0.14) \approx 0.7$

$\sigma = \sqrt{npq} = \sqrt{(6)(0.86)(0.14)} \approx 0.8$

On average, 5.2 out of every 6 patients have a successful surgery. The standard deviation is 0.8, so most samples of 6 surgeries would differ from the mean by at most 0.8 surgery.

5. $\mu = 5$

(a) $P(5) \approx \dfrac{(5)^5(2.71828)^{-5}}{5!} \approx 0.175$

(b) $P(x < 5) = P(0) + P(1) + P(2) + P(3) + P(4) \approx 0.007 + 0.034 + 0.084 + 0.140 + 0.175 \approx 0.440$

(c) $P(0) \approx \dfrac{(5)^0(2.71828)^{-5}}{0!} \approx 0.007$

6. $p = 0.56$

(a) $P(4) \approx (0.56)(0.44)^3 \approx 0.048$

(b) $P(2^{\text{nd}} \text{ or } 3^{\text{rd}}) = P(2) + P(3) \approx (0.56)(0.44)^1 + (0.56)(0.44)^2 \approx 0.355$

(c) $P(x > 3) = 1 - P(x \leq 3) = 1 - \left(P(1) + P(2) + P(3)\right)$
$\approx 1 - \left((0.56)(0.44)^0 + (0.56)(0.44)^1 + (0.56)(0.44)^2\right) \approx 0.085$

7. Event (a) is unusual because its probability is less than 0.05.

Normal Probability Distributions

5.1 INTRODUCTION TO NORMAL DISTRIBUTIONS AND THE STANDARD NORMAL DISTRIBUTION

5.1 TRY IT YOURSELF SOLUTIONS

1. (a) A: $x=45$, B: $x=60$, C: $x=45$ (B has the greatest mean.)

(b) Curve C is more spread out, so curve C has the greatest standard deviation.

2. Because a normal curve is symmetric about the mean, the estimated mean is: $\mu=300$
Inflection points are approximately: 263 and 337
Because the inflection points are one standard deviation from the mean, $\sigma = 37$

3. (1) 0.0143 (2) 0.98500

4.

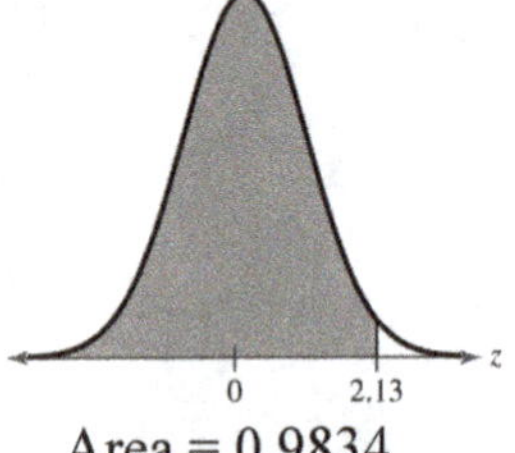

Area = 0.9834

5.

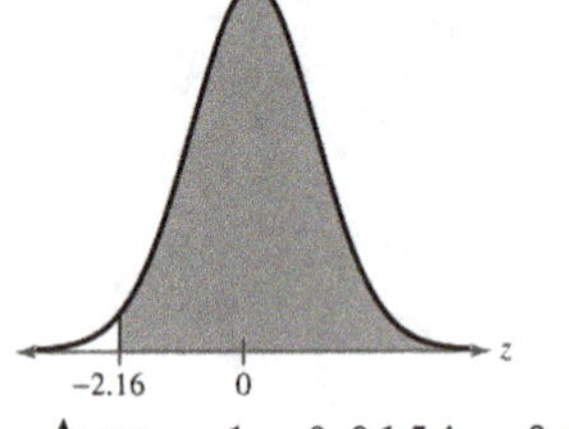

Area $= 1 - 0.0154 = 0.9846$

6.

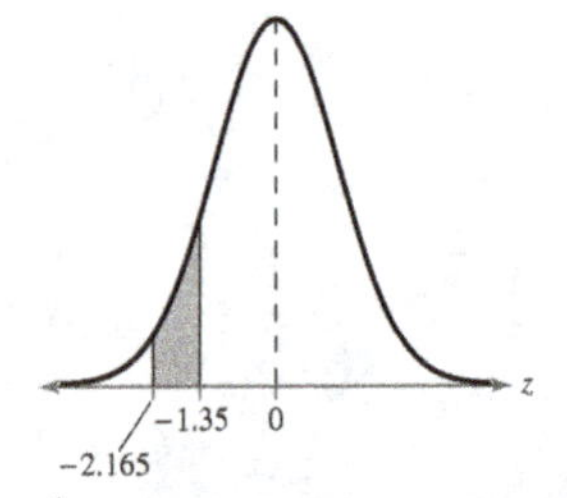

Area$= 0.0885 - 0.0152 = 0.0733$

5.1 EXERCISE SOLUTIONS

1. Answers will vary.

3. 1

5. Answers will vary.
Similarities: The two curves will have the same line of symmetry.

Differences: The curve with the larger standard deviation will be more spread out than the curve with the smaller standard deviation.

7. $\mu = 0,\ \sigma = 1$

9. "The" standard normal distribution is used to describe one specific normal distribution ($\mu = 0,\ \sigma = 1$). "A" normal distribution is used to describe a normal distribution with any mean and standard deviation.

11. No, the graph crosses the x-axis.

13. Yes, the graph fulfills the properties of the normal distribution. $\mu \approx 18.5,\ \sigma \approx 2$

15. No, the graph is skewed to the right.

17. (Area left of $z = 1.3) = 0.9032$

19. (Area right of $z = 2) = 1 -$ (Area leftt of $z = 2) = 1 - 0.9772 = 0.0228$

21. (Area left of $z = 1.2) -$ (Area left of $z = -0.7) = 0.8849 - 0.2420 = 0.6429$

23. 0.5675

25. 0.0050

27. $1 - 0.2578 = 0.7422$

29. $1 - 0.3613 = 0.6387$

31. $0.9979 - 0.5000 = 0.4979$

33. $0.9394 - 0.0606 = 0.8788$

35. $0.1003 + 0.1003 = 0.2006$

37. (a) **Life Spans of Tires**

Frequency

27,660 32,353 37,046 41,739 46,432

Distance (in miles)

It is reasonable to assume that the life spans are normally distributed because the histogram is symmetric and bell-shaped.

(b) $\bar{x} = 37{,}234.7;\ s = 6259.2$

(c) The sample mean of 37,234.7 hours is less than the claimed mean, so, on average, the tires in the sample lasted for a shorter time. The sample standard deviation of 6259.2 is greater than the claimed standard deviation, so the tires in the sample had greater variation in life span than the manufacturer's claim.

39. (a) $x = 162 \Rightarrow z = \dfrac{x-\mu}{\sigma} = \dfrac{162-150}{8.75} \approx 1.37$

$x = 168 \Rightarrow z = \dfrac{x-\mu}{\sigma} = \dfrac{168-150}{8.75} \approx 2.06$

$$x=155 \Rightarrow z=\frac{x-\mu}{\sigma}=\frac{155-150}{8.75}\approx 0.57$$

$$x=138 \Rightarrow z=\frac{x-\mu}{\sigma}=\frac{138-150}{8.75}\approx -1.37$$

(b) $x=168$ is unusual because its corresponding z-score (2.06) lies more than 2 standard deviations from the mean.

41. 0.9750

43. $1-0.0168=0.9832$

45. $0.8413-0.1587=0.6826$

47. $P(z<1.16)=0.8770$

49. Using technology: $P(z>2.175)=1-P(z<2.175)=1-0.9852=0.0148$

51. $P(-0.89<z<0)=0.5-0.1867=0.3133$

53. $P(-1.78<z<1.78)=0.9625-0.0375=0.9250$

55. $P(z<-2.58 \text{ or } z>2.58)=2(0.0049)=0.0098$

57.

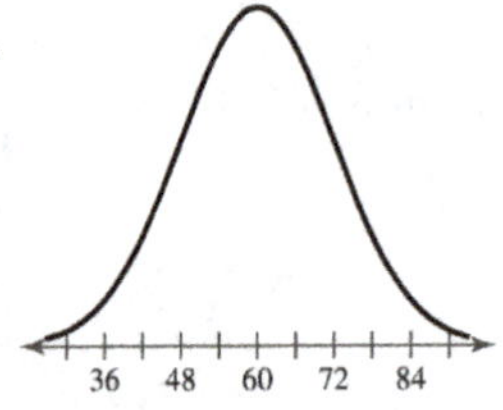

The normal distribution curve is centered at its mean (60) and has 2 points of inflection (48 and 72) representing $\mu \pm \sigma$.

59. The area under the curve is $(b-a)\left(\frac{1}{b-a}\right)=\frac{b-a}{b-a}=1$.

(Because a < b, you do not have to worry about division by 0.)

5.2 NORMAL DISTRIBUTIONS: FINDING PROBABILITIES

5.2 TRY IT YOURSELF SOLUTIONS

1.

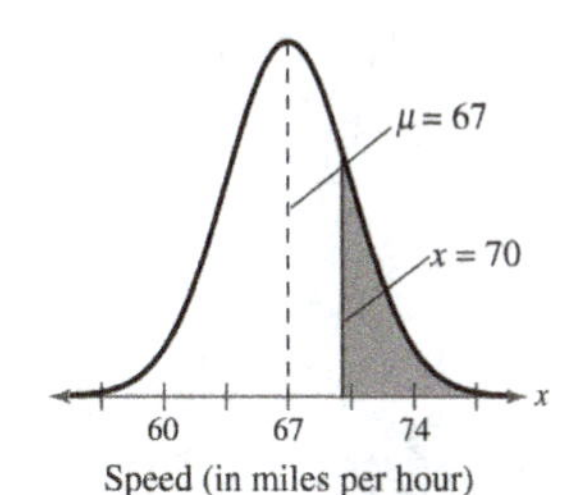

$z = \dfrac{x-\mu}{\sigma} = \dfrac{70-67}{3.5} \approx 0.857$

$P(z > 0.857) = 1 - P(z < 0.857) \approx 1 - 0.8043 \approx 0.1957$

The probability that a randomly selected vehicle is violating the speed limit of 70 miles per hour is 0.1957.

2. $z = \dfrac{x-\mu}{\sigma} = \dfrac{31-43}{12} = -1$

$z = \dfrac{x-\mu}{\sigma} = \dfrac{58-43}{12} = 1.25$

$P(31 < x < 58) = P(-1 < z < 1.25) = P(z < 1.25) - P(z < -1)$

$= 0.8944 - 0.1587 = 0.7357$

When 200 shoppers enter the store, you would expect $200(0.7357) = 147.14$ or about 147 shoppers to be in the store between 31 and 58 minutes.

3. $P(100 < x < 150) = P(0.12 < z < 2.12) = 0.9830 - 0.5478 = 0.4352$

The probability that a randomly selected person's triglyceride level is between 100 and 150 is 0.4352.

5.2 EXERCISE SOLUTIONS

1. $P(x < 170) = P(z < -0.2) = 0.4207$

3. $P(x > 182) = P(z > 0.4) = 1 - 0.6554 = 0.3446$

5. $P(160 < x < 170) = P(-0.7 < z < -0.2) = 0.4207 - 0.2420 = 0.1787$

7. (a) Using technology: $P(x < 4) \approx P(z < -0.6364) = 0.2623$

(b) Using technology: $P(4 < x < 6) = P(-0.6364 < z < 0.2727)$

$= 0.6075 - 0.2623 = 0.3452$

(c) Using technology: $P(x>8)=P(z>1.1818)=1-P(z<1.1818)=1-0.8814=0.1186$
None of these events are unusual because their probabilities are greater than 0.05.

9. (a) Using technology: $P(x<489)=P(z<-1.0377)=0.1497$

(b) Using technology: $P(495<x<507)=P(-0.4717<z<0.6604)=0.7455-0.3186=0.4269$

(c) Using technology: $P(x>522)=P(z>2.0755)=1-P(z<2.0755)=1-0.9810=0.0190$
The event in part (c) is unusual because its probability is less than 0.05.

11. (a) Using technology: $P(x<70)=P(z<-2.5)=0.0062$

(b) Using technology: $P(90<x<120)=P(-0.8333<z<1.6667)$
$=0.95221-0.20233=0.7499$

(c) Using technology: $P(x>140)=P(z>3.3333)=1-P(z<3.3333)=1-0.9996=0.0004$

13. Using technology: $P(750<x<1000)=P(-1.7254<z<-0.4301)$
$=0.3336-0.0422=0.2914$

15. Using technology: $P(285<x<294)=P(1.8<z<2.7)=0.99653-0.96407=0.0325$

17. (a) Using technology: $P(x<1300)=P(z<1.1244)=0.8696 \Rightarrow 86.96\%$

(b) Using technology: $P(x>1100)=P(z>0.0881)=1-P(z<0.0881)=1-0.5351=0.4649$
$(1000)(0.4649)=464.9 \Rightarrow 465$ scores

19. (a) Using technology: $P(x<290)=P(z<2.3)=0.9893 \Rightarrow 98.93\%$

(b) Using technology:
$P(260<x<300)=P(-0.7<z<3.3)=0.99952-0.24196=0.7576 \Rightarrow 75.76\%$

(c) Using technology: $P(x>287)=P(z>2)=1-P(z<2)=1-0.9772=0.0228$
$(250)(0.0228)=5.7 \Rightarrow 6$ mothers

21. Out of control, because the 10th observation is more than 3 standard deviations beyond the mean.

23. Out of control, because there are nine consecutive points below the mean, and two out of three consecutive points more than 2 standard deviations from the mean.

5.3 NORMAL DISTRIBUTIONS: FINDING VALUES

5.3 TRY IT YOURSELF SOLUTIONS

1. (1)

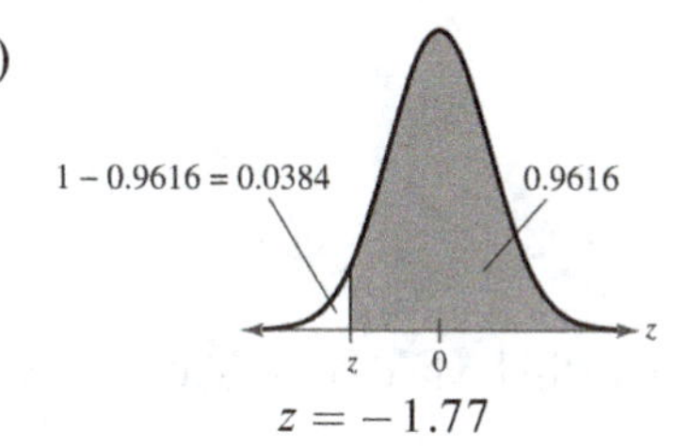

$z = -1.77$

(2)

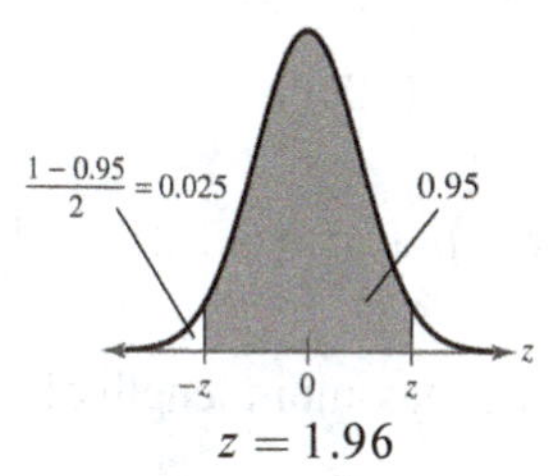

$z = 1.96$

2. (1)

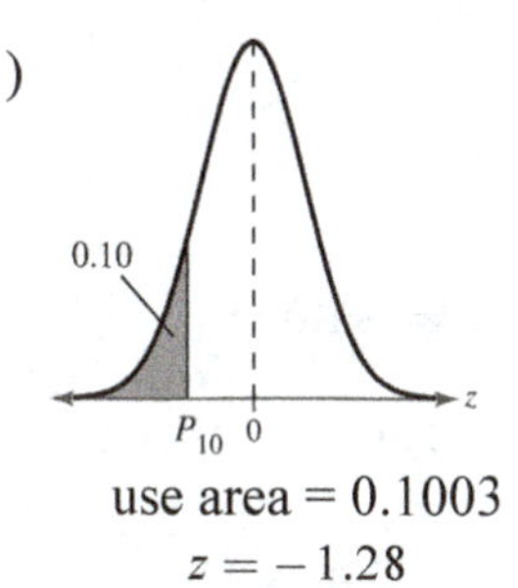

use area = 0.1003
$z = -1.28$

(2)

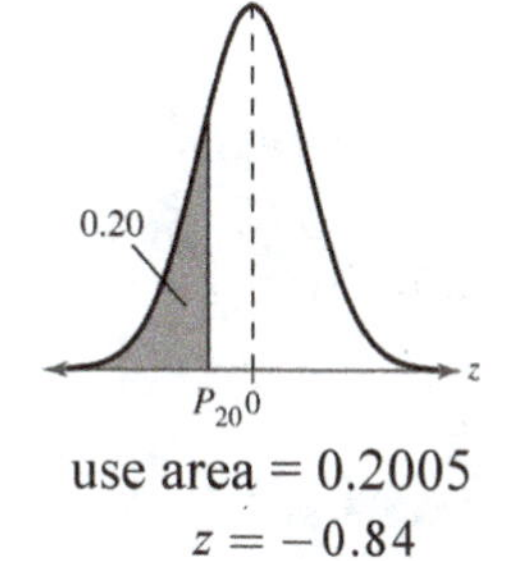

use area = 0.2005
$z = -0.84$

(3)

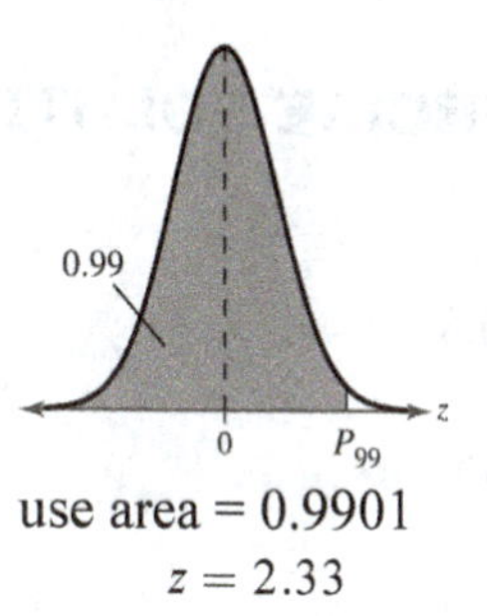

use area = 0.9901
$z = 2.33$

3. $\mu = 52,\ \sigma = 15$

(1) $z = -2.33 \Rightarrow x = \mu + z\sigma = 52 + (-2.33)(15) = 17.05$

(2) $z = 3 \Rightarrow x = \mu + z\sigma = 52 + (3)(15) = 97$

(3) $z = 0.58 \Rightarrow x = \mu + z\sigma = 52 + (0.58)(15) = 60.70$

17.05 pounds is below the mean, 60.7 pounds and 98.5 pounds are above the mean.

4. (a)

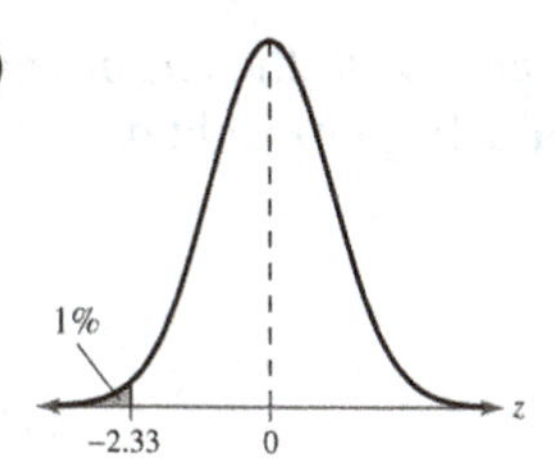

(b) $z = -2.33$

(c) $x = \mu + z\sigma = 129 + (-2.33)(5.18) \approx 116.93$

(d) So, the longest braking distance one of these cars could have and still be in the bottom 1% is about 117 feet.

5. (a)

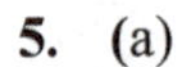

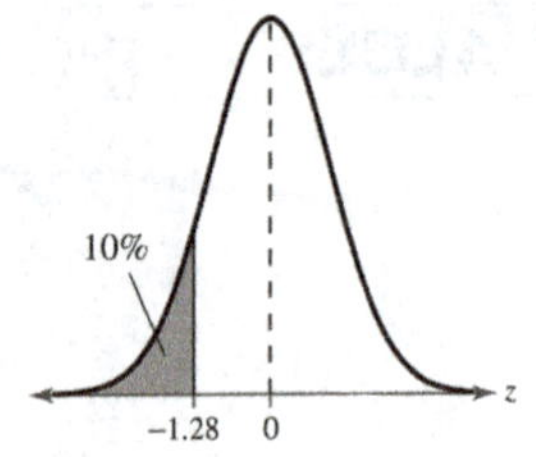

(b) $z = -1.28$

(c) $x = \mu + z\sigma = 11.2 + (-1.28)(2.1) = 8.512$

(d) The maximum length of time an employee could have worked and still be laid off is about 8.5 years.

5.3 EXERCISE SOLUTIONS

1. $z = -0.81$

3. $z = 0.45$

5. $z = -1.645$

7. $z = 1.555$

9. $z = -1.04$

11. $z = 1.175$

13. $z = -0.67$

15. $z = 1.34$

17. $z = -0.38$

19. $z = -0.58$

21. $z = \pm 1.645$

23. The z-score has an area of 0.0119 to its left.
$z = -1.18$

25. The z-score has an area of 0.637 to its right and has an area of $1 - 0.637 = 0.363$ to its left.
$z = -0.35$

27. The z-score has an area of 0.02275 to its left.
$z = -2.00$

29. Since the area between the z-scores is 0.80, then the combined area remaining (to the left of -z and to the right of z) is $1 - 0.80 = 0.20$. The positive z-score that has an area of 0.10 to its right has an area $1 - 0.10 = 0.90$ to its left.
$z = 1.28$

31. (a) 95th percentile $\Rightarrow$ Area = 0.95 $\Rightarrow z \approx 1.6449$ (using technology)
$x = \mu + z\sigma = 64.2 + (1.6449)(2.9) \approx 68.97$ inches

(b) 43rd percentile $\Rightarrow$ Area = 0.43 $\Rightarrow z \approx -0.1764$ (using technology)
$x = \mu + z\sigma = 64.2 + (-0.1764)(2.9) \approx 63.69$ inches

(c) 1st quartile $\Rightarrow$ Area = 0.25 $\Rightarrow z \approx -0.6745$ (using technology)
$x = \mu + z\sigma = 64.2 + (-0.6745)(2.9) \approx 62.24$ inches

33. (a) 5th percentile $\Rightarrow$ Area = 0.05 $\Rightarrow z \approx -1.64485$ (using technology)
$x = \mu + z\sigma = 2277 + (-1.64485)(584.2) \approx 1316.08$ kilowatt-hours

(b) 17th percentile $\Rightarrow$ Area = 0.17 $\Rightarrow z \approx -0.954165$ (using technology)
$x = \mu + z\sigma = 2277 + (-0.954165)(584.2) \approx 1719.58$ kilowatt-hours

(c) 3rd quartile $\Rightarrow$ Area = 0.75 $\Rightarrow\Rightarrow z \approx 0.6745$ (using technology)
$x = \mu + z\sigma = 2277 + (0.6745)(584.2) \approx 2671.04$ kilowatt-hours

35. (a) Top 5% $\Rightarrow$ Area = 0.95 $\Rightarrow z \approx 1.6449$ (using technology)
$x = \mu + z\sigma = 3.36 + (1.6449)(0.18) \approx 3.66$

(b) Middle 50% $\Rightarrow$ Area = 0.25 to 0.75 $\Rightarrow z \approx \pm 0.6745$ (using technology)
$x = \mu + z\sigma = 3.36 + (\pm 0.6745)(0.18) \approx 3.24$ to 3.48

37. (a) Upper 25% $\Rightarrow$ Area = 0.75 $\Rightarrow z \approx 0.6745$ (using technology)
$x = \mu + z\sigma = 5.4 + (0.6745)(0.4) \approx 5.67$ millions of cells per microliter

(b) Bottom 15% $\Rightarrow$ Area = 0.15 $\Rightarrow z \approx -1.0364$ (using technology)
$x = \mu + z\sigma = 5.4 + (-1.0364)(0.4) \approx 4.99$ millions of cells per microliter

39. Upper 4.5% $\Rightarrow$ Area = 0.955 $\Rightarrow z \approx 1.6954$ (using technology)
$x = \mu + z\sigma = 32 + (1.6954)(0.36) \approx 32.61$ ounces

41. Top 1% $\Rightarrow$ Area = 0.99 $\Rightarrow z \approx 2.3264$ (using technology)
$x = \mu + z\sigma \Rightarrow 8 = \mu + (2.3264)(0.03) \Rightarrow \mu \approx 7.93$ ounces

5.4 SAMPLING DISTRIBUTIONS AND THE CENTRAL LIMIT THEOREM

5.4 TRY IT YOURSELF SOLUTIONS

1.

Sample	Mean
1, 1, 1	1
1, 1, 3	1.67
1, 1, 5	2.33
1, 3, 1	1.67
1, 3, 3	2.33
1, 3, 5	3
1, 5, 1	2.33
1, 5, 3	3
1, 5, 5	3.67

Sample	Mean
3, 1, 1	1.67
3, 1, 3	2.33
3, 1, 5	3
3, 3, 1	2.33
3, 3, 3	3
3, 3, 5	3.67
3, 5, 1	3
3, 5, 3	3.67
3, 5, 5	4.33

Sample	Mean
5, 1, 1	2.33
5, 1, 3	3
5, 1, 5	3.67
5, 3, 1	3
5, 3, 3	3.67
5, 3, 5	4.33
5, 5, 1	3.67
5, 5, 3	4.33
5, 5, 5	5

$\bar{x}$	f	**Probability**
1	1	0.03704
1.67	3	0.11111
2.33	6	0.22222
3	7	0.25926
3.67	6	0.22222
4.33	3	0.11111
5	1	0.03704
	$n = 27$	$\sum P(x) = 1$

$\mu_{\bar{x}} = 3,\ \sigma_{\bar{x}}^2 \approx 0.889,\ \sigma_{\bar{x}} \approx 0.943$

$\mu_{\bar{x}} = \mu = 3,$

$$\sigma_{\bar{x}}^2 = \frac{\sigma^2}{n} = \frac{8/3}{3} \approx 0.889,$$

$$\sigma_{\bar{x}} = \frac{\sigma}{\sqrt{n}} = \frac{\sqrt{8/3}}{\sqrt{3}} \approx 0.943$$

2. $\mu_{\bar{x}} = \mu = 6.8,\ \sigma_{\bar{x}} = \dfrac{\sigma}{\sqrt{n}} = \dfrac{1.4}{\sqrt{64}} \approx 0.18$

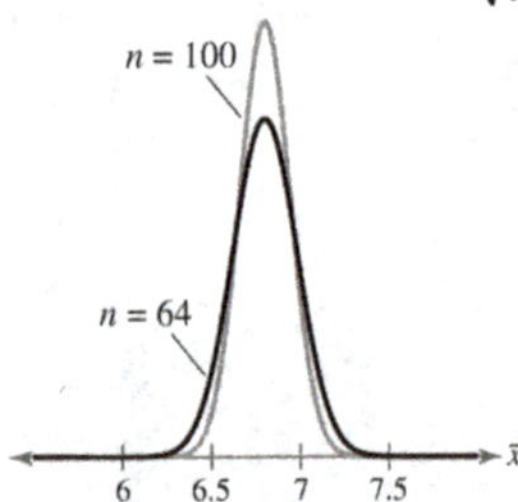

With a smaller sample size, the mean stays the same but the standard deviation increases.

3. $\mu_{\bar{x}} = \mu = 3.5,\ \sigma_{\bar{x}} = \dfrac{\sigma}{\sqrt{n}} = \dfrac{0.2}{\sqrt{16}} = 0.05$

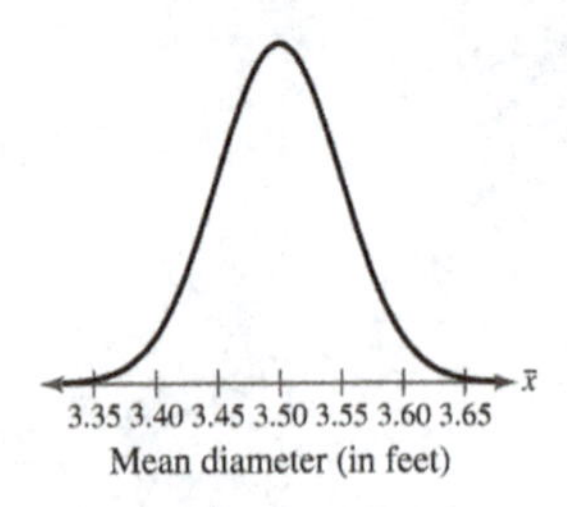

4. $\mu_{\bar{x}} = \mu = 20.7,\ \sigma_{\bar{x}} = \dfrac{\sigma}{\sqrt{n}} = \dfrac{6.5}{\sqrt{100}} = 0.65$

$$\bar{x} = 19.4:\ z = \frac{\bar{x} - \mu}{\frac{\sigma}{\sqrt{n}}} = \frac{19.4 - 20.7}{\frac{6.5}{\sqrt{100}}} = -\frac{1.3}{0.65} = -2$$

$$\bar{x}=22.5:\ z=\frac{\bar{x}-\mu}{\frac{\sigma}{\sqrt{n}}}=\frac{22.5-20.7}{\frac{6.5}{\sqrt{100}}}=\frac{1.8}{0.65}\approx 2.77$$

$$P(19.4<\bar{x}<22.5)=P(-2<z<2.77)\approx 0.9972-0.0228=0.9744$$

Of the samples of 100 drivers ages 16 to 19, about 97.44% will drive a mean distance each day that is between 19.4 and 22.5 miles.

5. $\mu_{\bar{x}}=\mu=235,500,\ \sigma_{\bar{x}}=\frac{\sigma}{\sqrt{n}}=\frac{50,000}{\sqrt{12}}\approx 14,433.76$

$$\bar{x}=225,000:\ z=\frac{\bar{x}-\mu}{\frac{\sigma}{\sqrt{n}}}=\frac{225,000-235,500}{\frac{50,000}{\sqrt{12}}}\approx\frac{-10,500}{14,433.76}\approx -0.73$$

$$P(\bar{x}>225,000)=P(z>-0.73)=1-P(z<-0.73)=1-0.2327=0.7673$$

About 77% of samples of 12 single-family houses will have a mean sales price greater than $225,000.

6. $x=200:\ z=\frac{x-\mu}{\sigma}=\frac{200-190}{48}\approx 0.21$

$$\bar{x}=200:\ z=\frac{\bar{x}-\mu}{\frac{\sigma}{\sqrt{n}}}=\frac{200-190}{\frac{48}{\sqrt{10}}}\approx\frac{10}{15.18}\approx 0.66$$

$P(z<0.21)=0.5832$

$P(z<0.66)=0.7454$

There is about a 58% chance that an LCD computer monitor will cost less than $200. There is about a 75% chance that the mean of a sample of 10 LCD computer monitors is less than $200.

5.4 EXERCISE SOLUTIONS

1. $\mu_{\bar{x}}=\mu=150$

$\sigma_{\bar{x}}=\frac{\sigma}{\sqrt{n}}=\frac{25}{\sqrt{50}}\approx 3.536$

3. $\mu_{\bar{x}}=\mu=790$

$\sigma_{\bar{x}}=\frac{\sigma}{\sqrt{n}}=\frac{48}{\sqrt{250}}\approx 3.036$

5. False. As the size of a sample increases, the mean of the distribution of sample means does not change.

7. False. A sampling distribution is normal if either $n\geq 30$ or the population is normal.

9. (c) Because $\mu_{\bar{x}}=16.5,\ \sigma_{\bar{x}}=\frac{\sigma}{\sqrt{n}}=\frac{11.9}{\sqrt{100}}=1.19$, and the graph approximates a normal curve.

11. (a) The mean and the standard deviation of the population (64, 48, 19, 79, 56) are:
$\mu=53.2,\ \sigma\approx 19.9$

(b)

Sample	Mean
19, 19	19
19, 48	33.5
19, 56	37.5
19, 64	41.5
19, 79	49
48, 19	33.5
48, 48	48
48, 56	52
48, 64	56

Sample	Mean
48, 79	63.5
56, 19	37.5
56, 48	52
56, 56	56
56, 64	60
56, 79	67.5
64, 19	41.5
64, 48	56
64, 56	60

Sample	Mean
64, 64	64
64, 79	71.5
79, 19	49
79, 48	63.5
79, 56	67.5
79, 64	71.5
79, 79	79

(c) $\mu_{\bar{x}} = 53.2,\ \sigma_{\bar{x}} \approx \dfrac{19.9}{\sqrt{2}} \approx 14.1$

The means are equal, but the standard deviation of the sampling distribution is smaller.

13. (a) The mean and the standard deviation of the population (350, 399, 418) are:
$\mu = 389,\ \sigma \approx 28.65$

(b)

Sample	Mean
350, 350, 350	350
350, 350, 399	366.33
350, 350, 418	372.67
350, 399, 350	366.33
350, 399, 399	382.67
350, 399, 418	389
350, 418, 350	372.67
350, 418, 399	389
350, 418, 418	395.33

Sample	Mean
399, 350, 350	366.33
399, 350, 399	382.67
399, 350, 418	389
399, 399, 350	382.67
399, 399, 399	399
399, 418, 350	389
399, 418, 399	405.33
399, 418, 418	411.67
418, 350, 350	372.67

Sample	Mean
418, 350, 399	389
418, 350, 418	395.33
418, 399, 350	389
418, 399, 399	405.33
418, 399, 418	411.67
418, 418, 350	395.33
418, 418, 399	411.67
418, 418, 418	418

(c) $\mu_{\bar{x}} = 389,\ \sigma_{\bar{x}} \approx \dfrac{28.65}{\sqrt{3}} \approx 16.54$

The means are equal, but the standard deviation of the sampling distribution is smaller.

15. $z = \dfrac{\bar{x} - \mu}{\frac{\sigma}{\sqrt{n}}} = \dfrac{24.3 - 24}{\frac{1.25}{\sqrt{64}}} \approx \dfrac{0.3}{0.156} \approx 1.92$

$P(\bar{x} < 24.3) = P(z < 1.92) = 0.9726$

The probability is not unusual because it is greater than 0.05.

17. $z = \dfrac{\bar{x} - \mu}{\frac{\sigma}{\sqrt{n}}} = \dfrac{551 - 550}{\frac{3.7}{\sqrt{45}}} \approx \dfrac{1}{0.552} \approx 1.8130$

$P(\bar{x} > 551) = P(z > 1.8130) = 1 - P(z < 1.8130) \approx 1 - 0.9651 = 0.0349$ (using technology)

The probability is unusual because it is less than 0.05.

19. $\mu_{\bar{x}} = 495$

$$\sigma_{\bar{x}} = \frac{\sigma}{\sqrt{n}} = \frac{120}{\sqrt{20}} \approx 26.83$$

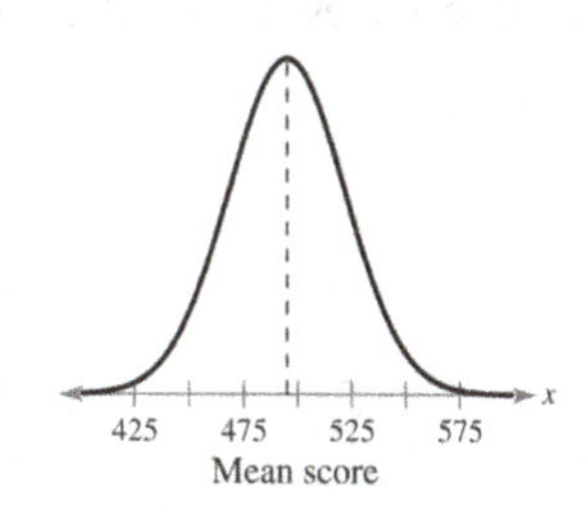

21. $\mu_{\bar{x}} = 23$

$$\sigma_{\bar{x}} = \frac{\sigma}{\sqrt{n}} = \frac{1.3}{\sqrt{25}} = 0.26$$

21 22 23 24 25
x
Mean temperature
(in degrees Celsius)

23. $\mu_{\bar{x}} = 1.64$

$$\sigma_{\bar{x}} = \frac{\sigma}{\sqrt{n}} = \frac{2.89}{\sqrt{12}} \approx 0.83$$

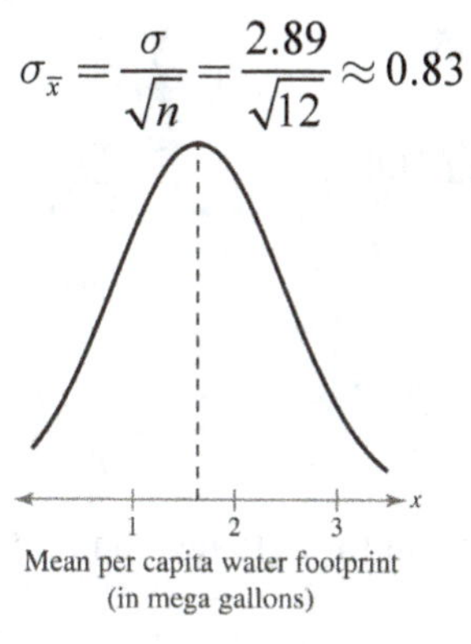

25. $\mu_{\bar{x}} = 132{,}000$

$$\sigma_{\bar{x}} = \frac{\sigma}{\sqrt{n}} = \frac{18{,}000}{\sqrt{35}} \approx 3042.56$$

130,000 135,000
x
Mean salary (in dollars)

27. $n = 40$: $\mu_{\bar{x}} = 495$, $\sigma_{\bar{x}} = \frac{\sigma}{\sqrt{n}} = \frac{120}{\sqrt{40}} \approx 18.97$

$n = 60$: $\mu_{\bar{x}} = 495$, $\sigma_{\bar{x}} = \frac{\sigma}{\sqrt{n}} = \frac{120}{\sqrt{60}} \approx 15.49$

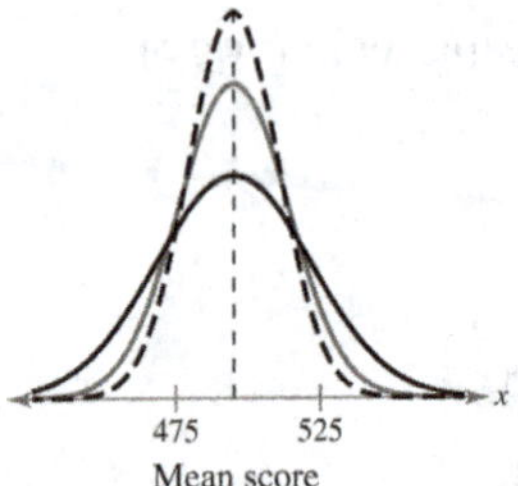

As the sample size increases, the standard deviation of the sample means decreases, while the mean of the sample means remains constant.

29. $z=\dfrac{\overline{x}-\mu}{\dfrac{\sigma}{\sqrt{n}}}=\dfrac{200-456}{\dfrac{1215}{\sqrt{32}}}\approx-1.1919$

$z=\dfrac{\overline{x}-\mu}{\dfrac{\sigma}{\sqrt{n}}}=\dfrac{500-456}{\dfrac{1215}{\sqrt{32}}}\approx0.2049$

$P(200<\overline{x}<500)=P(-1.1919<z<0.2049)\approx0.5812-0.1167=0.4645$
(using technology)
About 46% of samples of 32 years will have a mean gain between 200 and 500.

31. $z=\dfrac{\overline{x}-\mu}{\dfrac{\sigma}{\sqrt{n}}}=\dfrac{2.6-2.25}{\dfrac{1.30}{\sqrt{30}}}\approx1.4746$

$P(\overline{x}>2.6)=P(z>1.4746)\approx1-0.9298=0.0702$ (using technology)
About 7% of samples of 30 Chinese cities will have a mean childhood asthma rate greater than 2.6%.

33. $z=\dfrac{x-\mu}{\sigma}=\dfrac{3.2-2.25}{1.30}\approx0.7308$

$P(x<3.2)=P(z<0.7308)=0.7675$ (using technology)

$z=\dfrac{\overline{x}-\mu}{\dfrac{\sigma}{\sqrt{n}}}=\dfrac{3.2-2.25}{\dfrac{1.30}{\sqrt{10}}}\approx2.3109$

$P(\overline{x}<3.2)=P(z<2.3109)\approx0.9896$ (using technology)
It is more likely to select a sample of 10 cities with a mean childhood asthma prevalence less than 3.2% because the sample of 10 has a higher probability.

35. $z=\dfrac{\overline{x}-\mu}{\dfrac{\sigma}{\sqrt{n}}}=\dfrac{127.9-128}{\dfrac{0.20}{\sqrt{40}}}\approx\dfrac{-0.1}{0.0316}\approx-3.1623$

$P(\overline{x}<127.9)=P(z<-3.1623)\approx0.0008$ (using technology)
Yes, it is very unlikely that you would have randomly sampled 40 cans with a mean less than or equal to 127.9 ounces because it is more than 3 standard deviations from the mean of the sample means.

37. (a) $\mu = 96;\ \sigma = 0.5$

$$z = \frac{x - \mu}{\sigma} = \frac{96.25 - 96}{0.5} = 0.5$$

$P(x > 96.25) = P(z > 0.5) = 1 - P(z < 0.5) \approx 1 - 0.6915 = 0.3085$
(using technology)

(b) $$z = \frac{\bar{x} - \mu}{\frac{\sigma}{\sqrt{n}}} = \frac{96.25 - 96}{\frac{0.5}{\sqrt{40}}} \approx \frac{0.25}{0.079} \approx 3.1623$$

$P(\bar{x} > 96.25) = P(z > 3.1623) = 1 - P(z < 3.1623) \approx 1 - 0.9992 = 0.0008$
(using technology)

(c) Although there is about a 31% chance that a board cut by the machine will have a length greater than 96.25 inches, there is less than a 1% chance that the mean of a sample of 40 boards cut by the machine will have a length greater than 96.25 inches. Because there is less than a 1% chance that the mean of a sample of 40 boards will have a length greater than 96.25 inches, this is an unusual event.

39. Use the finite correction factor because $n = 55 > 0.05(1000) = 50$.

$$z = \frac{\bar{x} - \mu}{\frac{\sigma}{\sqrt{n}}\sqrt{\frac{N-n}{N-1}}} = \frac{50 - 47.12}{\frac{48.24}{\sqrt{55}}\sqrt{\frac{1000-55}{1000-1}}} \approx 0.4552$$

$P(\bar{x} < 50) \approx P(z < 0.4552) \approx 0.6755$
(using technology)

41. $$\mu = p = 0.63;\ \sigma = \sqrt{\frac{pq}{n}} = \sqrt{\frac{(0.63)(0.37)}{105}} \approx 0.0471$$

$$z = \frac{\hat{p} - p}{\sqrt{\frac{pq}{n}}} = \frac{0.55 - 0.63}{\sqrt{\frac{0.63(0.37)}{105}}} = \frac{-0.08}{0.0471} = -1.6979$$

$P(\hat{p} < 0.55) = P(z < -1.6979) = 0.0448$
(using technology)
The probability that less than 55% of a sample of 105 residents are in favor of building a new high school is about 4.5%. Because the probability is less than 0.05, this is an unusual event.

5.5 NORMAL APPROXIMATIONS TO BINOMIAL DISTRIBUTIONS

5.5 TRY IT YOURSELF SOLUTIONS

1. $n = 100,\ p = 0.29,\ q = 0.71$
$np = 100(0.29) = 29,\ nq = 100(0.71) = 71$

Because $np \geq 5$ and $nq \geq 5$, the normal distribution can be used.

$\mu = np = 100(0.29) = 29$

$\sigma = \sqrt{npq} = \sqrt{100(0.29)(0.71)} \approx 4.54$

2. (1) The discrete midpoint values are 57, 58, …, 83.
The corresponding interval is $56.5 < x < 83.5$.
The normal probability distribution is $P(56.5 < x < 83.5)$.

(2) The discrete midpoint values are …, 52, 53, 54.
The corresponding interval is $x < 54.5$.
The normal probability distribution is $P(x < 54.5)$.

3. $\mu = np = 100(0.29) = 29$

$\sigma = \sqrt{npq} = \sqrt{100(0.29)(0.71)} \approx 4.54$

$P(x > 30.5)$

$$z = \frac{x-\mu}{\sigma} \approx \frac{30.5-29}{4.54} \approx 0.330$$

$P(x > 30.5) = P(z > 0.330) = 1 - P(z < 0.330) = 1 - 0.6293 = 0.3707$

4. $P(x < 80.5)$

$$z = \frac{x-\mu}{\sigma} \approx \frac{80.5-94}{\sqrt{200(0.47)(0.53)}} \approx -1.91$$

$P(x < 80.5) = P(z < -1.91) = 0.0281$

5. $n = 75, \ p = 0.32$

$np = 75(0.32) = 24 \geq 5$ and $nq = 75(0.68) = 51 \geq 5$

The normal distribution can be used.

$\mu = np = 75(0.32) = 24$

$\sigma = \sqrt{npq} = \sqrt{75(0.32)(0.68)} \approx 4.04$

$P(14.5 < x < 15.5)$

$$z = \frac{x-\mu}{\sigma} = \frac{14.5-24}{4.04} \approx -2.35$$

$$z = \frac{x-\mu}{\sigma} = \frac{15.5-24}{4.04} \approx -2.10$$

$P(z < -2.35) = 0.0094$

$P(z < -2.10) = 0.0179$

$P(-2.35 < z < -2.10) = 0.0179 - 0.0094 = 0.0085$

5.5 EXERCISE SOLUTIONS

1. $np = (24)(0.85) = 20.4 \geq 5$

$nq = (24)(0.15) = 3.6 < 5$

Cannot use normal distribution.

3. $np = (18)(0.90) = 16.2 \geq 5$

$nq = (18)(0.10) = 1.8 < 5$

Cannot use normal distribution.

5. a

7. c

9. The probability of getting fewer than 25 successes; $P(x < 24.5)$

11. The probability of getting exactly 33 successes; $P(32.5 < x < 33.5)$

13. The probability of getting at most 150 successes; $P(x < 150.5)$

15. Binomial: $P(5 \leq x \leq 7) = P(x = 5) + P(x = 6) + P(x = 7)$

$$= {}_{16}C_5(0.4)^5(0.6)^{11} + {}_{16}C_6(0.4)^6(0.6)^{10} + {}_{16}C_7(0.4)^7(0.6)^9$$

$$\approx 0.549$$

Normal: $\mu = np = 16(0.4) = 6.4$, $\sigma = \sqrt{npq} = \sqrt{16(0.4)(0.6)} \approx 1.9596$

$$z = \frac{x - \mu}{\sigma} \approx \frac{4.5 - 6.4}{1.9596} \approx -0.9696$$

$$z = \frac{x - \mu}{\sigma} \approx \frac{7.5 - 6.4}{1.9596} \approx 0.5613$$

$$P(5 \leq x \leq 7) \approx P(4.5 \leq x \leq 7.5) = P(-0.9696 \leq z \leq 0.5613)$$

$$\approx 0.71272 - 0.16613 = 0.5466$$

(using technology)

The results are about the same.

17. $n = 30$, $p = 0.31$, $q = 0.69$

$np = 30(0.31) = 9.3 \geq 5$, $nq = 30(0.69) = 20.7 \geq 5$

Can use normal distribution.

$\mu = np = (30)(0.31) = 9.3$

$\sigma = \sqrt{npq} = \sqrt{(30)(0.31)(0.69)} \approx 2.533$

19. $n = 100$, $p = 0.41$

$np = 100(0.41) = 41 \geq 5$, $nq = 100(0.59) = 59 \geq 5$

$\mu = np = 100(0.41) = 41$, $\sigma = \sqrt{npq} = \sqrt{100(0.41)(0.59)} \approx 4.918$

Can use normal distribution.

(a) $z = \dfrac{x - \mu}{\sigma} \approx \dfrac{39.5 - 41}{4.918} \approx -0.3050$

$z = \dfrac{x - \mu}{\sigma} \approx \dfrac{40.5 - 41}{4.918} \approx -0.1017$

$$P(x=40)=P(39.5<x<40.5)$$
$$\approx P(-0.3050<z<-0.1017) \text{ (using technology)}$$
$$\approx 0.4595-0.3802\approx 0.0793$$

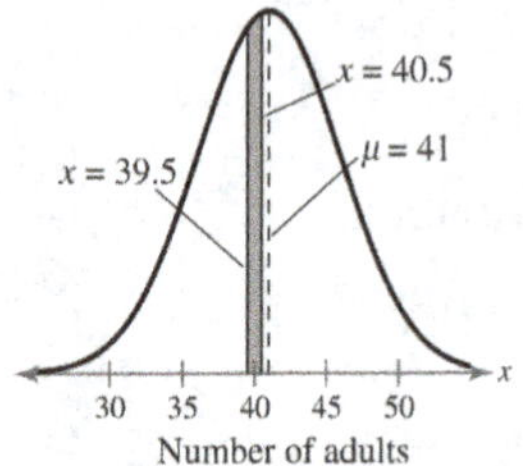

(b) $P(x\geq 40)=P(x>39.5)$
$$=P(z>0-0.3050)=1-P(z<-0.3050) \text{ (using technology)}$$
$$\approx 1-0.3802=0.6198$$

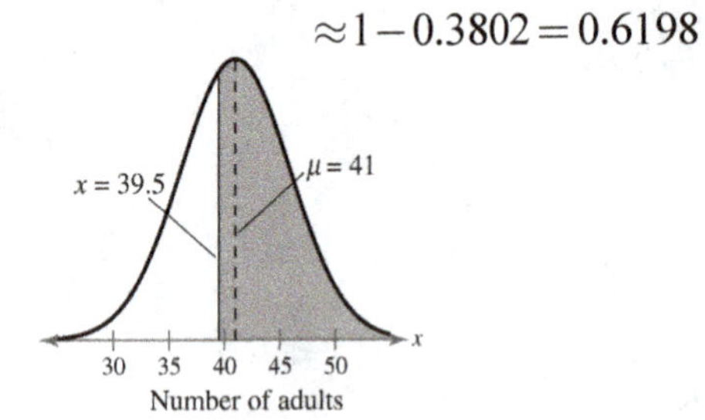

(c) $P(x<40)=P(x<39.5)=P(z<-0.3050)\approx 0.3802$

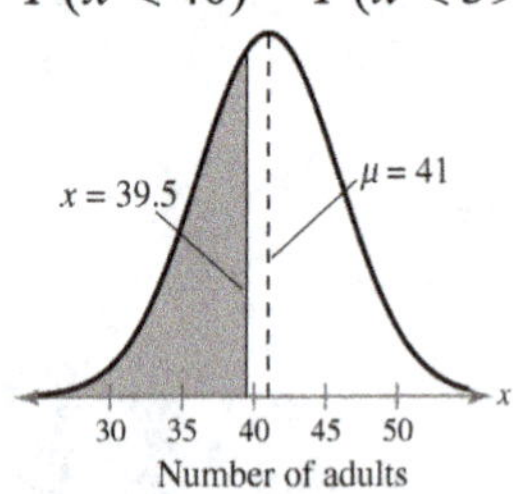

No unusual events because all of the probabilities are greater than 0.05.

21. $n=150,\ p=0.28$

$np=150(0.28)=42\geq 5,\ nq=150(0.72)=108\geq 5$

$\mu=np=150(0.28)=42,\ \sigma=\sqrt{npq}=\sqrt{150(0.28)(0.72)}\approx 5.499$

Can use normal distribution.

(a) $z=\dfrac{x-\mu}{\sigma}\approx\dfrac{40.5-42}{5.499}\approx -0.2728$

$P(x\leq 40)\approx P(x<40.5)=P(z<-0.2728)\approx 0.3925$ (using technology)

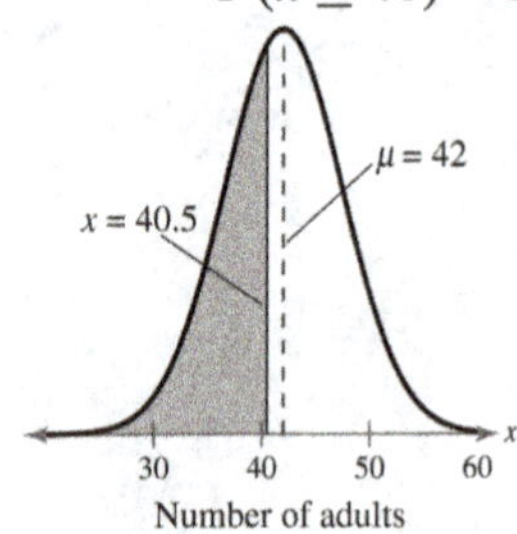

(b) $z = \dfrac{x-\mu}{\sigma} \approx \dfrac{50.5-42}{5.499} \approx 1.5457$

$$\begin{aligned} P(x > 50) &\approx P(x > 50.5) \\ &\approx P(z > 1.5457) = 1 - P(z < 1.5457) \\ &\approx 1 - 0.9389 \approx 0.0611 \end{aligned} \text{(using technology)}$$

$\mu = 42$

$x = 50.5$

30 40 50 60

Number of adults

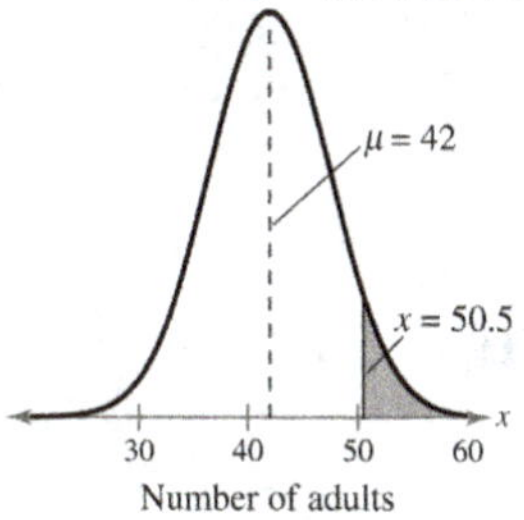

(c) $z = \dfrac{x-\mu}{\sigma} \approx \dfrac{19.5-42}{5.499} \approx -4.0917$

$z = \dfrac{x-\mu}{\sigma} \approx \dfrac{30.5-42}{5.499} \approx -2.0913$

$$\begin{aligned} P(20 \le x \le 30) &\approx P(19.5 < x < 30.5) \\ &\approx P(-4.0917 < z < -2.0913) \\ &\approx 0.01825 - 0.00002 \approx 0.0182 \end{aligned} \text{(using technology)}$$

$\mu = 42$

$x = 30.5$

$x = 19.5$

20 30 40 50 60

Number of adults

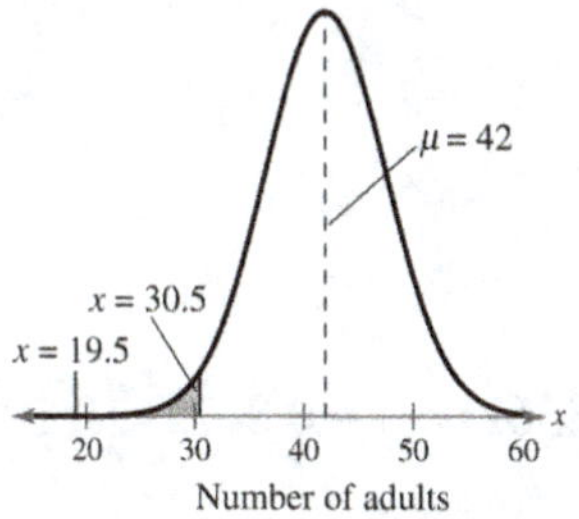

The event in part (c) is unusual because its probability is less than 0.05.

23. $n = 14,\ p = 0.33$

$np = 14(0.33) = 4.62 < 5,\ nq = 14(0.67) = 9.38 \ge 5$

Cannot use normal distribution because $np < 5$.

(a) $P(x = 8) = {}_{14}C_8 (0.33)^8 (0.67)^6 = 0.0382$

(b) $$\begin{aligned} P(x \le 4) &= P(x=0) + P(x=1) + P(x=2) + P(x=3) + P(x=4) \\ &= {}_{14}C_0 (0.33)^0 (0.67)^{14} + {}_{14}C_1 (0.33)^1 (0.67)^{13} + {}_{14}C_2 (0.33)^2 (0.67)^{12} \\ &\quad + {}_{14}C_3 (0.33)^3 (0.67)^{11} + {}_{14}C_4 (0.33)^4 (0.67)^{10} \\ &\approx 0.00367 + 0.02533 + 0.08109 + 0.15976 + 0.21639 \approx 0.4862 \end{aligned}$$

(c) $P(x<6)=P(x=0)+P(x=1)+P(x=2)+P(x=3)+P(x=4)+P(x=5)$

$$={}_{14}C_0(0.33)^0(0.67)^{14}+{}_{14}C_1(0.33)^1(0.67)^{13}+{}_{14}C_2(0.33)^2(0.67)^{12}$$
$$+{}_{14}C_3(0.33)^3(0.67)^{11}+{}_{14}C_4(0.33)^4(0.67)^{10}+{}_{14}C_5(0.33)^5(0.67)^9$$
$$\approx 0.00367+0.02533+0.08109+0.15976+0.21639+0.21316\approx 0.6994$$

The event is part (a) is unusual because its probability is less than 0.05.

25. $n=250,\ p=0.51,\ q=0.49$

$np=250(0.51)=127.5\geq 5,\ nq=250(0.49)=122.5\geq 5$

$\mu=np=250(0.51)=127.5,\ \sigma=\sqrt{npq}=\sqrt{250(0.51)(0.49)}\approx 7.9041$

Can use normal distribution.

(a) $z=\dfrac{x-\mu}{\sigma}\approx\dfrac{125.5-127.5}{7.9041}\approx -0.2530$

$P(x\leq 125)\approx P(x<125.5)\approx P(z<-0.2530)\approx 0.4001$ (using technology)

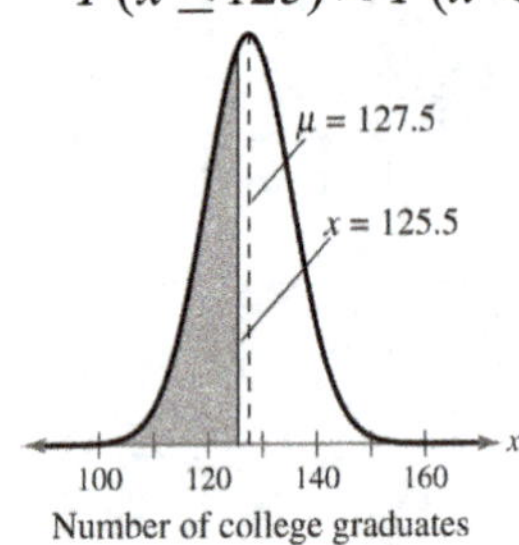

(b) $z=\dfrac{x-\mu}{\sigma}\approx\dfrac{134.5-127.5}{7.9041}\approx 0.88562$

$P(x\geq 135)\approx P(x>134.5)\approx P(z>0.88562)\approx 1-P(z<0.88562)\approx 1-0.8121\approx 0.1879$

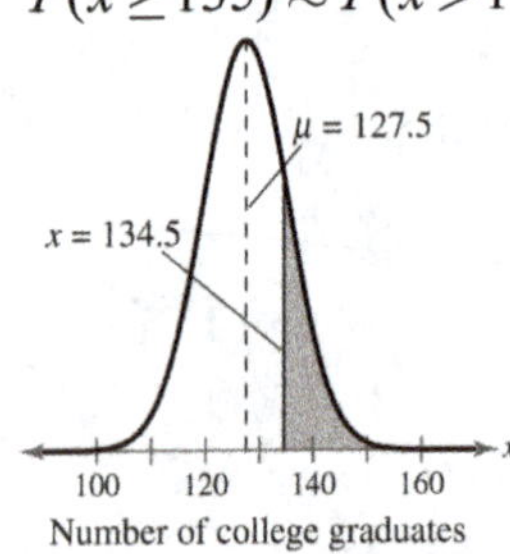

(c) $z=\dfrac{x-\mu}{\sigma}\approx\dfrac{99.5-127.5}{7.9041}\approx -3.5425$

$z=\dfrac{x-\mu}{\sigma}\approx\dfrac{125.5-127.5}{7.9041}\approx -0.2530$

$P(100<x<125)\approx P(99.5<x<125.5)$

$\approx P(-3.5425<z<-0.2530)\approx 0.4001-0.0002\approx 0.3999$ (using technology)

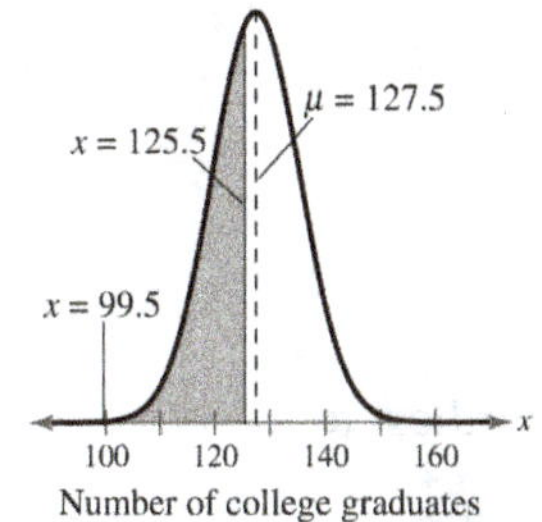

No unusual events because all of the probabilities are greater than 0.05.

27. (a) $n=700,\ p=0.033$

$np=700(0.033)=23.1\geq 5,\ nq=700(0.967)=676.9\geq 5$

Can use normal distribution.

$\mu=np=700(0.033)=23.1,\quad \sigma=\sqrt{npq}=\sqrt{700(0.033)(0.967)}\approx 4.7263$

$P(x\geq 30)\approx P(x>29.5)$

$\approx P(z>1.35413)=1-P(z<1.35413)$ (using technology)

$\approx 1-0.9122\approx 0.0878$

(b) $n=750,\ p=0.033$

$np=750(0.033)=24.75\geq 5,\ nq=750(0.967)=725.25\geq 5$

Can use normal distribution.

$\mu=np=750(0.033)=24.75,\quad \sigma=\sqrt{npq}=\sqrt{750(0.033)(0.967)}\approx 4.8922$

$P(x\geq 30)\approx P(x>29.5)$

$\approx P(z>0.97094)=1-P(z<0.97094)$ (using technology)

$\approx 1-0.8342\approx 0.1658$

(c) $n=1000,\ p=0.033$

$np=1000(0.033)=33\geq 5,\ nq=1000(0.967)=967\geq 5$

Can use normal distribution.

$\mu=np=1000(0.033)=33,\quad \sigma=\sqrt{npq}=\sqrt{1000(0.033)(0.967)}\approx 5.64898$

$P(x\geq 30)\approx P(x>29.5)$

$\approx P(z>-0.61958)=1-P(z<-0.61958)$ (using technology)

$\approx 1-0.2678\approx 0.7322$

29. $n=250,\ p=0.70,\ \mu=np=250(0.70)=175,\ \sigma=\sqrt{npq}=\sqrt{250(0.70)(0.30)}\approx 7.2457$

60% say no $\rightarrow n=250(0.60)=150$ say no while 100 say yes.

$$z=\frac{x-\mu}{\sigma}\approx\frac{100.5-175}{7.2457}\approx -10.28$$

$P(\text{less than or equal to 100 say yes})=P(x\leq 100)\approx P(x<100.5)\approx P(z<-10.28)\approx 0$

It is highly unlikely that 60% responded no. Answers will vary.

31. $n=100,\ p=0.75,\ \mu=np=100(0.75)=75,\ \sigma=\sqrt{npq}=\sqrt{100(0.75)(0.25)}\approx 4.3301$

$$z=\frac{x-\mu}{\sigma}\approx\frac{69.5-75}{4.3301}\approx -1.2702$$

$P(\text{reject claim})=P(x<70)\approx P(x<69.5)\approx P(z<-1.2702)\approx 0.1020$

CHAPTER 5 REVIEW EXERCISE SOLUTIONS

1. $\mu = 15,\ \sigma = 3$

3. Curve *B* has the greatest mean because its line of symmetry occurs the farthest to the right.

5. 0.6772

7. 0.6293

9. $1 - 0.2843 = 0.7157$

11. 0.00236 (using technology)

13. $0.5 - 0.0505 = 0.4995$

15. $0.95637 - 0.51994 \approx 0.4364$ (using technology)

17. $0.0668 + 0.0668 = 0.1336$

19. $x = 17 \rightarrow z = \dfrac{x-\mu}{\sigma} = \dfrac{17-21.3}{6.5} \approx -0.66$

$x = 29 \rightarrow z = \dfrac{x-\mu}{\sigma} = \dfrac{29-21.3}{6.5} \approx 1.18$

$x = 8 \rightarrow z = \dfrac{x-\mu}{\sigma} = \dfrac{8-21.3}{6.5} \approx -2.05$

$x = 23 \rightarrow z = \dfrac{x-\mu}{\sigma} = \dfrac{23-21.3}{6.5} \approx 0.26$

21. $P(z < 1.28) = 0.8997$

23. $P(-2.15 < x < 1.55) = 0.93943 - 0.01578 \approx 0.9237$ (using technology)

25. $P(z < -2.50 \text{ or } z > 2.50) = 2(0.0062) = 0.0124$

27. $z = \dfrac{x-\mu}{\sigma} = \dfrac{84-74}{8} = 1.25$

$P(x < 84) = P(z < 1.25) = 0.8944$

29. $z = \dfrac{x-\mu}{\sigma} = \dfrac{80-74}{8} = 0.75$

$P(x > 80) = P(z > 0.75) = 1 - P(z < 0.75) = 1 - 0.7734 = 0.2266$

31. $z = \dfrac{x-\mu}{\sigma} = \dfrac{60-74}{8} = -1.75$

$z = \dfrac{x-\mu}{\sigma} = \dfrac{70-74}{8} = -0.5$

$P(60 < x < 70) = P(-1.75 < z < -0.5) \approx 0.30854 - 0.04001 \approx 0.2685$ (using technology)

33. (a) $z = \dfrac{x-\mu}{\sigma} = \dfrac{12.3-14.7}{11.5} \approx -0.2087$

$P(x < 12.3) \approx P(z < -0.2087) \approx 0.4173$ (using technology)

(b) $z = \dfrac{x-\mu}{\sigma} = \dfrac{15.4-14.7}{11.5} \approx 0.0609$

$z = \dfrac{x-\mu}{\sigma} = \dfrac{19.6-14.7}{11.5} \approx 0.4261$

$P(15.4 < x < 19.6) \approx P(0.0609 < z < 0.4261) \approx 0.6650 - 0.5243 \approx 0.1407$
(using technology)

(c) $z = \dfrac{x-\mu}{\sigma} = \dfrac{17.7-14.7}{11.5} \approx 0.2609$

$P(x > 17.7) \approx P(z > 0.2609) = 1 - P(z < 0.2609) \approx 1 - 0.6029 = 0.3971$ (using technology)

35. No unusual events because all of the probabilities are greater than 0.05.

37. $z = -0.07$

39. $z = 2.457$ (using technology)

41. $z = 1.04$

43. 30.5% to right $\Rightarrow$ 69.5% to left, $z = 0.51$

45. $x = \mu + z\sigma = 132 + (-2.75)(4.53) \approx 119.54$ feet

47. 90th percentile $\Rightarrow$ Area $= 0.90 \Rightarrow z = 1.2816$

$x = \mu + z\sigma = 132 + (1.2816)(4.53) \approx 137.81$ feet

49. Top 15% $\Rightarrow$ Area $= 0.85 \Rightarrow z = 1.036$ (using technology)

$x = \mu + z\sigma = 132 + (1.036)(4.53) \approx 136.70$ feet

51. (a) $\mu = 1.5$, $\sigma \approx 1.118$

(b)

Sample	Mean
0, 0	0
0, 1	0.5
0, 2	1
0, 3	1.5
1, 0	0.5
1, 1	1
1, 2	1.5
1, 3	2

Sample	Mean
2, 0	1
2, 1	1.5
2, 2	2
2, 3	2.5
3, 0	1.5
3, 1	2
3, 2	2.5
3, 3	3

(c) $\mu_{\bar{x}} = 1.5$, $\sigma_{\bar{x}} \approx \dfrac{1.118}{\sqrt{2}} \approx 0.791$

The means are equal, but the standard deviation of the sampling distribution is smaller.

53. $\mu_{\bar{x}} = 471.5,\ \sigma_{\bar{x}} = \dfrac{\sigma}{\sqrt{n}} = \dfrac{187.9}{\sqrt{35}} \approx 31.761$

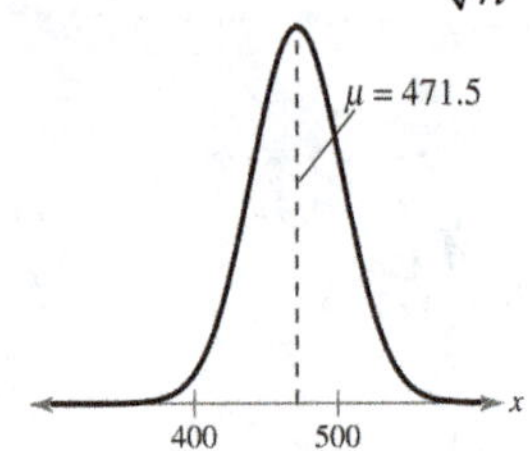

55. (a) $z = \dfrac{\bar{x}-\mu}{\frac{\sigma}{\sqrt{n}}} = \dfrac{12.3-14.7}{\frac{11.5}{\sqrt{2}}} \approx -0.29514$

$P(\bar{x} < 12.3) \approx P(z < -0.29514) \approx 0.3839$ (using technology)

(b) $z = \dfrac{\bar{x}-\mu}{\frac{\sigma}{\sqrt{n}}} = \dfrac{15.4-14.7}{\frac{11.5}{\sqrt{2}}} \approx 0.0861$

$z = \dfrac{\bar{x}-\mu}{\frac{\sigma}{\sqrt{n}}} = \dfrac{19.6-14.7}{\frac{11.5}{\sqrt{2}}} \approx 0.6026$

$P(15.4 < \bar{x} < 19.6) \approx P(0.0861 < z < 0.6026) \approx 0.7266 - 0.5343 \approx 0.1923$
(using technology)

(c) $z = \dfrac{\bar{x}-\mu}{\frac{\sigma}{\sqrt{n}}} = \dfrac{17.7-14.7}{\frac{11.5}{\sqrt{2}}} \approx 0.3689$

$P(\bar{x} > 17.7) = P(z > 0.3689) = 1 - P(z < 0.3689) \approx 1 - 0.6439 \approx 0.3561$ (using technology)

The probabilities in parts (a) and (c) are smaller, and the probability in part (b) is larger. This is to be expected because the standard error of the sample mean is smaller.

57. (a) $z = \dfrac{\bar{x}-\mu}{\frac{\sigma}{\sqrt{n}}} = \dfrac{21.6-20.8}{\frac{5.6}{\sqrt{36}}} \approx 0.8571$

$P(\bar{x} < 21.6) = P(z < 0.8571) = 0.8043$

(b) $z = \dfrac{\bar{x}-\mu}{\frac{\sigma}{\sqrt{n}}} = \dfrac{19.8-20.8}{\frac{5.6}{\sqrt{36}}} \approx -1.0714$

$P(\bar{x} > 19.8) = P(z > -1.0714) \approx 1 - 0.1420 = 0.8580$

(c) $z = \dfrac{\overline{x} - \mu}{\frac{\sigma}{\sqrt{n}}} = \dfrac{20.5 - 20.8}{\frac{5.6}{\sqrt{36}}} \approx -0.3214$

$z = \dfrac{\overline{x} - \mu}{\frac{\sigma}{\sqrt{n}}} = \dfrac{21.5 - 20.8}{\frac{5.6}{\sqrt{36}}} = 0.75$

$P(20.5 < \overline{x} < 21.5) = P(-0.3214 < z < 0.75) \approx 0.7734 - 0.3740 \approx 0.3994$ (using technology)

59. (a) $z = \dfrac{\overline{x} - \mu}{\frac{\sigma}{\sqrt{n}}} = \dfrac{60,000 - 61,000}{\frac{11,000}{\sqrt{45}}} \approx -0.6098$

$P(\overline{x} < 60,000) \approx P(z < -0.6098) \approx 0.2710$ (using technology)

(b) $z = \dfrac{\overline{x} - \mu}{\frac{\sigma}{\sqrt{n}}} = \dfrac{63,000 - 61,000}{\frac{11,000}{\sqrt{45}}} \approx 1.2197$

$P(\overline{x} > 63,000) \approx P(z > 1.2197) \approx 1 - 0.8887 \approx 0.1113$ (using technology)

61. $n = 20,\ p = 0.75,\ q = 0.25$

$np = 20(0.75) = 15 \geq 5,\ nq = 20(0.25) = 5 \geq 5$

Can use the normal distribution.

$\mu = np = 20(0.75) = 15,\ \sigma = \sqrt{npq} = \sqrt{20(0.75)(0.25)} \approx 1.936$

63. The probability of getting at least 25 successes; $P(x \geq 25) \approx P(x > 24.5)$

65. The probability of getting exactly 45 successes; $P(x = 45) \approx P(44.5 < x < 45.5)$

67. The probability of getting less than 60 successes; $P(x < 60) \approx P(x < 59.5)$

69. $n = 70,\ p = 0.32,\ q = 0.68$

$np = 70(0.32) = 22.4 \geq 5,\ nq = 70(0.68) = 47.6 \geq 5$

Can use the normal distribution.

$\mu = np = 70(0.32) = 22.4\ \sigma = \sqrt{npq} = \sqrt{70(0.32)(0.68)} \approx 3.9028$

(a) $z = \dfrac{x - \mu}{\sigma} \approx \dfrac{15.5 - 22.4}{3.9028} \approx -1.7680$

$P(x \leq 15) \approx P(x < 15.5) = P(z < -1.7680) \approx 0.0385$ (using technology)

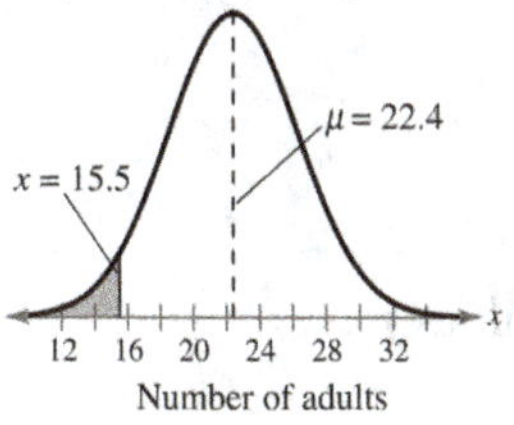

$x = 15.5$ $\mu = 22.4$

12 16 20 24 28 32 x

Number of adults

(b) $z=\dfrac{x-\mu}{\sigma}\approx\dfrac{24.5-22.4}{3.9028}\approx 0.5381$

$z=\dfrac{x-\mu}{\sigma}\approx\dfrac{25.5-22.4}{3.9028}\approx 0.7943$

$P(x=25)\approx P(24.5<x<25.5)\approx P(0.5381<z<0.7943)\approx 0.7865-0.7047\approx 0.0818$ (using technology)

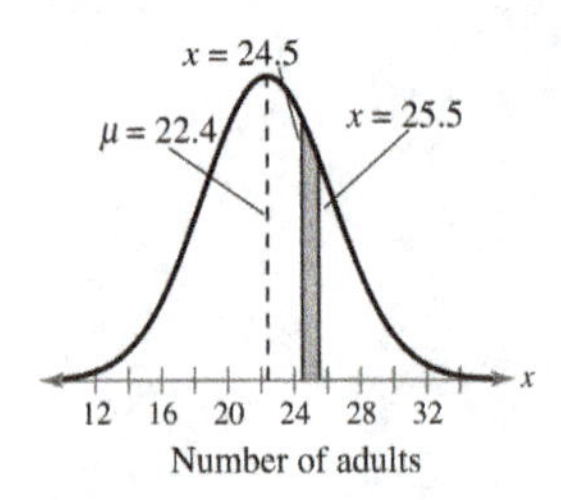

(c) $z=\dfrac{x-\mu}{\sigma}\approx\dfrac{30.5-22.4}{3.9028}\approx 2.0754$

$P(x>30)\approx P(x>30.5)\approx P(z>2.0754)\approx 1-0.9810\approx 0.0190$ (using technology)

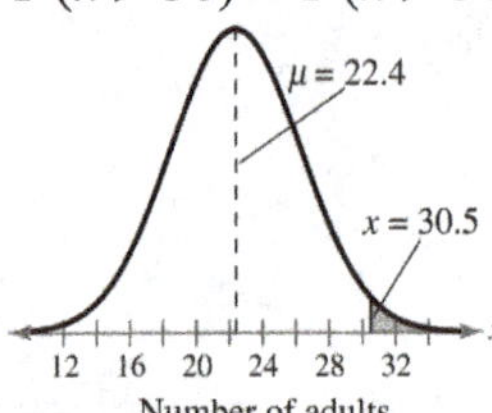

The events in parts (a) and (c) are unusual because their probabilities are less than 0.05.

CHAPTER 5 QUIZ SOLUTIONS

1. (a) $P(z>-1.68)=1-0.0465=0.9535$

(b) $P(z<2.23)=0.9871$

(c) $P(-0.47<z<0.47)\approx 0.6808-0.3192\approx 0.3616$ (using technology)

(d) $P(z<-1.992 \text{ or } z>-0.665)\approx 0.0232+(1-0.2530)=0.7702$ (using technology)

2. (a) $z=\dfrac{x-\mu}{\sigma}=\dfrac{5.97-9.2}{1.62}\approx -1.9938$

$P(x<5.97)\approx P(z<-1.9938)\approx 0.0231$ (using technology)

(b) $z=\dfrac{x-\mu}{\sigma}=\dfrac{40.5-87}{19}\approx -2.4474$

$P(x>40.5)\approx P(z>-2.4474)\approx 1-0.0072\approx 0.9928$ (using technology)

(c) $z = \dfrac{x-\mu}{\sigma} = \dfrac{5.36-5.5}{0.08} = -1.75$

$z = \dfrac{x-\mu}{\sigma} = \dfrac{5.64-5.5}{0.08} = 1.75$

$P(5.36 < x < 5.64) = P(-1.75 < z < 1.75) \approx 0.95994 - 0.04006 \approx 0.9199$ (using technology)

(d) $z = \dfrac{x-\mu}{\sigma} = \dfrac{19.6-18.5}{4.25} \approx 0.2588$

$z = \dfrac{x-\mu}{\sigma} = \dfrac{26.1-18.5}{4.25} \approx 1.7882$

$P(19.6 < x < 26.1) \approx P(0.2588 < z < 1.7882) \approx 0.96313 - 0.60211 \approx 0.3610$
(using technology)

3. $z = \dfrac{x-\mu}{\sigma} = \dfrac{125-100}{15} \approx 1.6667$

$P(x > 125) \approx P(z > 1.6667) \approx 1 - 0.9522 \approx 0.0478$ (using technology)

Yes, the event is unusual because its probability is less than 0.05.

4. $z = \dfrac{x-\mu}{\sigma} = \dfrac{95-100}{15} \approx -0.3333$

$z = \dfrac{x-\mu}{\sigma} = \dfrac{105-100}{15} \approx 0.3333$

$P(95 < x < 105) \approx P(-0.3333 < z < 0.3333) \approx 0.63056 - 0.36944 \approx 0.2611$ (using technology)

No, the event is not unusual because its probability is greater than 0.05.

5. $z = \dfrac{x-\mu}{\sigma} = \dfrac{112-100}{15} = 0.80$

$P(x > 112) = P(z > 0.80) = 1 - .7881 = 0.2119 \rightarrow 21.19\%$ (using technology)

6. $z = \dfrac{x-\mu}{\sigma} = \dfrac{90-100}{15} \approx -0.6667$

$P(x < 90) = P(z < -0.6667) = 0.2525$ (using technology)

$(2000)(0.2525) = 505$ people

7. Top 5% $\rightarrow z \approx 1.6449$

$\mu + z\sigma = 100 + (1.6449)(15) \approx 124.7 \rightarrow 125$

8. Bottom 10% $\rightarrow z \approx -1.2816$

$\mu + z\sigma = 100 + (-1.2816)(15) \approx 80.8 \rightarrow 80$

(Because you are finding the highest score that would still place you in the bottom 10%, round down, because if you rounded up you would be outside the bottom 10%.)

9. $z = \dfrac{\bar{x} - \mu}{\frac{\sigma}{\sqrt{n}}} = \dfrac{105 - 100}{\frac{15}{\sqrt{60}}} \approx \dfrac{5}{1.9365} \approx 2.5820$

$P(\bar{x} > 105) \approx P(z > 2.5820) \approx 1 - 0.9951 \approx 0.0049$ (using technology)

About 0.5% of samples of 60 students will have a mean IQ score greater than 105. This is a very unusual event.

10. $z = \dfrac{x - \mu}{\sigma} = \dfrac{105 - 100}{15} \approx 0.3333$

$P(x > 105) \approx P(z > 0.3333) \approx 1 - 0.6306 \approx 0.3694$ (using technology)

$z = \dfrac{\bar{x} - \mu}{\frac{\sigma}{\sqrt{n}}} = \dfrac{105 - 100}{\frac{15}{\sqrt{15}}} \approx \dfrac{5}{3.873} \approx 1.2910$

$P(\bar{x} > 105) \approx P(z > 1.2910) \approx 1 - 0.9016 \approx 0.0984$ (using technology)

You are more likely to select one student with a test score greater than 105 because the standard error of the mean is less than the standard deviation.

11. $n = 250,\ p = 0.16,\ q = 0.84$

$np = 250(0.16) = 40 \geq 5$, $nq = 250(0.84) = 210 \geq 5$

Can use normal distribution.

$\mu = np = 250(0.16) = 40,\ \sigma = \sqrt{npq} = \sqrt{250(0.16)(0.84)} \approx 5.797$

12. (a) $z = \dfrac{x - \mu}{\sigma} \approx \dfrac{40.5 - 40}{5.797} \approx 0.08625$

$P(x \leq 40) \approx P(x < 40.5) \approx P(z < 0.08625) \approx 0.5344$ (using technology)

(b) $z = \dfrac{x - \mu}{\sigma} \approx \dfrac{44.5 - 40}{5.797} \approx 0.7763$

$P(x < 45) \approx P(x < 44.5) \approx P(z < 0.7763) \approx 0.7812$ (using technology)

(c) $z = \dfrac{x - \mu}{\sigma} \approx \dfrac{47.5 - 40}{5.797} \approx 1.2938$

$z = \dfrac{x - \mu}{\sigma} \approx \dfrac{48.5 - 40}{5.797} \approx 1.4663$

$P(x = 48) \approx P(47.5 < x < 48.5) \approx P(1.2938 < z < 1.4663) \approx 0.9287 - 0.9021 \approx 0.0266$ (using technology)

The event in part (c) is unusual because its probability is less than 0.05.

CUMULATIVE REVIEW, CHAPTERS 3-5

1. (a) $np = 30(0.61) = 18.3 \geq 5,\ nq = 30(0.39) = 11.7 \geq 5$

Can use normal distribution.

(b) $\mu = np = 30(0.61) = 18.3,\ \sigma = \sqrt{npq} = \sqrt{30(0.61)(0.39)} \approx 2.6715$

$z = \dfrac{x-\mu}{\sigma} \approx \dfrac{14.5-18.3}{2.6715} \approx -1.4224$

$P(x \le 14) \approx P(x < 14.5) \approx P(z < -1.4224) \approx 0.0775$ (using technology)

(c) $z = \dfrac{x-\mu}{\sigma} \approx \dfrac{13.5-18.3}{2.6715} \approx -1.7967$

$P(x = 14) \approx P(13.5 < x < 14.5) \approx P(-1.7967 < z < -1.4224) \approx 0.07746 - 0.03619 \approx 0.0413$ (using technology)

Yes, because the probability is less than 0.05.

3.

x	$P(x)$	$xP(x)$	$x-\mu$	$(x-\mu)^2$	$(x-\mu)^2 P(x)$
0	0.113	0.000	-2.446	5.9829	0.6761
1	0.188	0.188	-1.446	2.0909	0.3931
2	0.188	0.376	-0.446	0.1989	0.0374
3	0.288	0.864	0.554	0.3069	0.0884
4	0.150	0.600	1.554	2.4149	0.3622
5	0.038	0.190	2.554	6.5229	0.2479
6	0.038	0.228	3.554	12.6309	0.4800
		$\sum xP(x) = 2.446$			$\sum (x-\mu)^2 P(x) \approx 2.285$

(a) $\mu = \sum xP(x) \approx 2.45$

(b) $\sigma^2 = \sum (x-\mu)^2 P(x) \approx 2.29$

(c) $\sigma = \sqrt{\sigma^2} \approx 1.51$

(d) $E(x) = \mu \approx 2.45$

The number of fouls per game for Garrett Temple is about 2.45 fouls. The standard deviation is 1.5, so most of Temple's games differ from the mean by no more than about 1 or 2 fouls.

5. (a) $(16)(15)(14)(13) = 43{,}680$

(b) $\dfrac{(7)(6)(5)(4)}{(16)(15)(14)(13)} = \dfrac{840}{43{,}680} \approx 0.0192$

7. 0.0010

9. $0.9984 - 0.500 = 0.4984$

11. $0.5478 + (1 - 0.9573) = 0.5905$

13. $p = \dfrac{1}{200} = 0.005$

(a) $P(x=5)=(0.005)(0.995)^4 \approx 0.0049$

(b) $P(x \leq 3) \approx (0.005)(0.995)^0 + (0.005)(0.995)^1 + (0.005)(0.995)^2$
$\approx 0.005 + 0.004975 + 0.004950 \approx 0.0149$

(c) $P(x > 20) = 1 - P(x \leq 20) \approx 1 - 0.0954 = 0.9046$

15. (a) $\mu_{\bar{x}} = 70$

$$\sigma_{\bar{x}} = \frac{\sigma}{\sqrt{n}} = \frac{1.2}{\sqrt{40}} \approx 0.1897$$

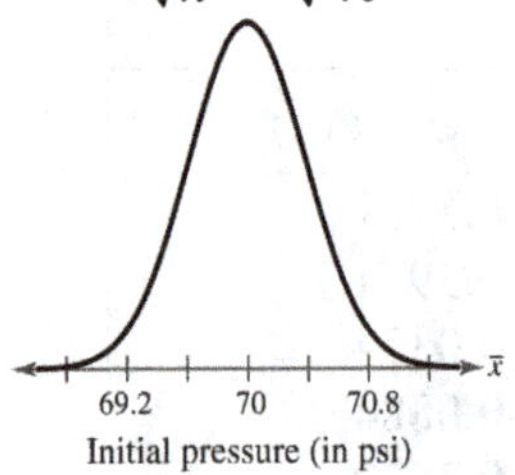

(b) $P(\bar{x} \leq 69) = P\left(z \leq \frac{69-70}{\frac{1.2}{\sqrt{15}}}\right) \approx P(z < -3.2275) \approx 0.0006$ (using technology)

17. (a) ${}_{12}C_4 = 495$

(b) $\frac{(1)(1)(1)(1)}{{}_{12}C_4} = 0.0020$

CHAPTER 6

Confidence Intervals

6.1 CONFIDENCE INTERVALS FOR THE MEAN (LARGE SAMPLES)

6.1 TRY IT YOURSELF SOLUTIONS

1. $\bar{x} = \frac{\sum x}{n} = \frac{626}{30} \approx 20.9$

Another point estimate for the population mean is about 20.9.

2. $z_c = 1.96,\ n = 30,\ \sigma = 2.3$

$E = z_c \frac{\sigma}{\sqrt{n}} \approx 1.96 \frac{2.3}{\sqrt{30}} \approx 0.8$

You are 95% confident that the margin of error for the population mean is about 0.8 hour.

3. $\bar{x} \approx 20.9,\ E \approx 0.8$

$\bar{x} - E \approx 20.9 - 0.8 = 20.1$

$\bar{x} + E \approx 20.9 + 0.8 = 21.7$

(20.1, 21.7)

This confidence interval is wider than the one found in Example 3.

4. 75% CI: (20.632, 21.468) $\Rightarrow$ (20.6, 21.5)

85% CI: (20.526, 21.574) $\Rightarrow$ (20.5, 21.6)

90% CI: (20.452, 21.648) $\Rightarrow$ (20.5, 21.7)

As the confidence level increases, so does the width of the interval.

5. $n = 30,\ \bar{x} = 22.9,\ \sigma = 1.5,\ z_c = 1.645$

$E = z_c \frac{\sigma}{\sqrt{n}} = 1.645 \frac{1.5}{\sqrt{30}} \approx 0.5$

$\bar{x} - E \approx 22.9 - 0.5 = 22.4$

$\bar{x} + E \approx 22.9 + 0.5 = 23.4$

(22.4, 23.4)

With 90% confidence, you can say that the mean age of the students is between 22.4 and 23.4 years. Because of the larger sample size, the confidence interval is slightly narrower.

6. $z_c = 1.96,\ E = 0.75,\ \sigma \approx 2.3$

$$n = \left(\frac{z_c \sigma}{E}\right)^2 \approx \left(\frac{1.96 \cdot 2.3}{0.75}\right)^2 \approx 36.13 \to 37$$

Because of the larger margin of error, the sample size needed is smaller.

6.1 EXERCISE SOLUTIONS

1. You are more likely to be correct using an interval estimate because it is unlikely that a point estimate will exactly equal the population mean.

3. d; As the level of confidence increases, z_c increases, causing wider intervals.

5. 1.28

7. 1.15

9. $\bar{x} - \mu = 3.8 - 4.27 = -0.47$

11. $\bar{x} - \mu = 26.43 - 24.67 = 1.76$

13. $E = z_c \frac{\sigma}{\sqrt{n}} = 1.96 \frac{5.2}{\sqrt{30}} \approx 1.861$

15. $E = z_c \frac{\sigma}{\sqrt{n}} = 1.28 \frac{1.3}{\sqrt{75}} \approx 0.192$

17. Because $c = 0.88$ is the lowest level of confidence, the interval associated with it will be the narrowest. Thus, this matches (c).

19. Because $c = 0.95$ is the third lowest level of confidence, the interval associated with it will be the third narrowest. Thus, this matches (b).

21. $\bar{x} \pm z_c \frac{\sigma}{\sqrt{n}} = 12.3 \pm 1.645 \frac{1.5}{\sqrt{50}} \approx 12.3 \pm 0.349 \approx (12.0,\ 12.6)$

23. $\bar{x} \pm z_c \frac{\sigma}{\sqrt{n}} = 10.5 \pm 2.575 \frac{2.14}{\sqrt{45}} \approx 10.5 \pm 0.821 \approx (9.7,\ 11.3)$

25. $(12.0, 14.8) \Rightarrow \bar{x} = \frac{14.8 + 12.0}{2} = 13.4,\ E = 14.8 - 13.4 = 1.4$

27. $(1.71, 2.05) \Rightarrow \bar{x} = \frac{2.05 + 1.71}{2} = 1.88,\ E = 2.05 - 1.88 = 0.17$

29. $c = 0.90 \Rightarrow z_c = 1.645$

$$n = \left(\frac{z_c \sigma}{E}\right)^2 = \left(\frac{(1.645)(6.8)}{1}\right)^2 \approx 125.13 \Rightarrow 126$$

31. $c = 0.80 \Rightarrow z_c = 1.28$

$$n = \left(\frac{z_c \sigma}{E}\right)^2 = \left(\frac{(1.28)(4.1)}{2}\right)^2 \approx 6.89 \Rightarrow 7$$

33. $(26.2, 30.1) \Rightarrow 2E = 30.1 - 26.2 = 3.9 \Rightarrow E = 1.95$ and $\bar{x} = 26.2 + E = 26.2 + 1.95 = 28.15$

35. 90% CI: $\bar{x} \pm z_c \frac{\sigma}{\sqrt{n}} = 1368.48 \pm 1.645 \frac{202.60}{\sqrt{48}} \approx 1368.48 \pm 48.104 \approx (1320.4,\ 1416.6)$

95% CI: $\bar{x} \pm z_c \frac{\sigma}{\sqrt{n}} = 1368.48 \pm 1.96 \frac{202.60}{\sqrt{48}} \approx 1368.48 \pm 57.316 \approx (1311.2,\ 1425.8)$

With 90% confidence, you can say that the population mean price is between $1320.40 and $1416.60.
With 95% confidence, you can say that the population mean price is between $1311.20 and $1425.80.
The 95% CI is wider.

37. $\bar{x} \pm z_c \frac{\sigma}{\sqrt{n}} = 84.13 \pm 1.645 \frac{14.56}{\sqrt{64}} \approx 84.13 \pm 2.994 \approx (81.14,\ 87.12)$

$\bar{x} \pm z_c \frac{\sigma}{\sqrt{n}} = 84.13 \pm 1.96 \frac{14.56}{\sqrt{64}} \approx 84.13 \pm 3.567 \approx (80.56,\ 87.70)$

With 90% confidence, you can say that the population mean temperature is between 81.14 °F and 87.12 °F.
With 95% confidence, you can say that the population mean temperature is between 80.56 °F and 87.70 °F.
The 95% CI is wider.

39. No; The margin of error is large ($E = 48.1$).

41. No; The right endpoint of the 95% CI is 87.70.

43. (a) An increase in the level of confidence will widen the confidence interval and the less certain you can be about a point estimate.

(b) An increase in the sample size will narrow the confidence interval because it decreases the standard error.

(c) An increase in the population standard deviation will widen the confidence interval because small standard deviations produce more precise intervals, which are smaller.

44. Answers will vary.

45. $\bar{x} = \frac{\sum x}{n} = \frac{686}{30} \approx 22.87$

90% CI: $\bar{x} \pm z_c \frac{\sigma}{\sqrt{n}} = 22.87 \pm 1.645 \frac{6.5}{\sqrt{30}} \approx 22.87 \pm 1.95 \approx (20.9,\ 24.8)$

99% CI: $\bar{x} \pm z_c \frac{\sigma}{\sqrt{n}} = 22.87 \pm 2.575 \frac{6.5}{\sqrt{30}} \approx 22.87 \pm 3.06 \approx (19.8,\ 25.9)$

With 90% confidence, you can say that the population mean length of time is between 20.9 and 24.8 minutes.
With 99% confidence, you can say that the population mean length of time is between 19.8 and 25.9 minutes.
The 99% CI is wider.

47. $n=\left(\frac{z_c\sigma}{E}\right)^2=\left(\frac{1.96\cdot 4.8}{1}\right)^2\approx 88.510\to 89$

49. (a) $n=\left(\frac{z_c\sigma}{E}\right)^2=\left(\frac{1.96\cdot 3.10}{0.75}\right)^2\approx 65.6\to 66$ servings

(b) 95% CI: $\bar{x}\pm z_c\frac{\sigma}{\sqrt{n}}=29\pm 1.96\frac{3.10}{\sqrt{66}}\approx 29\pm 0.748\approx(28.252,\ 29.748)$

No. The 95% CI is (28.252, 29.748). If the population mean is within 3% of the sample mean, then it falls outside the CI.

Yes. If the population mean is within 0.3% of the sample mean, then it falls within the CI.

51. (a) $n=\left(\frac{z_c\sigma}{E}\right)^2=\left(\frac{1.645\cdot 0.75}{0.5}\right)^2\approx 6.1\to 7$ cans

(b) 90% CI: $\bar{x}\pm z_c\frac{\sigma}{\sqrt{n}}=127.75\pm 1.645\frac{0.75}{\sqrt{8}}\approx 127.75\pm 0.44\approx(127.3,\ 128.2)$

Yes; The 90% CI is (127.3, 128.2) and 128 ounces falls within that interval.

53. (a) $n=\left(\frac{z_c\sigma}{E}\right)^2=\left(\frac{2.575\cdot 0.5}{0.15}\right)^2\approx 73.7\to 74$ balls

(b) 99% CI: $\bar{x}\pm z_c\frac{\sigma}{\sqrt{n}}=27.5\pm 2.575\frac{0.5}{\sqrt{84}}\approx 27.5\pm 0.140\approx(27.360,\ 27.640)$

Yes; The 99% CI is (27.360, 27.640) and there are amounts less than 27.6 inches that fall within that interval.

55. *Sample answer:* A 99% CI may not be practical to use in all situations. It may produce a CI so wide that is has no practical application.

57. (a) $\sqrt{\frac{N-n}{N-1}}=\sqrt{\frac{1000-500}{1000-1}}\approx 0.707$

(b) $\sqrt{\frac{N-n}{N-1}}=\sqrt{\frac{1000-100}{1000-1}}\approx 0.949$

(c) $\sqrt{\frac{N-n}{N-1}}=\sqrt{\frac{1000-75}{1000-1}}\approx 0.962$

(d) $\sqrt{\frac{N-n}{N-1}}=\sqrt{\frac{1000-50}{1000-1}}\approx 0.975$

(e) $\sqrt{\frac{N-n}{N-1}}=\sqrt{\frac{100-50}{100-1}}\approx 0.711$

(f) $\sqrt{\frac{N-n}{N-1}}=\sqrt{\frac{400-50}{400-1}}\approx 0.937$

(g) $\sqrt{\frac{N-n}{N-1}}=\sqrt{\frac{700-50}{700-1}}\approx 0.964$

(h) $\sqrt{\frac{N-n}{N-1}}=\sqrt{\frac{1200-50}{1200-1}}\approx 0.979$

The finite population correction factor approaches 1 as the population size increases and the sample size remains the same.
The finite population correction factor approaches 1 as the population size increases and the sample size remains the same.

59. *Sample answer:*

$E = \dfrac{z_c \sigma}{\sqrt{n}}$ Write original equation.

$E\sqrt{n} = z_c \sigma$ Multiply each side by $\sqrt{n}$.

$\sqrt{n} = \dfrac{z_c \sigma}{E}$ Divide each side by E.

$n = \left(\dfrac{z_c \sigma}{E}\right)^2$ Square each side.

6.2 CONFIDENCE INTERVALS FOR THE MEAN (SMALL SAMPLES)

6.2 TRY IT YOURSELF SOLUTIONS

1. $\text{d.f.} = n - 1 = 22 - 1 = 21$

90% confidence level

$t_c = 1.721$

For a t-distribution curve with 21 degrees of freedom, 90% of the area under the curve lies between $t = \pm 1.721$.

2. $\text{d.f.} = n - 1 = 16 - 1 = 15$

90% CI: $t_c = 1.753$

$$E = t_c \frac{s}{\sqrt{n}} = 1.753 \frac{10}{\sqrt{16}} \approx 4.4$$

90% CI: $\bar{x} \pm E \approx 162 \pm 4.4 = (157.6,\ 166.4)$

With 90% confidence, you can say that the population mean temperature of coffee sold is between 157.6°F and 166.4°F.

99% CI: $t_c = 2.947$

$$E = t_c \frac{s}{\sqrt{n}} = 2.947 \frac{10}{\sqrt{16}} \approx 7.4$$

99% CI: $\bar{x} \pm E \approx 162 \pm 7.4 = (154.6,\ 169.4)$

With 99% confidence, you can say that the population mean temperature of coffee sold is between 154.6°F and 169.4°F.

3. $\text{d.f.} = n - 1 = 36 - 1 = 35$

90% CI: $t_c = 1.690$

$E = t_c \frac{s}{\sqrt{n}} = 1.690 \frac{2.39}{\sqrt{36}} \approx 0.67$

90% CI: $\overline{x} \pm E \approx 9.75 \pm 0.67 = (9.08,\ 10.42)$

With 90% confidence, you can say that the population mean number of days the car model sits on the lot is between 9.08 and 10.42 days.

95% CI: $t_c = 2.030$

$E = t_c \frac{s}{\sqrt{n}} = 2.030 \frac{2.39}{\sqrt{36}} \approx 0.81$

95% CI: $\overline{x} \pm E \approx 9.75 \pm 0.81 = (8.94,\ 10.56)$

With 95% confidence, you can say that the population mean number of days the car model sits on the lot is between 8.94 and 10.56 days.

The 90% confidence interval is slightly narrower.

4. Is σ known? No

Is $n \geq 30$? No

Is the population normally distributed? Yes

Use the t-distribution because σ is not known and the population is normally distributed.

6.2 EXERCISE SOLUTIONS

1. $t_c = 1.833$

3. $t_c = 2.947$

5. $E = t_c \frac{s}{\sqrt{n}} = 2.131 \frac{5}{\sqrt{16}} \approx 2.664$

7. $E = t_c \frac{s}{\sqrt{n}} = 1.691 \frac{2.4}{\sqrt{35}} \approx 0.686$

9. $\overline{x} \pm t_c \frac{s}{\sqrt{n}} = 12.5 \pm 2.015 \frac{2.0}{\sqrt{6}} \approx 12.5 \pm 1.645 \approx (10.9,\ 14.1)$

11. $\overline{x} \pm t_c \frac{s}{\sqrt{n}} = 4.3 \pm 2.650 \frac{0.34}{\sqrt{14}} \approx 4.3 \pm 0.241 \approx (4.1,\ 4.5)$

13. $(14.7,\ 22.1) \Rightarrow \overline{x} = \frac{14.7 + 22.1}{2} = 18.4 \Rightarrow E = 22.1 - 18.4 = 3.7$

15. $(64.6,\ 83.6) \Rightarrow \overline{x} = \frac{64.6 + 83.6}{2} = 74.1 \Rightarrow E = 83.6 - 74.1 = 9.5$

17. $E = t_c \frac{s}{\sqrt{n}} = 2.365 \frac{7.2}{\sqrt{8}} \approx 6.0$

$\overline{x} \pm E \approx 35.5 \pm 6.0 \approx (29.5,\ 41.5)$

With 95% confidence, you can say that the population mean commute time to work is between 29.5 and 41.5 minutes.

19. $E = t_c \frac{s}{\sqrt{n}} = 2.365 \frac{184.00}{\sqrt{8}} \approx 153.85$

$\bar{x} \pm E \approx 526.50 \pm 153.85 \approx (372.65, 680.35)$

With 95% confidence, you can say that the population mean cell phone price is between \$372.65 and \$680.35.

21. $E = z_c \frac{\sigma}{\sqrt{n}} = 1.96 \frac{9.3}{\sqrt{8}} \approx 6.4$

$\bar{x} \pm E \approx 35.5 \pm 6.4 \approx (29.1,\ 41.9)$

With 95% confidence, you can say that the population mean commute time to work is between 29.1 and 41.9 minutes. This confidence interval is slightly wider than the one found in Exercise 17.

23. Yes, \$431.61 falls between \$372.65 and \$680.35.

25. (a) $\bar{x} \approx 1185$

(b) $s \approx 168.1$

(c) $\bar{x} \pm t_c \frac{s}{\sqrt{n}} \approx 1185 \pm 3.106 \frac{168.1}{\sqrt{12}} \approx 1185 \pm 150.72 \approx (1034.3,\ 1335.7)$

27. (a) $\bar{x} \approx 7.49$

(b) $s \approx 1.64$

(c) $\bar{x} \pm t_c \frac{s}{\sqrt{n}} \approx 7.49 \pm 2.947 \frac{1.64}{\sqrt{16}} \approx 7.49 \pm 1.21 \approx (6.28,\ 8.70)$

29. No, 1020 does not fall between 1034.3 and 1335.7.

31. (a) $\bar{x} \approx 68,757.94$

(b) $s \approx 15,834.18$

(c) Use a t-distribution because σ unknown and $n \geq 30$.

$\bar{x} \pm t_c \frac{s}{\sqrt{n}} = 68,757.94 \pm 2.453 \frac{15,834.18}{\sqrt{32}} \approx 68,757.94 \pm 6866.227 \approx (61,892,\ 75,624)$

33. Yes, \$72,000 falls between \$61,892 and \$75,624.

35. Use a t-distribution because σ unknown and $n \geq 30$.

$\bar{x} \pm t_c \frac{s}{\sqrt{n}} = 27.7 \pm 2.009 \frac{6.12}{\sqrt{50}} \approx 27.7 \pm 1.74 \approx (26.0,\ 29.4)$

With 95% confidence, you can say that the population mean BMI is between 26.0 and 29.4.

37. Neither distribution can be used because $n < 30$ and the mileages are not normally distributed.

39. No; Half the sample mean is $\frac{4.36}{2} = 2.18\%$, which falls outside the confidence interval.

41. $n = 25, \bar{x} = 56.0, s = 0.25$

$$t = \frac{\bar{x} - \mu}{\frac{s}{\sqrt{n}}} = \frac{56.0 - 55.5}{\frac{0.25}{\sqrt{25}}} = 10$$

$-t_{0.99} = -2.797,\ t_{0.99} = 2.797$

They are not making good tennis balls because for this sample the t-value is $t = 10$, which is not between $-t_{0.99} = -2.797$ and $t_{0.99} = 2.797$.

6.3 CONFIDENCE INTERVALS FOR POPULATION PROPORTIONS

6.3 TRY IT YOURSELF SOLUTIONS

1. $x = 717,\ n = 717 + 1338 + 1769 + 956 = 4780$

$$\hat{p} = \frac{x}{n} = \frac{717}{4780} = 0.15 = 15\%$$

2. $\hat{p} = 0.15,\ \hat{q} = 1 - 0.15 = 0.85$

$n\hat{p} = (4780)(0.15) = 717 > 5$

$n\hat{q} = (4790)(0.85) = 4063 > 5$

Distribution of $\hat{p}$ is approximately normal.

$z_c = 1.645$

$$E = z_c\sqrt{\frac{\hat{p}\hat{q}}{n}} = 1.645\sqrt{\frac{0.15 \cdot 0.85}{4780}} \approx 0.01$$

$\hat{p} \pm E \approx 0.15 \pm 0.01 \approx (0.14,\ 0.16)$

3. $n = 800,\ \hat{p} = 0.50$

$\hat{q} = 1 - \hat{p} = 1 - 0.50 = 0.50$

$n\hat{p} = 800 \cdot 0.50 = 400 > 5$

$n\hat{q} = 800 \cdot 0.50 = 400 > 5$

Distribution of $\hat{p}$ is approximately normal.

$$z_c = 2.575,\ E = z_c\sqrt{\frac{\hat{p}\hat{q}}{n}} = 2.575\sqrt{\frac{0.50(0.50)}{800}} \approx 0.046$$

$\hat{p} \pm E \approx 0.50 \pm 0.046 = (0.454,\ 0.546)$

4. (1) $\hat{p}=0.5,\ \hat{q}=0.5,\ z_c=1.645,\ E=0.02$

$$n=\hat{p}\hat{q}\left(\frac{z_c}{E}\right)^2=(0.5)(0.5)\left(\frac{1.645}{0.02}\right)^2\approx 1691.266\rightarrow 1692$$

At least 1692 people should be included in the sample.

(2) $\hat{p}=0.063,\ \hat{q}=0.937,\ z_c=1.645,\ E=0.02$

$$n=\hat{p}\hat{q}\left(\frac{z_c}{E}\right)^2=(0.063)(0.937)\left(\frac{1.645}{0.02}\right)^2\approx 399.3\rightarrow 400$$

At least 400 people should be included in the sample.

6.3 EXERCISE SOLUTIONS

1. False. To estimate the value of p, the population proportion of successes, use the point estimate $\hat{p}=\frac{x}{n}$.

3. $\hat{p}=\frac{x}{n}=\frac{62}{1040}\approx 0.060,\ \hat{q}=1-\hat{p}\approx 0.940$

5. $\hat{p}=\frac{x}{n}=\frac{1310}{2016}\approx 0.650,\ \hat{q}=1-\hat{p}\approx 0.350$

7. $(0.905,\ 0.933)\rightarrow \hat{p}=\frac{0.905+0.933}{2}=0.919\Rightarrow E=0.933-0.919=0.014$

9. $(0.512,\ 0.596)\rightarrow \hat{p}=\frac{0.512+0.596}{2}=0.554\Rightarrow E=0.596-0.554=0.042$

11. $\hat{p}=\frac{x}{n}=\frac{1322}{2241}\approx 0.590,\ \hat{q}=1-\hat{p}\approx 0.410$

90% CI: $\hat{p}\pm z_c\sqrt{\frac{\hat{p}\hat{q}}{n}}\approx 0.590\pm 1.645\sqrt{\frac{(0.590)(0.410)}{2241}}\approx 0.590\pm 0.017\approx (0.573,\ 0.607)$

95% CI: $\hat{p}\pm z_c\sqrt{\frac{\hat{p}\hat{q}}{n}}\approx 0.590\pm 1.96\sqrt{\frac{(0.590)(0.410)}{2241}}\approx 0.590\pm 0.020\approx (0.570,\ 0.610)$

With 90% confidence, you can say that the population proportion of U.S. adults who say they have made a New Year's resolution is between 57.3% and 60.7%. With 95% confidence, you can say it is between 57.0% and 61.0%. The 95% confidence interval is slightly wider.

13. $\hat{p}=\frac{x}{n}=\frac{700}{1000}=0.7,\ \hat{q}=1-\hat{p}=0.3$

$$\hat{p}\pm z_c\sqrt{\frac{\hat{p}\hat{q}}{n}}\approx 0.7\pm 2.575\sqrt{\frac{(0.7)(0.3)}{1000}}\approx 0.7\pm 0.037\approx (0.663,\ 0.737)$$

With 99% confidence, you can say that the population proportion of U.S. adults who say they think police officers should be required to wear body cameras while on duty is between 66.3% and 73.7%.

15. $\hat{p} = \frac{x}{n} = \frac{49,311}{1,626,773} \approx 0.0303, \hat{q} = 1 - \hat{p} = 0.9697$

$$\hat{p} \pm z_c\sqrt{\frac{\hat{p}\hat{q}}{n}} \approx 0.0303 \pm 1.96\sqrt{\frac{(0.0303)(0.9697)}{1,626,773}} \approx 0.0303 \pm 0.0003 \approx (0.030,\ 0.031)$$

17. (a) $n = \hat{p}\hat{q}\left(\frac{z_c}{E}\right)^2 = 0.5 \cdot 0.5\left(\frac{1.96}{0.04}\right)^2 \approx 600.25 \rightarrow 601$ adults

(b) $n = \hat{p}\hat{q}\left(\frac{z_c}{E}\right)^2 = 0.25 \cdot 0.75\left(\frac{1.96}{0.04}\right)^2 \approx 450.2 \rightarrow 451$ adults

(c) Having an estimate of the population proportion reduces the minimum sample size needed.

19. (a) $n = \hat{p}\hat{q}\left(\frac{z_c}{E}\right)^2 = 0.5 \cdot 0.5\left(\frac{1.645}{0.03}\right)^2 \approx 751.67 \rightarrow 752$ adults

(b) $n = \hat{p}\hat{q}\left(\frac{z_c}{E}\right)^2 = 0.11 \cdot 0.89\left(\frac{1.645}{0.03}\right)^2 \approx 294.4 \rightarrow 295$ adults

(c) Having an estimate of the population proportion reduces the minimum sample size needed.

21. 90% CI $\approx (0.573,\ 0.607)$
95% CI $\approx (0.570,\ 0.610)$
Yes; 0.59 falls within both confidence intervals.

23. No; The minimum sample size needed is 451 adults.

25. United States:

$\hat{p} = 0.32,\ \hat{q} = 0.68,\ n = 1003$

$$\hat{p} \pm z_c\sqrt{\frac{\hat{p}\hat{q}}{n}} = 0.32 \pm 2.575\sqrt{\frac{(0.32)(0.68)}{1003}} \approx 0.32 \pm 0.038 \approx (0.282,\ 0.358)$$

Canada:

$\hat{p} = 0.21,\ \hat{q} = 0.79,\ n = 1020$

$$\hat{p} \pm z_c\sqrt{\frac{\hat{p}\hat{q}}{n}} = 0.21 \pm 2.575\sqrt{\frac{(0.21)(0.79)}{1020}} \approx 0.21 \pm 0.033 \approx (0.177,\ 0.243)$$

France:

$\hat{p} = 0.25,\ \hat{q} = 0.75,\ n = 999$

$$\hat{p} \pm z_c\sqrt{\frac{\hat{p}\hat{q}}{n}} = 0.25 \pm 2.575\sqrt{\frac{(0.25)(0.75)}{999}} \approx 0.25 \pm 0.035 \approx (0.215,\ 0.285)$$

Japan:

$\hat{p}=0.50,\ \hat{q}=0.50,\ n=1000$

$$\hat{p}\pm z_c\sqrt{\frac{\hat{p}\hat{q}}{n}}=0.50\pm 2.575\sqrt{\frac{(0.50)(0.50)}{1000}}\approx 0.50\pm 0.041\approx (0.459,\ 0.541)$$

Australia:

$\hat{p}=0.13,\ \hat{q}=0.87,\ n=1000$

$$\hat{p}\pm z_c\sqrt{\frac{\hat{p}\hat{q}}{n}}=0.13\pm 2.575\sqrt{\frac{(0.13)(0.87)}{1000}}\approx 0.13\pm 0.027\approx (0.103,\ 0.157)$$

27. (a) Expect to stay at first employer for 3 or more years:

$\hat{p}=0.69,\ \hat{q}=1-\hat{p}=0.31$

$$\hat{p}\pm z_c\sqrt{\frac{\hat{p}\hat{q}}{n}}=0.69\pm 1.96\sqrt{\frac{(0.69)(0.31)}{2000}}\approx 0.69\pm 0.020\approx (0.670,\ 0.710)$$

Completed an apprenticeship or internship:

$\hat{p}=0.68,\ \hat{q}=1-\hat{p}=0.32$

$$\hat{p}\pm z_c\sqrt{\frac{\hat{p}\hat{q}}{n}}=0.68\pm 1.96\sqrt{\frac{(0.68)(0.32)}{2000}}\approx 0.68\pm 0.020\approx (0.660,\ 0.700)$$

Employed in field of study:

$\hat{p}=0.65,\ \hat{q}=1-\hat{p}=0.35$

$$\hat{p}\pm z_c\sqrt{\frac{\hat{p}\hat{q}}{n}}=0.65\pm 1.96\sqrt{\frac{(0.65)(0.35)}{2000}}\approx 0.65\pm 0.021\approx (0.629,\ 0.671)$$

Feel underemployed:

$\hat{p}=0.51,\ \hat{q}=1-\hat{p}=0.49$

$$\hat{p}\pm z_c\sqrt{\frac{\hat{p}\hat{q}}{n}}=0.51\pm 1.96\sqrt{\frac{(0.51)(0.49)}{2000}}\approx 0.51\pm 0.022\approx (0.488,\ 0.532)$$

Prefer to work for a large company:

$\hat{p}=0.14,\ \hat{q}=1-\hat{p}=0.86$

$$\hat{p}\pm z_c\sqrt{\frac{\hat{p}\hat{q}}{n}}=0.14\pm 1.96\sqrt{\frac{(0.14)(0.86)}{2000}}\approx 0.14\pm 0.015\approx (0.125,\ 0.155)$$

(b) Expect to stay at first employer for 3 or more years:

$\hat{p}=0.69,\ \hat{q}=1-\hat{p}=0.31$

$$\hat{p}\pm z_c\sqrt{\frac{\hat{p}\hat{q}}{n}}=0.69\pm 2.575\sqrt{\frac{(0.69)(0.31)}{2000}}\approx 0.69\pm 0.027\approx (0.663,\ 0.717)$$

Completed an apprenticeship or internship:

$\hat{p}=0.68,\ \hat{q}=1-\hat{p}=0.32$

$$\hat{p}\pm z_{\alpha/2}\sqrt{\frac{\hat{p}\hat{q}}{n}}=0.68\pm 2.575\sqrt{\frac{(0.68)(0.32)}{2000}}\approx 0.68\pm 0.027\approx (0.653,\ 0.707)$$

Employed in field of study:

$\hat{p}=0.65,\ \hat{q}=1-\hat{p}=0.35$

$$\hat{p}\pm z_c\sqrt{\frac{\hat{p}\hat{q}}{n}}=0.68\pm 2.575\sqrt{\frac{(0.68)(0.32)}{2000}}\approx 0.68\pm 0.027\approx (0.653,\ 0.707)$$

Feel underemployed:

$\hat{p} = 0.51,\ \hat{q} = 1 - \hat{p} = 0.49$

$$\hat{p} \pm z_c\sqrt{\frac{\hat{p}\hat{q}}{n}} = 0.51 \pm 2.575\sqrt{\frac{(0.51)(0.49)}{2000}} \approx 0.51 \pm 0.029 \approx (0.481,\ 0.539)$$

Prefer to work for a large company:

$\hat{p} = 0.14,\ \hat{q} = 1 - \hat{p} = 0.86$

$$\hat{p} \pm z_c\sqrt{\frac{\hat{p}\hat{q}}{n}} = 0.14 \pm 2.575\sqrt{\frac{(0.14)(0.86)}{2000}} \approx 0.14 \pm 0.020 \approx (0.120,\ 0.160)$$

29. $70\% \pm 3.4\% \to (66.6\%,\ 73.4\%) \to (0.666,\ 0.734)$

$$E = z_c\sqrt{\frac{\hat{p}\hat{q}}{n}} \to z_c = E\sqrt{\frac{n}{\hat{p}\hat{q}}} = 0.034\sqrt{\frac{1003}{(0.70)(0.30)}} \approx\to z_c = 2.35$$

$P(-2.35 < z < 2.35) = 0.9906 - 0.0094 = 0.9812$

$(0.666,\ 0.734)$ is approximately a 98.1% CI.

31. $71\% \pm 3\% \to (68\%,\ 74\%) \to (0.68,\ 0.74)$

$$E = z_c\sqrt{\frac{\hat{p}\hat{q}}{n}} \to z_c = E\sqrt{\frac{n}{\hat{p}\hat{q}}} = 0.03\sqrt{\frac{1000}{(0.71)(0.29)}} \approx\to z_c = 2.09$$

$P(-2.09 < z < 2.09) = 0.9817 - 0.0183 = 0.9634$

$(0.68,\ 0.74)$ is approximately a 96.3% CI.

33. $47\% \pm 2\% \to (45\%,\ 49\%) \to (0.45,\ 0.49)$

$$E = z_c\sqrt{\frac{\hat{p}\hat{q}}{n}} \to z_c = E\sqrt{\frac{n}{\hat{p}\hat{q}}} = 0.02\sqrt{\frac{3539}{(0.47)(0.53)}} \approx\to z_c = 2.38$$

$P(-2.38 < z < 2.38) = 0.9913 - 0.0087 = 0.9826$

$(0.45,\ 0.49)$ is approximately a 98.3% CI.

$53\% \pm 2\% \to (51\%,\ 55\%) \to (0.51,\ 0.55)$

$$E = z_c\sqrt{\frac{\hat{p}\hat{q}}{n}} \to z_c = E\sqrt{\frac{n}{\hat{p}\hat{q}}} = 0.02\sqrt{\frac{3539}{(0.53)(0.47)}} \approx\to z_c = 2.38$$

$P(-2.38 < z < 2.38) = 0.9913 - 0.0087 = 0.9826$

$(0.51,\ 0.55)$ is approximately a 98.3% CI.

35. If $n\hat{p} < 5$ or $n\hat{q} < 5$, the sampling distribution of $\hat{p}$ may not be normally distributed, so z_c cannot be used to calculate the confidence interval.

37.

$\hat{p}$	$\hat{q}=1-\hat{p}$	$\hat{p}\hat{q}$
0.0	1.0	0.00
0.1	0.9	0.09
0.2	0.8	0.16
0.3	0.7	0.21
0.4	0.6	0.24
0.5	0.5	0.25
0.6	0.4	0.24
0.7	0.3	0.21
0.8	0.2	0.16
0.9	0.1	0.09
1.0	0.0	0.00

$\hat{p}$	$\hat{q}=1-\hat{p}$	$\hat{p}\hat{q}$
0.45	0.55	0.2475
0.46	0.54	0.2484
0.47	0.53	0.2491
0.48	0.52	0.2496
0.49	0.51	0.2499
0.50	0.50	0.2500
0.51	0.49	0.2499
0.52	0.48	0.2496
0.53	0.47	0.2491
0.54	0.46	0.2484
0.55	0.45	0.2475

$\hat{p}=0.5$ gives the maximum value of $\hat{p}\hat{q}$.

6.4 CONFIDENCE INTERVALS FOR VARIANCE AND STANDARD DEVIATION

6.4 TRY IT YOURSELF SOLUTIONS

1. $\text{d.f.}=n-1=29$

level of confidence = 0.90

Area to the right of χ_R^2 is 0.05.

Area to the right of χ_L^2 is 0.95.

$\chi_R^2=42.557$, $\chi_L^2=17.708$

For a chi-square distribution curve with 29 degrees of freedom, 90% of the area under the curve lies between 17.708 and 42.557.

2. 90% CI: $\chi_R^2=42.557$, $\chi_L^2=17.708$

95% CI: $\chi_R^2=45.722$, $\chi_L^2=16.047$

$$90\%\text{ CI for } \sigma^2: \left(\frac{(n-1)s^2}{\chi_R^2}, \frac{(n-1)s^2}{\chi_L^2}\right)=\left(\frac{29\cdot(1.2)^2}{42.557}, \frac{29\cdot(1.2)^2}{17.708}\right)\approx(0.98,\ 2.36)$$

$$95\%\text{ CI for } \sigma^2: \left(\frac{(n-1)s^2}{\chi_R^2}, \frac{(n-1)s^2}{\chi_L^2}\right)=\left(\frac{29\cdot(1.2)^2}{45.722}, \frac{29\cdot(1.2)^2}{16.047}\right)\approx(0.91,\ 2.60)$$

$$90\%\text{ CI for } \sigma: \left(\sqrt{0.981}, \sqrt{2.358}\right)\approx(0.99,\ 1.54)$$

$$95\%\text{ CI for } \sigma: \left(\sqrt{0.913}, \sqrt{2.602}\right)\approx(0.96,\ 1.61)$$

With 90% confidence, you can say that the population variance is between 0.98 and 2.36 and that the population standard deviation is between 0.99 and 1.54. With 95% confidence, you can say that the population variance is between 0.91 and 2.60, and that the population standard deviation is between 0.96 and 1.61.

6.4 EXERCISE SOLUTIONS

1. Yes.

3. $\chi_R^2 = 14.067$, $\chi_L^2 = 2.167$ **5.** $\chi_R^2 = 32.852$, $\chi_L^2 = 8.907$ **7.** $\chi_R^2 = 52.336$, $\chi_L^2 = 13.121$

9. (a) $\left(\dfrac{(n-1)s^2}{\chi_R^2}, \dfrac{(n-1)s^2}{\chi_L^2}\right) \approx \left(\dfrac{29\cdot(11.56)}{45.722}, \dfrac{29\cdot(11.56)}{16.047}\right) \approx (7.33,\ 20.89)$

(b) $\left(\sqrt{7.3321}, \sqrt{20.8911}\right) \approx (2.71,\ 4.57)$

11. (a) $\left(\dfrac{(n-1)s^2}{\chi_R^2}, \dfrac{(n-1)s^2}{\chi_L^2}\right) \approx \left(\dfrac{17\cdot(35)^2}{27.587}, \dfrac{17\cdot(35)^2}{8.672}\right) \approx (755,\ 2401)$

(b) $\left(\sqrt{754.885}, \sqrt{2401.407}\right) \approx (27,\ 49)$

13. (a) $s^2 \approx 0.0756$

$\left(\dfrac{(n-1)s^2}{\chi_R^2}, \dfrac{(n-1)s^2}{\chi_L^2}\right) \approx \left(\dfrac{17\cdot(0.0756)}{30.191}, \dfrac{17\cdot(0.0756)}{7.564}\right) \approx (0.0426,\ 0.1699)$

(b) $\left(\sqrt{0.04257}, \sqrt{0.16991}\right) \approx (0.2063,\ 0.4122)$

With 95% confidence, you can say that the population variance is between 0.0426 and 0.1699, and the population standard deviation is between 0.2063 and 0.4122 inch.

15. (a) $s^2 \approx 362.98$

$\left(\dfrac{(n-1)s^2}{\chi_R^2}, \dfrac{(n-1)s^2}{\chi_L^2}\right) \approx \left(\dfrac{20\cdot(362.98)}{39.997}, \dfrac{20\cdot(362.98)}{7.434}\right) \approx (181.50,\ 976.54)$

(b) $\left(\sqrt{181.50}, \sqrt{976.54}\right) \approx (13.47,\ 31.25)$

With 99% confidence, you can say that the population variance is between 181.50 and 976.54, and the population standard deviation is between 13.47 and 31.25 thousand dollars.

17. (a) $\left(\dfrac{(n-1)s^2}{\chi_R^2}, \dfrac{(n-1)s^2}{\chi_L^2}\right) \approx \left(\dfrac{13\cdot(3.54)^2}{29.819}, \dfrac{13\cdot(3.54)^2}{3.565}\right) \approx (5.46,\ 45.70)$

(b) $\left(\sqrt{5.46}, \sqrt{45.70}\right) \approx (2.34,\ 6.76)$

With 99% confidence, you can say that the population variance is between 5.46 and 45.70, and the population standard deviation is between 2.34 and 6.76 days.

19. (a) $\left(\dfrac{(n-1)s^2}{\chi_R^2}, \dfrac{(n-1)s^2}{\chi_L^2}\right) \approx \left(\dfrac{18\cdot(15)^2}{31.526}, \dfrac{18\cdot(15)^2}{8.231}\right) \approx (128,\ 492)$

(b) $\left(\sqrt{128.465},\ \sqrt{492.042}\right) \approx (11,\ 22)$

With 95% confidence, you can say that the population variance is between 128 and 492, and the population standard deviation is between 11 and 22 grains per gallon.

21. (a) $\left(\dfrac{(n-1)s^2}{\chi_R^2},\ \dfrac{(n-1)s^2}{\chi_L^2}\right) \approx \left(\dfrac{17\cdot(0.25)^2}{24.769},\ \dfrac{17\cdot(0.25)^2}{10.085}\right) \approx (0.04,\ 0.11)$

(b) $\left(\sqrt{0.0429},\ \sqrt{0.1054}\right) \approx (0.21,\ 0.32)$

With 80% confidence, you can say that the population variance is between 0.04 and 0.11, and the population standard deviation is between 0.21 and 0.32 day.

23. (a) $\left(\dfrac{(n-1)s^2}{\chi_R^2}\cdot\dfrac{(n-1)s^2}{\chi_L^2}\right) \approx \left(\dfrac{(21)(3.6)^2}{38.932}\cdot\dfrac{(21)(3.6)^2}{8.897}\right) \approx (7.0,\ 30.6)$

(b) $\left(\sqrt{6.99},\ \sqrt{30.59}\right) \approx (2.6,\ 5.5)$

With 98% confidence, you can say that the population variance is between 7.0 and 30.6, and the population standard deviation is between 2.6 and 5.5 minutes.

25. 95% CI for $\sigma: (0.2063,\ 0.4122)$

Yes, because all of the values in the confidence interval are less than 0.5.

27. 80% CI for $\sigma: (0.21,\ 0.32)$

No, because 0.25 is contained in the confidence interval.

29. Answers will vary. *Sample answer:* Unlike a confidence interval for a population mean or proportion, a confidence interval for a population variance does not have a margin of error. The left and right endpoints must be calculated separately.

CHAPTER 6 REVIEW EXERCISE SOLUTIONS

1. (a) $\bar{x} \approx 103.5$

(b) $E = z_c \dfrac{\sigma}{\sqrt{n}} \approx 1.645 \dfrac{45}{\sqrt{40}} \approx 11.7$

3. (a) $\bar{x} \pm z_c \dfrac{\sigma}{\sqrt{n}} = 103.5 \pm 1.645 \dfrac{45}{\sqrt{40}} \approx 103.5 \pm 11.7 \approx (91.8,\ 115.2)$

With 90% confidence, you can say that the population mean waking time is between 91.8 and 115.2 minutes past 5:00 A.M.

(b) $\bar{x} \pm 0.10\bar{x} = 103.5 \pm 0.10(103.5) = 103.5 \pm 10.35 = (93.15,\ 113.85)$

Yes; If the population mean is within 10% of the sample mean, then it falls inside the confidence interval.

5. $(20.75,\ 24.10) \rightarrow \bar{x} = \dfrac{20.75 + 24.10}{2} = 22.425 \rightarrow E = 24.10 - 22.425 = 1.675$

7. $n = \left(\dfrac{z_c \sigma}{E}\right)^2 \approx \left(\dfrac{1.96 \cdot 45}{10}\right)^2 \approx 77.79 \Rightarrow 78$ people

9. $t_c = 1.383$ **11.** $t_c = 2.624$

13. (a) $E = t_c \dfrac{s}{\sqrt{n}} = 1.753 \dfrac{25.6}{\sqrt{16}} \approx 11.2$

(b) $\bar{x} \pm t_c \dfrac{s}{\sqrt{n}} \approx 72.1 \pm 11.2 \approx (60.9,\ 83.3)$

15. (a) $E = t_c \dfrac{s}{\sqrt{n}} = 2.718\left(\dfrac{0.9}{\sqrt{12}}\right) \approx 0.7$

(b) $\bar{x} \pm t_c \dfrac{s}{\sqrt{n}} \approx 6.8 \pm 0.7 \approx (6.1,\ 7.5)$

17. $\bar{x} \pm t_c \dfrac{s}{\sqrt{n}} = 165 \pm 1.690 \dfrac{67}{\sqrt{36}} \approx 165 \pm 18.87 \approx (146,\ 184)$

With 90% confidence, you can say that the population mean height is between 146 and 184 feet.

19. (a) $\hat{p} = \dfrac{x}{n} = \dfrac{745}{1035} \approx 0.720,\ \hat{q} = 1 - \hat{p} \approx 0.280$

(b) 90% CI: $\hat{p} \pm z_c \sqrt{\dfrac{\hat{p}\hat{q}}{n}} = 0.720 \pm 1.645 \sqrt{\dfrac{0.720 \cdot 0.280}{1035}} \approx 0.720 \pm 0.023 \approx (0.697,\ 0.743)$

95% CI: $\hat{p} \pm z_c \sqrt{\dfrac{\hat{p}\hat{q}}{n}} = 0.720 \pm 1.96 \sqrt{\dfrac{0.720 \cdot 0.280}{1035}} \approx 0.720 \pm 0.027 \approx (0.693,\ 0.747)$

(c) With 90% confidence, you can say that the population proportion of U.S. adults who say they want the U.S. to play a leading or major role in global affairs is between 69.7% and 74.3%. With 95% confidence, you can say it is between 69.3% and 74.7%. The 95% confidence interval is slightly wider.

21. (a) $\hat{p} = \dfrac{x}{n} = \dfrac{1167}{2202} \approx 0.530,\ \hat{q} = 1 - \hat{p} \approx 0.470$

(b) 90% CI: $\hat{p} \pm z_c \sqrt{\dfrac{\hat{p}\hat{q}}{n}} = 0.530 \pm 1.645 \sqrt{\dfrac{0.530 \cdot 0.470}{2202}} \approx 0.530 \pm 0.017 \approx (0.513,\ 0.547)$

95% CI: $\hat{p} \pm z_c \sqrt{\dfrac{\hat{p}\hat{q}}{n}} = 0.530 \pm 1.96 \sqrt{\dfrac{0.530 \cdot 0.470}{2202}} \approx 0.530 \pm 0.021 \approx (0.509,\ 0.551)$

(c) With 90% confidence, you can say that the population proportion of U.S. adults who think antibiotics are effective against viral infections is between 51.3% and 54.7%. With 95% confidence, you can say it is between 50.9% and 55.1%. The 95% confidence interval is slightly wider.

23. No; it falls outside both confidence intervals.

25. (a) $n = \hat{p}\hat{q}\left(\frac{z_c}{E}\right)^2 = 0.50 \cdot 0.50\left(\frac{1.96}{0.05}\right)^2 \approx 384.16 \rightarrow 385$ adults

(b) $n = \hat{p}\hat{q}\left(\frac{z_c}{E}\right)^2 = 0.32 \cdot 0.68\left(\frac{1.96}{0.05}\right)^2 \approx 334.4 \rightarrow 335$ adults

(c) Having an estimate of the population proportion reduces the minimum sample size needed.

27. $\chi_R^2 = 23.337$, $\chi_L^2 = 4.404$

29. $\chi_R^2 = 24.996$, $\chi_L^2 = 7.261$

31. $s^2 \approx 359.936$

(a) 95% CI for σ^2: $\left(\frac{(n-1)s^2}{\chi_R^2}, \frac{(n-1)s^2}{\chi_L^2}\right) \approx \left(\frac{12 \cdot (359.936)}{23.337}, \frac{12 \cdot (359.936)}{4.404}\right) \approx (185.1,\ 980.8)$

(b) 95% CI for σ: $\left(\sqrt{185.1}, \sqrt{980.8}\right) \approx (13.6,\ 31.3)$

With 95% confidence we can say that the population variance is between 185.1 and 980.8, and the population standard deviation is between 13.6 and 31.3 knots.

CHAPTER 6 QUIZ SOLUTIONS

1. (a) $\bar{x} = \frac{\sum x}{n} = \frac{77.94}{30} = 2.598$

(b) $E = z_c \frac{\sigma}{\sqrt{n}} \approx 1.96 \frac{0.344}{\sqrt{30}} \approx 0.123$

(c) $\bar{x} \pm z_c \frac{\sigma}{\sqrt{n}} \approx 2.598 \pm 0.123 \approx (2.475,\ 2.721)$

With 95% confidence, you can say that the population mean winning time is between 2.475 and 2.721 hours.

(d) No; It falls outside the confidence interval.

2. $n=\left(\frac{z_c\sigma}{E}\right)^2=\left(\frac{2.575\cdot 0.325}{0.13}\right)^2\approx 41.4\to 42$ champions

3. (a) $\bar{x}=6.61,\ s\approx 3.376$

(b) $\bar{x}\pm t_c\frac{s}{\sqrt{n}}\approx 6.61\pm 1.833\frac{3.376}{\sqrt{10}}\approx 6.61\pm 1.957\approx (4.65,\ 8.57)$

With 90% confidence you can say that the population mean amount of time is between 4.65 and 8.57 minutes.

(c) $\bar{x}\pm z_c\frac{\sigma}{\sqrt{n}}\approx 6.61\pm 1.645\frac{3.5}{\sqrt{10}}\approx 6.61\pm 1.82\approx (4.79,\ 8.43)$

With 90% confidence you can say that the population mean amount of time is between 4.79 and 8.43 minutes. This confidence interval is narrower than the one found in part (b).

4. $\bar{x}\pm t_c\frac{s}{\sqrt{n}}=133{,}326\pm 2.201\frac{36{,}729}{\sqrt{12}}\approx 133{,}326\pm 23{,}336\approx (109{,}990,\ 156{,}662)$

With 95% confidence you can say that the population mean annual earnings is between $109,990 and $156,662.

5. Yes. $131,935 falls within the confidence interval, (109,990, 156,662).

6. (a) $\hat{p}=\frac{x}{n}=\frac{753}{1018}\approx 0.740$

(b) $\hat{p}\pm z_c\sqrt{\frac{\hat{p}\hat{q}}{n}}\approx 0.740\pm 1.645\sqrt{\frac{0.740\cdot 0.260}{1018}}\approx 0.740\pm 0.023\approx (0.717,\ 0.763)$

With 90% confidence you can say that the population proportion of U.S. adults who say that the energy situation in the United States is very or fairly serious is between 71.7% and 76.3%.

(c) No. The values fall outside the confidence interval.

(d) $n=\hat{p}\hat{q}\left(\frac{z_c}{E}\right)^2\approx 0.740\cdot 0.260\left(\frac{2.575}{0.04}\right)^2\approx 797.3\to 798$ adults

7. (a) $\left(\frac{(n-1)s^2}{\chi_R^2},\ \frac{(n-1)s^2}{\chi_L^2}\right)=\left(\frac{9\cdot(3.38)^2}{19.023},\ \frac{9\cdot(3.38)^2}{2.700}\right)\approx (5.41,\ 38.08)$

(b) $\left(\sqrt{5.4050},\ \sqrt{38.0813}\right)\approx (2.32,\ 6.17)$

With 95% confidence you can say that the population standard deviation is between 2.32 and 6.17 minutes.

Hypothesis Testing with One Sample

CHAPTER 7

7.1 INTRODUCTION TO HYPOTHESIS TESTING

7.1 TRY IT YOURSELF SOLUTIONS

1. (1) The mean is not 74 months; $\mu \neq 74$.

$H_0: \mu = 74;\ H_a: \mu \neq 74$ (claim)

(2) The variance is less than or equal to 2.7; $\sigma^2 \leq 2.7$.

$\sigma^2 > 2.7$

$H_0: \sigma^2 \leq 2.7$ (claim); $H_a: \sigma^2 > 2.7$

(3) The proportion is more than 24%; $p > 0.24$.

$p \leq 0.24$

$p \leq 0.24;\ H_a: p > 0.24$ (claim)

2. $H_0: p \leq 0.01;\ H_a: p > 0.01$

A type I error will occur if the actual proportion is less than or equal to 0.01, but you reject H_0.

A type II error will occur if the actual proportion is greater than 0.01, but you fail to reject H_0.

A type II error is more serious because you would be misleading the consumer, possibly causing serious injury or death.

3. (1) H_0: The mean life of a certain type of automobile battery is 74 months.

H_a: The mean life of a certain type of automobile battery is not 74 months.

$H_0: \mu = 74;\ H_a: \mu \neq 74$

Two-tailed

$\frac{1}{2}$ *P*-value area $\frac{1}{2}$ *P*-value area

$-z$ z z

(2) H_0: The variance of the life of a manufacturer's home theater systems is less than or equal to 2.7.

H_a: The variance of the life of a manufacturer's home theater systems is greater than 2.7.

$H_0: \sigma^2 \leq 2.7; H_a: \sigma^2 > 2.7$

Right-tailed

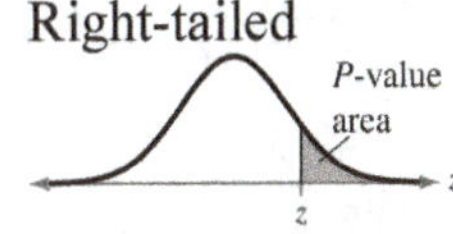

(3) H_0: The proportion of homeowners who feel their house is too small for their family is less than or equal to 24%.

H_a: The proportion of homeowners who feel their house is too small for their family is greater than 24%.

$H_0: p \leq 0.24$; $H_a: p > 0.24$
Right-tailed

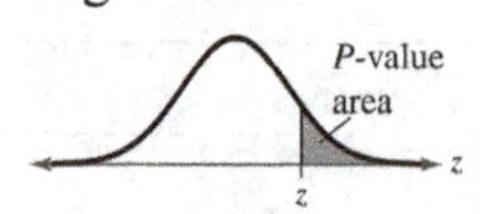

4. (1) There is enough evidence to support the claim that the mean life of a certain type of automobile battery is not 74 months.
There is not enough evidence to support the claim that the mean life of a certain type of automobile battery is not 74 months.

(2) There is enough evidence to support the realtor's claim that the proportion of homeowners who feel their house is too small for their family is more than 24%.
There is not enough evidence to support the realtor's claim that the proportion of homeowners who feel their house is too small for their family is more than 24%.

5. (1) $H_0: \mu \geq 650$; $H_a: \mu < 650$ (claim)
If you reject H_0, then you will support the claim that the mean repair cost per automobile is less than \$650.

(2) $H_0: \mu = 98.6$ (claim); $H_a: \mu \neq 98.6$
If you reject H_0, then you will reject the claim that the mean temperature is about 98.6°F.

7.1 EXERCISE SOLUTIONS

1. The two types of hypotheses used in a hypothesis test are the null hypothesis and the alternative hypothesis.
The alternative hypothesis is the complement of the null hypothesis.

3. You can reject the null hypothesis, or you can fail to reject the null hypothesis.

5. False. In a hypothesis test, you assume the null hypothesis is true.

7. True

9. False. A small *P*-value in a test will favor a rejection of the null hypothesis.

11. $H_0: \mu \leq 645$ (claim); $H_a: \mu > 645$

13. $H_0: \sigma = 5$; $H_a: \sigma \neq 5$ (claim)

15. $H_0: p \geq 0.45$; $H_a: p < 0.45$ (claim)

17. c; $H_0: \mu \leq 3$

19. b; $H_0: \mu = 3$

21. Right-tailed

23. Two-tailed

25. $\mu > 8$
$H_0: \mu \leq 8$; $H_a: \mu > 8$ (claim)

27. $\sigma \leq 320$
$H_0: \sigma \leq 320$ (claim); $H_a: \sigma > 320$

29. $p = 0.73$
$H_0: p = 0.73$ (claim); $H_a: p \neq 0.73$

31. A type I error will occur if the actual proportion of new customers who return to buy their next textbook is at least 0.60, but you reject $H_0: p \geq 0.60$.
A type II error will occur if the actual proportion of new customers who return to buy their next textbook is less than 0.60, but you fail to reject $H_0: p \geq 0.60$.

33. A type I error will occur if the actual standard deviation of the length of time to play a game is less than or equal to 12 minutes, but you reject $H_0: \sigma \leq 12$.
A type II error will occur of the actual standard deviation of the length of time to play a game is greater than 12 minutes, but you fail to reject $H_0: \sigma \leq 12$.

35. A type I error will occur when the actual proportion of applicants who become campus security officers is at most 0.25, but you reject H_0: $p \leq 0.25$.
A type II error will occur when the actual proportion of applicants who become campus security officers is greater than 0.25, but you fail to reject H_0: $p \leq 0.25$.

37. H_0: The proportion of homeowners who have a home security alarm is greater than or equal to 14%.
H_a: The proportion of homeowners who have a home security alarm is less than 14%.
H_0: $p \geq 0.14$; H_a: $p < 0.14$
Left-tailed because the alternative hypothesis contains $<$.

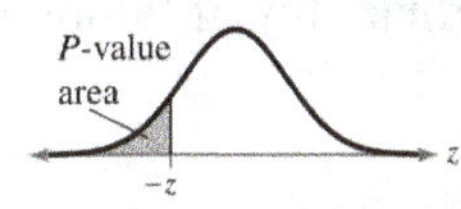

39. H_0: The standard deviation of the 18-hole scores for a golfer is greater than or equal to 2.1 strokes.
H_a: The standard deviation of the 18-hole scores for a golfer is less than 2.1 strokes.
$H_0: \sigma \geq 2.1$; $H_a: \sigma < 2.1$
Left-tailed because the alternative hypothesis contains $<$.

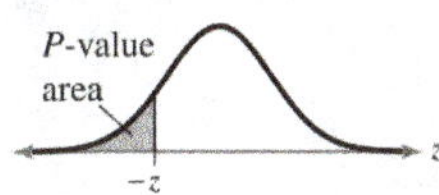

41. H_0: The mean graduation rate at a high school is less than or equal to 97%.
H_a: The mean graduation rate at a high school is greater than 97%.
$H_0: \mu \leq 97$; $H_a: \mu > 97$
Right-tailed because the alternative hypothesis contains $>$.

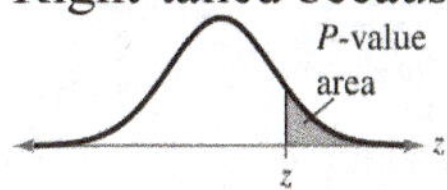

43. Alternative hypothesis

(a) There is enough evidence to support the scientist's claim that the mean incubation period for swan eggs is less than 40 days.

(b) There is not enough evidence to support the scientist's claim that the mean incubation period for swan eggs is less than 40 days.

45. Null hypothesis

(a) There is enough evidence to reject the claim that the standard deviation of the life of the lawn mower is at most 2.8 years.

(b) There is not enough evidence to reject the claim that the standard deviation of the life of the lawnmower is at most 2.8 years.

47. Null hypothesis

(a) There is enough evidence to reject the report's claim that at least 65% of individuals convicted of terrorism or terrorism-related offenses in the United States are foreign born.

(b) There is not enough evidence to reject the report's claim that at least 65% of individuals convicted of terrorism or terrorism-related offenses in the United States are foreign born.

49. $H_0: \mu \geq 60$; $H_a: \mu < 60$

51. (a) H_0: $\mu \geq 5$; H_a: $\mu < 5$

(b) H_0: $\mu \leq 5$; H_a: $\mu > 5$

53. If you decrease α, you are decreasing the probability that you reject H_0. Therefore, you are increasing the probability of failing to reject H_0. This could increase β, the probability of failing to reject H_0 when H_0 is false.

55. Yes; If the *P*-value is less than $\alpha = 0.05$, it is also less than $\alpha = 0.10$.

57. (a) Fail to reject H_0 because the CI includes values greater than 70.

(b) Reject H_0 because the CI is located below 70.

(c) Fail to reject H_0 because the CI includes values greater than 70.

59. (a) Reject H_0 because the CI is located to the right of 0.20.

(b) Fail to reject H_0 because the CI includes values less than 0.20.

(c) Fail to reject H_0 because the CI includes values less than 0.20.

7.2 HYPOTHESIS TESTING FOR THE MEAN (LARGE SAMPLES)

7.2 TRY IT YOURSELF SOLUTIONS

1. (1) Fail to reject H_0 because $0.0745 > 0.05$.

(2) Reject H_0 because $0.0745 < 0.10$.

2.

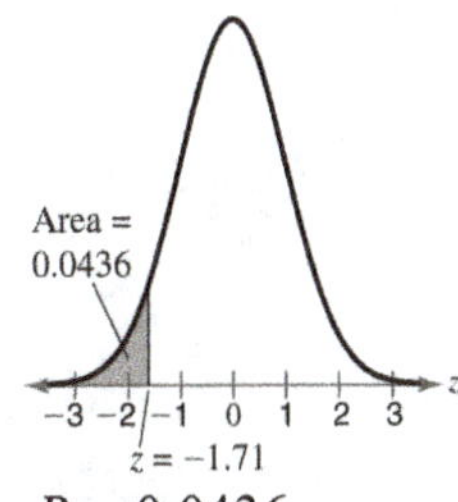

$P = 0.0436$

Reject H_0 because $P = 0.0436 < 0.05$.

3. Area to the left of $z = 1.64$ is 0.9495.

$P = 2(\text{area}) = 2(0.0505) = 0.1010$

Fail to reject H_0 because $P = 0.1010 > 0.10$.

4. The claim is "the mean speed is greater than 35 miles per hour."

$H_0: \mu \leq 35$; $H_\alpha: \mu > 35$ (claim)

$\alpha = 0.05$

$$z = \frac{\bar{x} - \mu}{\frac{\sigma}{\sqrt{n}}} = \frac{36 - 35}{\frac{4}{\sqrt{100}}} = 2.5$$

P-value = {Area right of $z = 2.50$ } $= 0.0062$

Reject H_0 because P-value $= 0.0062 < 0.05$.

There is enough evidence at the 5% level of significance to support the claim that the average speed is greater than 35 miles per hour.

5. The claim is "the mean time number of workdays missed due to illness or injury in the past 12 months is 3.5 days."

$H_0: \mu = 3.5$ (claim); $H_\alpha: \mu \neq 3.5$

$\alpha = 0.01$

$$z = \frac{\overline{x} - \mu}{\frac{\sigma}{\sqrt{n}}} = \frac{4 - 3.5}{\frac{1.5}{\sqrt{25}}} \approx 1.67$$

P-value = 2{Area to the right of $z = 1.67$ } $= 2(0.0475) = 0.0950$

Fail to reject H_0 because P-value$= 0.0950 > 0.01$.

There is not enough evidence at the 1% level of significance to support the claim that the mean number of workdays missed due to illness or injury in the past 12 months is 3.5 days.

6. $P = 0.0440 > 0.01 = \alpha$. Fail to reject H_0.

7.

Area = 0.1003

$z_0 = -1.28$

Rejection region: $z < -1.28$

8.

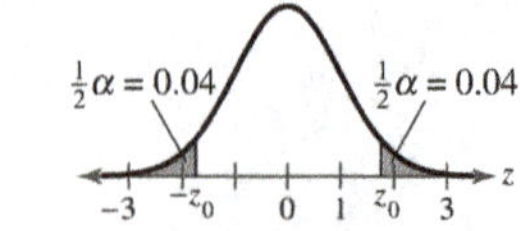

0.0401 and 0.9599

$-z_0 = -1.75$ and $z_0 = 1.75$

Rejection region: $z < -1.75$, $z > 1.75$

9. The claim is "the mean work day of the company's mechanical engineers is less than 8.5 hours."

$H_0: \mu \geq 8.5$; $H_a: \mu < 8.5$ (claim)

$\alpha = 0.01$

$z_0 = -2.33$; Rejection region: $z < -2.33$

$$z = \frac{x - \mu}{\frac{\sigma}{\sqrt{n}}} = \frac{8.2 - 8.5}{\frac{0.5}{\sqrt{25}}} = -3.00$$

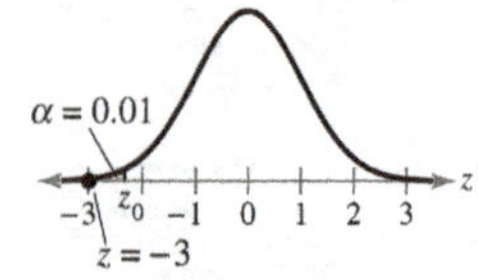

Because $-3.00 < -2.33$, reject H_0.

There is enough evidence at the 1% level of significance to support the claim that the mean work day is less than 8.5 hours.

10. $\alpha = 0.01$

$-z_0 = -2.575$, $z_0 = 2.575$; Rejection regions: $z < -2.575$, $z > 2.575$

Because $z \approx -1.98$ is not in the rejection region, we fail to reject H_0.

There is not enough evidence at the 1% level of significance to reject the claim that the mean cost of raising a child (age 2 and under) by married-couple families in the United States is $14,050.

7.2 EXERCISE SOLUTIONS

1. In the z-test using rejection region(s), the test statistic is compared with critical values. The z-test using a P-value compares the P-value with the level of significance α.

3. (a) Fail to reject H_0 because $P = 0.0461 > 0.01 = \alpha$.

(b) Reject H_0 because $P = 0.0461 < 0.05 = \alpha$.

(c) Reject H_0 because $P = 0.0461 < 0.10 = \alpha$.

5. (a) Fail to reject H_0 because $P = 0.1271 > 0.01 = \alpha$.

(b) Fail to reject H_0 because $P = 0.1271 > 0.05 = \alpha$.

(c) Fail to reject H_0 because $P = 0.1271 > 0.10 = \alpha$.

7. (a) Fail to reject H_0 because $P = 0.0838 > 0.01 = \alpha$.

(b) Reject H_0 because $P = 0.0838 > 0.05 = \alpha$.

(c) Reject H_0 because $P = 0.0838 < 0.10 = \alpha$.

9.

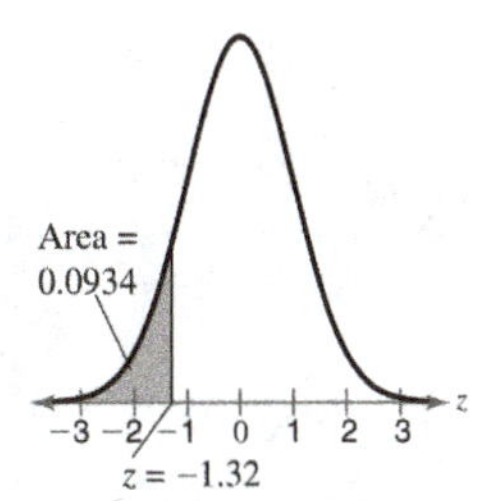

$P = 0.0934$; Reject H_0 because $P = 0.0934 < 0.10$.

11.

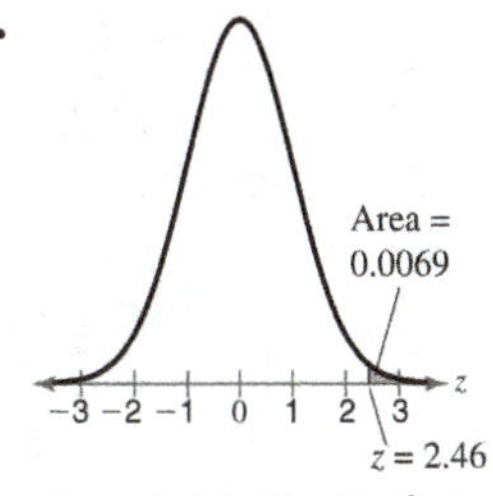

$P = 0.0069$; Reject H_0 because $P = 0.0069 < 0.01$.

13.

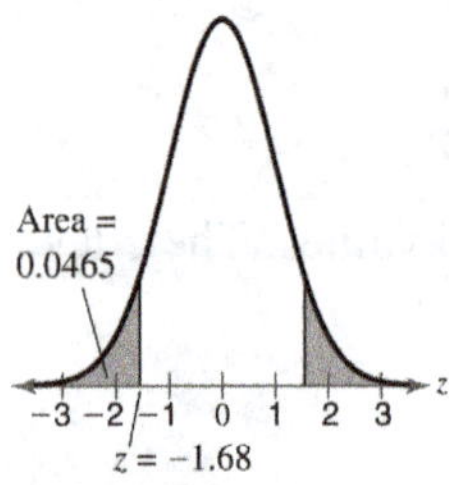

$P = 2(\text{Area}) = 2(0.0465) = 0.0930$; Fail to reject H_0 because $P = 0.0930 > 0.05$.

15. (a) $P = 0.0089$ (b) $P = 0.3050$

The larger P-value corresponds to the larger area.

17. Fail to reject H_0, $(P = 0.0628 > 0.05)$.

19. Critical value: $z_0 = -1.88$; Rejection region: $z < -1.88$

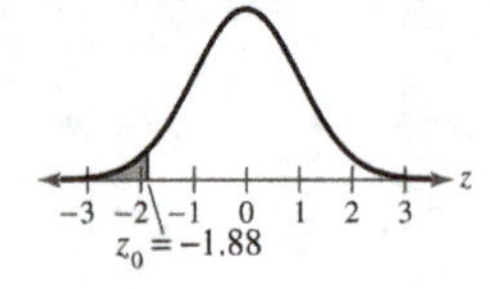

21. Critical value: $z_0 = 1.645$; Rejection region: $z > 1.645$

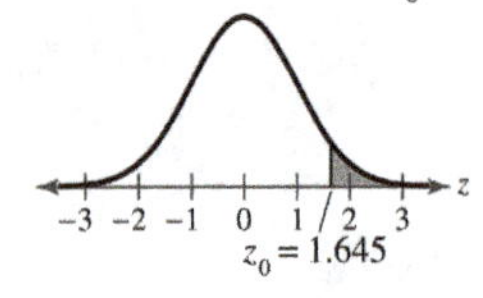

23. Critical values: $-z_0 = -2.33$, $z_0 = 2.33$; Rejection regions: $z < -2.33$, $z > 2.33$

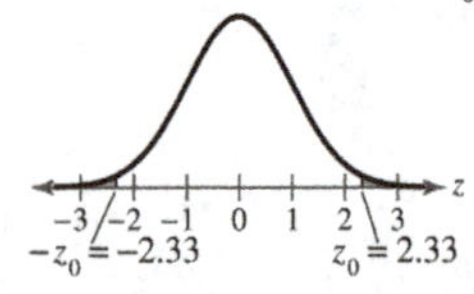

25. (a) Fail to reject H_0 because $z < 1.285$.

(b) Fail to reject H_0 because $z < 1.285$.

(c) Fail to reject H_0 because $z < 1.285$.

(d) Reject H_0 because $z > 1.285$.

27. $H_0: \mu = 40$ (claim); $H_a: \mu \neq 40$

$\alpha = 0.05 \rightarrow z_0 = \pm 1.96$

$$z = \frac{\overline{x} - \mu}{\frac{\sigma}{\sqrt{n}}} = \frac{39.2 - 40}{\frac{1.97}{\sqrt{25}}} \approx -2.030$$

Because $z = -2.03 < -1.96$, reject H_0. There is enough evidence at the 5% level of significance to reject the claim.

29. $H_0: \mu = 5880$; $H_a: \mu \neq 5880$ (claim)

$\alpha = 0.03 \rightarrow z_0 = \pm 2.17$

$$z = \frac{\overline{x} - \mu}{\frac{\sigma}{\sqrt{n}}} = \frac{5771 - 5880}{\frac{413}{\sqrt{67}}} \approx -2.16$$

Because $-2.17 < -2.16 < 2.17$, fail to reject H_0. There is not enough evidence at the 3% level of significance to support the claim.

31. (a) The claim is "the mean total score for the school's applicants is more than 499."
$H_0: \mu \leq 499$; $H_a: \mu > 499$ (claim)

(b) $$z = \frac{\overline{x} - \mu}{\frac{\sigma}{\sqrt{n}}} = \frac{502 - 499}{\frac{10.6}{\sqrt{100}}} \approx 2.83$$
Area = 0.9977

(c) P-value = {Area to right of $z = 2.83$ } = 0.0023

(d) Because $P = 0.0023 < 0.01 = \alpha$, reject H_0.

(e) There is enough evidence at the 1% level of significance to support the report's claim that the mean total score for the school's applicants is more than 499.

33. (a) The "the mean winning times for Boston Marathon women's open division champions is at least 2.68 hours."
$H_0: \mu \geq 2.68$ (claim); $H_a: \mu < 2.68$

(b) $$z = \frac{\overline{x} - \mu}{\frac{\sigma}{\sqrt{n}}} = \frac{2.60 - 2.68}{\frac{0.32}{\sqrt{30}}} \approx -1.37$$
Area = 0.0853

(c) P-value = {Area to the left of $z = -1.37$ } = 0.0853

(d) Because $P = 0.0853 > 0.05 = \alpha$, fail to reject H_0.

(e) There is not enough evidence at the 5% level of significance to reject the statistician's claim that the mean winning times for Boston Marathon women's open division champions is at least 2.68 hours.

35. (a) The claim is "the mean height of top-rated roller coasters is 160 feet."
$H_0: \mu = 160$ (claim); $H_a: \mu \neq 160$

(b) $\bar{x} \approx 164.61$

$$z = \frac{\bar{x} - \mu}{\frac{\sigma}{\sqrt{n}}} = \frac{164.61 - 160}{\frac{71.6}{\sqrt{36}}} \approx 0.3863 \approx 0.39$$

Area to the right of 0.3863= 0.3496

(c) P-value $= 2\{\text{Area to right of } z = 0.3863\} = 2(0.3496) = 0.6992$

(d) Because $P = 0.6992 > 0.05 = \alpha$, fail to reject H_0.

(e) There is not enough evidence at the 5% level of significance to reject the claim that the mean height of top-rated roller coasters is 160 feet.

37. (a) The claim is "the mean caffeine content per 12-ounce bottle of a population of caffeinated soft drinks is 37.7 milligrams."
$H_0: \mu = 37.7$ (claim); $H_a: \mu \neq 37.7$

(b) $-z_0 = -2.575,\ z_0 = 2.575$;
Rejection regions: $z < -2.575,\ z > 2.575$

(c) $$z = \frac{\bar{x} - \mu}{\frac{\sigma}{\sqrt{n}}} = \frac{36.4 - 37.7}{\frac{10.8}{\sqrt{36}}} \approx -0.72$$

(d) Because $-2.575 < z < 2.575$, fail to reject H_0.

(e) There is not enough evidence at the 1% level of significance to reject the consumer research organization's claim that the mean caffeine content per 12-ounce bottle of a population of caffeinated soft drinks is 37.7 milligrams.

39. (a) The claim is "the mean sodium content in one of its breakfast sandwiches is no more than 920 milligrams."
$H_0: \mu \leq 920$ (claim); $H_a: \mu > 920$

(b) $z_0 = 1.28$; Rejection region: $z > 1.28$

(c) $$z = \frac{\bar{x} - \mu}{\frac{\sigma}{\sqrt{n}}} = \frac{925 - 920}{\frac{18}{\sqrt{44}}} \approx 1.84$$

(d) Because $z > 1.28$, reject H_0.

(e) There is enough evidence at the 10% level of significance to reject the claim that the mean sodium content in one of their breakfast sandwiches is no more than 920 milligrams.

41. (a) The claim is "the mean life of a fluorescent lamp is at least 10,000 hours."
$H_0: \mu \geq 10{,}000$ (claim); $H_a: \mu < 10{,}000$

(b) $z_0 = -1.23$; Rejection region: $z < -1.23$

(c) $\bar{x} \approx 9580.9$

$$z = \frac{\bar{x} - \mu}{\frac{\sigma}{\sqrt{n}}} \approx \frac{9580.9 - 10{,}000}{\frac{1850}{\sqrt{32}}} \approx -1.28$$

(d) Because $z < -1.23$, reject H_0.

(e) There is enough evidence at the 11% level of significance to reject the lamp manufacturer's claim that the mean life of fluorescent lamps is at least 10,000 hours.

43. Outside; When the standardized test statistic is inside the rejection region, $P < \alpha$.

7.3 HYPOTHESIS TESTING FOR THE MEAN (SMALL SAMPLES)

7.3 TRY IT YOURSELF SOLUTIONS

1. d.f $= n - 1 = 14 - 1 = 13$ and $\alpha = 0.01$, One tail (left-tailed)

2. d.f $= n - 1 = 9 - 1 = 8$ and $\alpha = 0.10$, One tail (right-tailed)
$t_0 = 1.397$

3. d.f $= n - 1 = 16 - 1 = 15$ and $\alpha = 0.05$, Two tail
$-t_0 = -2.131,\ t_0 = 2.131$

4. The claim is "the mean age of a used car sold in the last 12 months is less than 4.1 years."
$H_0: \mu \geq 4.1$; $H_a: \mu < 4.1$ (claim)
$\alpha = 0.10$ and d.f. $= n - 1 = 25 - 1 = 24$
$t_0 = -1.318$; Rejection region: $t < -1.318$

$$t = \frac{\bar{x} - \mu}{\frac{s}{\sqrt{n}}} = \frac{3.7 - 4.1}{\frac{1.3}{\sqrt{25}}} \approx -1.54$$

Because $t < -1.318$, reject H_0.
There is enough evidence at the 10% level of significance to support the claim that the mean age of a used car sold in the last 12 months is less than 4.1 years.

5. The claim is "the mean conductivity of the river is 1890 milligrams per liter."
$H_0: \mu = 1890$ (claim); $H_a: \mu \neq 1890$
$\alpha = 0.01$ and d.f. $= n - 1 = 38$
$-t_0 = -2.712,\ t_0 = 2.712$; Rejection regions: $t < -2.712,\ t > 2.712$

$$t = \frac{\overline{x} - \mu}{\frac{s}{\sqrt{n}}} = \frac{2350 - 1890}{\frac{900}{\sqrt{39}}} \approx 3.192$$

Because $t > 2.712$, reject H_0.

There is enough evidence at the 1% level of significance to reject the company's claim that the mean conductivity of the river is 1890 milligrams per liter.

6. The claim is "the mean wait time is at most 18 minutes."

$H_0: \mu \leq 18$ minutes (claim); $H_a: \mu > 18$ minutes

$P - \text{value} = 0.9997$

$P - \text{value} = 0.9997 > 0.05 = \alpha$

Fail to reject H_0.

There is not enough evidence at the 5% level of significance to reject the office's claim that the mean wait time is at most 18 minutes.

7.3 EXERCISE SOLUTIONS

1. Identify the level of significance α and the degrees of freedom, $\text{d.f.} = n - 1$. Find the critical value(s) using the t-distribution table in the row with $n - 1$ d.f. If the hypothesis test is:

(1) left-tailed, use the "One Tail, α" column with a negative sign.

(2) right-tailed, use the "One Tail, α" column with a positive sign.

(3) two-tailed, use the "Two Tail, α," column with a negative and a positive sign.

3. Critical value: $t_0 = -1.328$; Rejection region: $t < -1.328$

5. Critical value: $t_0 = 1.717$; Rejection region: $t > 1.717$

7. Critical values: $t_0 = -2.056, t_0 = 2.056$; Rejection regions: $t < -2.056, t > 2.056$

9. (a) Fail to reject H_0 because $t > -2.086$.

(b) Fail to reject H_0 because $t > -2.086$.

(c) Fail to reject H_0 because $t > -2.086$.

(d) Reject H_0 because $t < -2.086$.

11. (a) Reject H_0 because $t < -1.725$.

(b) Fail to reject H_0 because $-1.725 < t < 1.725$.

(c) Reject H_0 because $t > 1.725$.

13. $H_0: \mu = 15$ (claim); $H_a: \mu \neq 15$

$\alpha = 0.01$ and d.f. $= n - 1 = 35$

$t_0 = \pm 2.724$

$$t = \frac{\bar{x} - \mu}{\frac{s}{\sqrt{n}}} = \frac{13.9 - 15}{\frac{3.23}{\sqrt{36}}} \approx -2.043$$

Because $-2.724 < t < 2.724$, fail to reject H_0. There is not enough evidence at the 1% level of significance to reject the claim.

15. $H_0: \mu \geq 8000$ (claim); $H_a: \mu < 8000$

$\alpha = 0.01$ and d.f. $= n - 1 = 24$

$t_0 = -2.492$

$$t = \frac{\bar{x} - \mu}{\frac{s}{\sqrt{n}}} = \frac{7700 - 8000}{\frac{450}{\sqrt{25}}} \approx -3.333$$

Because $t < -2.492$, reject H_0. There is enough evidence at the 1% level of significance to reject the claim.

17. $H_0: \mu \geq 4915$; $H_a: \mu < 4915$ (claim)

$\alpha = 0.02$ and d.f. $= n - 1 = 50$

$t_0 = -2.109$

$$t = \frac{\bar{x} - \mu}{\frac{s}{\sqrt{n}}} = \frac{5017 - 4915}{\frac{5613}{\sqrt{51}}} \approx 0.13$$

Because $t > -2.109$, fail to reject H_0. There is not enough evidence at the 2% level of significance to support the claim.

19. (a) The claim is "the mean price of a three-year-old sports utility vehicle (in good condition) is \$20,000."

$H_0: \mu = 20{,}000$ (claim); $H_a: \mu \neq 20{,}000$

(b) $-t_0 = -2.080$, $t_0 = 2.080$; Rejection region: $t < -2.080$, $t > 2.080$

(c) $$t = \frac{\bar{x} - \mu}{\frac{s}{\sqrt{n}}} = \frac{20{,}640 - 20{,}000}{\frac{1990}{\sqrt{22}}} \approx 1.508$$

(d) Because $-2.080 < t < 2.080$, fail to reject H_0.

(e) There is not enough evidence at the 5% level of significance to reject the claim that the mean price of a three-year-old sports utility vehicle (in good condition) is \$20,000.

21. (a) The claim is "the mean credit card debt by state is greater than \$5500 per person."

$H_0: \mu \leq 5500$; $H_a: \mu > 5500$ (claim)

(b) $t_0 = 1.699$; Rejection region: $t > 1.699$

(c) $t = \dfrac{\bar{x} - \mu}{\frac{s}{\sqrt{n}}} = \dfrac{5594 - 5500}{\frac{597}{\sqrt{30}}} \approx 0.86$

(d) Because $t < 1.699$, fail to reject H_0.

(e) There is not enough evidence at the 5% level of significance to support the credit reporting agency's claim that the mean credit card debt by state is greater than $5500 per person.

23. (a) The claim is "the mean amount of carbon monoxide in the air in U.S. cities is less than 2.34 parts per million."
$H_0: \mu \geq 2.34$; $H_a: \mu < 2.34$ (claim)

(b) $t_0 = -1.295$; Rejection region: $t < -1.295$

(c) $t = \dfrac{\bar{x} - \mu}{\frac{s}{\sqrt{n}}} = \dfrac{2.37 - 2.34}{\frac{2.11}{\sqrt{64}}} \approx 0.11$

(d) Because $t > -1.295$, fail to reject H_0.

(e) There is not enough evidence at the 10% level of significance to support the claim that the mean amount of carbon monoxide in the air in U.S. cities is less than 2.34 parts per million.

25. (a) The claim is "the mean annual salary for senior-level product engineers is $98,000."
$H_0: \mu = 98{,}000$ (claim); $H_a: \mu \neq 98{,}000$

(b) $-t_0 = -2.131$, $t_0 = 2.131$; Rejection region: $t < -2.131$, $t > 2.131$

(c) $\bar{x} \approx 92{,}068.9$, $s \approx 12{,}671.6$
$t = \dfrac{\bar{x} - \mu}{\frac{s}{\sqrt{n}}} \approx \dfrac{92{,}068.9 - 98{,}000}{\frac{12{,}671.6}{\sqrt{16}}} \approx -1.87$

(d) Because $-2.131 < t < 2.131$, fail to reject H_0.

(e) There is not enough evidence at the 5% level of significance to reject the employment information service's claim that the mean annual salary for senior-level product engineers is $98,000.

27. (a) The claim is "the mean minimum time it takes for a sedan to travel a quarter mile is greater than 14.7 seconds."
$H_0: \mu \leq 14.7$; $H_a: \mu > 14.7$ (claim)

(b) $t = \dfrac{\overline{x} - \mu}{\dfrac{s}{\sqrt{n}}} = \dfrac{15.4 - 14.7}{\dfrac{2.10}{\sqrt{22}}} \approx 1.56$

P-value = {Area to right of $t = 1.56$ } $= 0.0664$

(c) Because $P < 0.10 = \alpha$, reject H_0.

(d) There is enough evidence at the 10% level of significance to support the consumer group's claim that the mean minimum time it takes for a sedan to travel a quarter mile is greater than 14.7 seconds.

29. (a) The claim is "the mean class size for full-time faculty is fewer than 32 students."
$H_0: \mu \geq 32$; $H_a: \mu < 32$ (claim)

(b) $\overline{x} \approx 30.167$ $\quad s \approx 4.004$

$t = \dfrac{\overline{x} - \mu}{\dfrac{s}{\sqrt{n}}} = \dfrac{30.167 - 32}{\dfrac{4.004}{\sqrt{18}}} \approx -1.942$

P-value = {Area to left of $t = -1.942$ } $= 0.0344$

(c) Because $P < 0.05 = \alpha$, reject H_0.

(d) There is enough evidence at the 5% level of significance to support the claim that the mean class size for full-time faculty is fewer than 32 students.

31. Because σ is unknown, $n < 30$, the sample is random, and the gas mileage is normally distributed, use the t-distribution.
$H_0: \mu \geq 23$ (claim); $H_a: \mu < 23$

$t = \dfrac{\overline{x} - \mu}{\dfrac{s}{\sqrt{n}}} \approx \dfrac{22 - 23}{\dfrac{4}{\sqrt{5}}} \approx -0.559$

P-value = {Area left of $t = t = -0.559$ } $= 0.3030$
Because $P > 0.05 = \alpha$, fail to reject H_0. There is not enough evidence at the 5% level of significance to reject the claim that the mean gas mileage for the luxury sedan is at least 23 miles per gallon.

33. More likely; The tails of the t-distribution curve are thicker than those of a standard normal distribution curve. So, if you incorrectly use a standard normal sampling distribution instead of a t-sampling distribution, the area under the curve at the tails will be smaller than what it would be for the t-test, meaning the critical value(s) will lie closer to the mean. This makes it more likely for the test statistic to be in the rejection region(s). This result is the same regardless of whether the test is left-tailed, right-tailed, or two-tailed; in each case, the tail thickness affects the location of the critical value(s).

7.4 HYPOTHESIS TESTING FOR PROPORTIONS

7.4 TRY IT YOURSELF SOLUTIONS

1. $np = (150)(0.90) = 135 > 5,\ nq = (150)(0.10) = 15 > 5$

The claim is "more than 90% of U.S. adults have access to a smartphone."

$H_0: p \leq 0.90;\ H_a: p > 0.90$ (claim)

$\alpha = 0.01$

$z_0 = 2.33$; Rejection region: $z > 2.33$

$$z = \frac{\hat{p} - p}{\sqrt{\frac{pq}{n}}} = \frac{0.87 - 0.90}{\sqrt{\frac{(0.90)(0.10)}{150}}} \approx -1.22$$

Fail to reject H_0.

There is not enough evidence at the 1% level of significance to support the claim that more than 90% of U.S. adults have access to a smartphone.

2. $np = (1768)(0.67) \approx 1185 > 5,\ nq = (1768)(0.33) \approx 583 > 5$

The claim is "67% of U.S. adults believe that doctors prescribing antibiotics for viral infections for which antibiotics are not effective is a significant cause of drug-resistant superbugs."

$H_0: p = 0.67$ (claim); $H_a: p \neq 0.67$

$\alpha = 0.10$

$-z_0 = -1.645,\ z_0 = 1.645$; Rejection region: $z < -1.645,\ z > 1.645$

$$z = \frac{\hat{p} - p}{\sqrt{\frac{pq}{n}}} = \frac{\left(\frac{1150}{1768}\right) - 0.67}{\sqrt{\frac{(0.67)(0.33)}{1768}}} \approx -1.75$$

Reject H_0.

There is enough evidence at the 10% level of significance to reject the claim that 67% of U.S. adults believe that doctors prescribing antibiotics for viral infections for which antibiotics are not effective is a significant cause of drug-resistant superbugs.

7.4 EXERCISE SOLUTIONS

1. If $np \geq 5$ and $nq \geq 5$, the normal distribution can be used.

3. $np = (40)(0.12) = 4.8 < 5$

$nq = (40)(0.88) = 35.2 > 5$

Cannot use normal distribution because $np < 5$.

5. $np = (500)(0.15) = 75 > 5$

$nq = (500)(0.85) = 425 > 5 \rightarrow$ use normal distribution

$H_0: p = 0.15$; $H_a: p \neq 0.15$ (claim)

$-z_0 = -1.96$, $z_0 = 1.96$; Rejection region: $z < -1.96$, $z > 1.96$

$$z = \frac{\hat{p} - p}{\sqrt{\frac{pq}{n}}} = \frac{0.12 - 0.15}{\sqrt{\frac{(0.15)(0.85)}{500}}} \approx -1.88$$

Fail to reject H_0. There is not enough evidence at the 5% level of significance to support the claim.

7. (a) The claim is "less than 80% of U.S. adults think that healthy children should be required to be vaccinated."

$H_0: p \geq 0.80$; $H_a: p < 0.80$ (claim)

(b) $z_0 = -1.645$; Rejection region: $z < -1.645$

(c) $$z = \frac{\hat{p} - p}{\sqrt{\frac{pq}{n}}} = \frac{0.82 - 0.80}{\sqrt{\frac{(0.80)(0.20)}{200}}} \approx 0.707$$

(d) Because $z > -1.645$, fail to reject H_0.

(e) There is not enough evidence at the 5% level of significance to support the medical researcher's claim that less than 80% of U.S. adults think that healthy children should be required to be vaccinated.

9. (a) The claim is "at most 3% of working college students are employed as teachers or teaching assistants."

$H_0: p \leq 0.03$ (claim); $H_a: p > 0.03$

(b) $z_0 = 2.33$; Rejection region: $z > 2.33$

(c) $$z = \frac{\hat{p} - p}{\sqrt{\frac{pq}{n}}} = \frac{0.04 - 0.03}{\sqrt{\frac{(0.03)(0.97)}{200}}} \approx 0.83$$

(d) Because $z < 2.33$, fail to reject H_0.

(e) There is not enough evidence at the 1% level of significance to reject the education researcher's claim that at most 3% of working college students are employed as teachers or teaching assistants.

11. (a) The claim is "85% of Americans think they are unlikely to contract the Zika virus."

$H_0: p = 0.85$ (claim); $H_a: p \neq 0.85$

(b) $-z_0 = -1.96,\ z_0 = 1.96$; Rejection region: $z < -1.96,\ z > 1.96$

(c) $$z = \frac{\hat{p} - p}{\sqrt{\frac{pq}{n}}} = \frac{\left(\frac{225}{250}\right) - 0.85}{\sqrt{\frac{(0.85)(0.15)}{250}}} \approx 2.21$$

(d) Because $z > 1.96$, reject H_0.

(e) There is enough evidence at the 5% level of significance to reject the medical researcher's claim that 85% of Americans think they are unlikely to contract the Zika virus.

13. (a) The claim is "27% of U.S. adults would travel into space on a commercial flight if they could afford it."
$H_0: p = 0.27$ (claim); $H_a: p \neq 0.27$

(b) $$z = \frac{\hat{p} - p}{\sqrt{\frac{pq}{n}}} = \frac{0.30 - 0.27}{\sqrt{\frac{(0.27)(0.73)}{1000}}} \approx 2.14$$
P-value = 2{Area to right of $z = 2.14$} $= 2(0.0162) \approx 0.03$

(c) Because $P < 0.05 = \alpha$, reject H_0.

(d) There is enough evidence at the 5% level of significance to reject the research center's claim that 27% of U.S. adults would travel into space on a commercial flight if they could afford it.

15. (a) The claim is "less than 67% of U.S. households own a pet."
$H_0: p \geq 0.67$; $H_a: p < 0.67$ (claim)

(b) $$z = \frac{\hat{p} - p}{\sqrt{\frac{pq}{n}}} = \frac{\left(\frac{390}{600}\right) - 0.67}{\sqrt{\frac{(0.67)(0.33)}{600}}} \approx -1.04$$
P-value = Area to left of $z = -1.04 \approx 0.15$

(c) Because $P > 0.10 = \alpha$, fail to reject H_0.

(d) There is not enough evidence at the 10% level of significance to support the humane society's claim that less than 67% of U.S. households own a pet.

17. $H_0: p \geq 0.63$ (claim); $H_a: p < 0.63$
$z_0 = -1.645$: Rejection region: $z < -1.645$

$$z=\frac{\hat{p}-p}{\sqrt{\frac{pq}{n}}}=\frac{0.59-0.63}{\sqrt{\frac{(0.63)(0.37)}{100}}}\approx-1.45$$

Because $z>-1.645$, fail to reject H_0. There is not enough evidence at the 5% level of significance to reject the claim that at least 63% of adults make an effort to live in ways that help protect the environment some of the time.

19. (a) The claim is "less than 80% of U.S. adults think that healthy children should be required to be vaccinated."

$H_0: p\geq 0.80$; $H_a: p<0.80$ (claim)

(b) $z_0=-1.645$; Rejection region: $z<-1.645$

(c) $z=\frac{x-np}{\sqrt{npq}}=\frac{164-200(0.80)}{\sqrt{200(0.80)(0.20)}}\approx 0.707$

(d) Because $z>-1.645$, fail to reject H_0.

(e) There is not enough evidence at the 5% level of significance to support the medical researcher's claim that less than 80% of U.S. adults think that healthy children should be required to be vaccinated.

The results are the same.

7.5 HYPOTHESIS TESTING FOR VARIANCE AND STANDARD DEVIATION

7.5 TRY IT YOURSELF SOLUTIONS

1. d.f. $=17$, $\alpha=0.01$

$\chi_0^2=33.409$

2. d.f. $=29$, $\alpha=0.05$

$\chi_0^2=17.708$

3. d.f. $=50$, $\alpha=0.01$

$\chi_R^2=79.490$

$\chi_L^2=27.991$

4. The claim is "the variance of the amount of sports drink in a 12-ounce bottle is no more than 0.40."

$H_0: \sigma^2\leq 0.40$ (claim); $H_a: \sigma^2>0.40$

$\alpha=0.01$ and d.f. $=n-1=30$

$\chi_0^2 = 50.892$; Rejection region: $\chi^2 > 50.892$

$$\chi^2 = \frac{(n-1)s^2}{\sigma^2} = \frac{(30)(0.75)}{0.40} = 56.250$$

Because $\chi^2 > 50.892$, reject H_0.

There is enough evidence at the 1% level of significance to reject the claim that the variance of the amount of sports drink in a 12-ounce bottle is no more than 0.40.

5. The claim is "the standard deviation of the lengths of response times is less than 3.7 minutes."

$H_0: \sigma \geq 3.7$; $H_a: \sigma < 3.7$ (claim)

$\alpha = 0.05$ and $\text{d.f.} = n-1 = 8$

$\chi_0^2 = 2.733$; Rejection region: $\chi^2 < 2.733$

$$\chi^2 = \frac{(n-1)s^2}{\sigma^2} = \frac{(8)(3.0)^2}{(3.7)^2} \approx 5.259$$

Because $\chi^2 > 2.733$, fail to reject H_0.

There is not enough evidence at the 5% level of significance to support the claim that the standard deviation of the lengths of response times is less than 3.7 minutes.

6. The claim is "the variance of the weight losses is 25.5."

$H_0: \sigma^2 = 25.5$ (claim); $H_a: \sigma^2 \neq 25.5$

$\alpha = 0.10$ and $\text{d.f.} = n-1 = 12$

$\chi_L^2 = 5.226$ and $\chi_R^2 = 21.026$; Rejection region: $\chi^2 > 21.026$, $\chi^2 < 5.226$

$$\chi^2 = \frac{(n-1)s^2}{\sigma^2} = \frac{(12)(10.8)}{25.5} \approx 5.082$$

Because $\chi^2 < 5.226$, reject H_0.

There is enough evidence at the 10% level of significance to reject the claim that the variance of the weight losses of users is 25.5.

7.5 EXERCISE SOLUTIONS

1. Specify the level of significance α. Determine the degrees of freedom. Determine the critical values using the χ^2-distribution. For a right-tailed test, use the value that corresponds to d.f. and α. For a left-tailed test, use the value that corresponds to d.f. and $1-\alpha$. For a two-tailed test, use the value that corresponds to d.f. and $\frac{1}{2}\alpha$, and d.f. and $1-\frac{1}{2}\alpha$.

3. The requirement of a normal distribution is more important when testing a standard deviation than when testing a mean. When the population is not normal, the results of the chi-square test can be misleading because the chi-square test is not as robust as the tests for the population mean.

5. Critical value: $\chi_0^2 = 38.885$; Rejection region: $\chi^2 > 38.885$

7. Critical value: $\chi_0^2 = 0.872$; Rejection region: $\chi^2 < 0.872$

9. Critical values: $\chi_L^2 = 60.391,\ \chi_R^2 = 101.879$; Rejection regions: $\chi^2 < 60.391,\ \chi^2 > 101.879$

11. Critical value: $\chi_0^2 = 49.588$; Rejection region: $\chi^2 > 49.588$

13. (a) Fail to reject H_0 because $\chi^2 < 6.251$.

(b) Fail to reject H_0 because $\chi^2 < 6.251$.

(c) Fail to reject H_0 because $\chi^2 < 6.251$.

(d) Reject H_0 because $\chi^2 > 6.251$.

15. $H_0: \sigma^2 = 0.52$ (claim); $H_a: \sigma^2 \neq 0.52$

$\chi_L^2 = 7.564,\ \chi_R^2 = 30.191$; Rejection regions: $\chi^2 < 7.564,\ \chi^2 > 30.191$

$$\chi^2 = \frac{(n-1)s^2}{\sigma^2} = \frac{(17)(0.508)}{(0.52)} \approx 16.608$$

Because $7.564 < \chi^2 < 30.191$, fail to reject H_0. There is not enough evidence at the 5% level of significance to reject the claim.

17. $H_0: \sigma^2 \leq 17.6$ (claim); $H_a: \sigma^2 > 17.6$

$\chi_0^2 = 63.691$; Rejection region: $\chi^2 > 63.691$

$$\chi^2 = \frac{(n-1)s^2}{\sigma^2} = \frac{(40)(28.33)}{17.6} \approx 64.39$$

Because $\chi^2 > 63.691$, reject H_0. There is enough evidence at the 1% level of significance to reject the claim.

19. $H_0: \sigma^2 = 32.8$; $H_a: \sigma^2 \neq 32.8$ (claim)

$\chi_L^2 = 77.929,\ \chi_R^2 = 124.342$; Rejection regions: $\chi^2 < 77.929,\ \chi^2 > 124.342$

$$\chi^2 = \frac{(n-1)s^2}{\sigma^2} = \frac{(100)(40.9)}{(32.8)} \approx 124.7$$

Because $\chi^2 > 124.342$, Reject H_0. There is enough evidence at the 10% level of significance to support the claim.

21. $H_0: \sigma \geq 40$; $H_a: \sigma < 40$ (claim)

$\chi_0^2 = 3.053$; Rejection region: $\chi^2 < 3.053$

$$\chi^2 = \frac{(n-1)s^2}{\sigma^2} = \frac{(11)(40.8)^2}{(40)^2} \approx 11.444$$

Because $\chi^2 > 3.053$, fail to reject H_0. There is not enough evidence at the 1% level of significance to support the claim.

23. (a) The claim is "the variance of the diameters in a certain tire model is 8.6."
$H_0: \sigma^2 = 8.6$ (claim); $H_a: \sigma^2 \neq 8.6$

(b) $\chi_L^2 = 1.735$, $\chi_R^2 = 23.589$; Rejection regions: $\chi^2 < 1.735$, $\chi^2 > 23.589$

(c) $\chi^2 = \dfrac{(n-1)s^2}{\sigma^2} = \dfrac{(9)(4.3)}{8.6} = 4.5$

(d) Because $1.735 < \chi^2 < 23.589$, fail to reject H_0.

(e) There is not enough evidence at the 1% level of significance to reject the claim that the variance of the diameters in a certain tire model is 8.6.

25. (a) The claim is "the standard deviation for grade 12 students on a mathematics assessment test is less than 35 points."
$H_0: \sigma \geq 35$; $H_a: \sigma < 35$ (claim)

(b) $\chi_0^2 = 18.114$; Rejection region: $\chi^2 < 18.114$

(c) $\chi^2 = \dfrac{(n-1)s^2}{\sigma^2} = \dfrac{(27)(34)^2}{(35)^2} \approx 25.48$

(d) Because $\chi^2 > 18.114$, fail to reject H_0.

(e) There is not enough evidence at the 10% level of significance to support the school administrator's claim that the standard deviation for grade 12 students on a mathematics assessment test is less than 35 points.

27. (a) The claim is "the standard deviation of waiting times experienced by patients is no more than 0.5 minute."
$H_0: \sigma \leq 0.5$ (claim); $H_a: \sigma > 0.5$

(b) $\chi_0^2 = 33.196$; Rejection region: $\chi^2 > 33.196$

(c) $\chi^2 = \dfrac{(n-1)s^2}{\sigma^2} = \dfrac{(24)(0.7)^2}{(0.5)^2} \approx 47.04$

(d) Because $\chi^2 > 33.196$, reject H_0.

(e) There is enough evidence at the 10% level of significance to reject the claim that the standard deviation of waiting times experienced by patients is no more than 0.5 minute.

29. (a) The claim is "the standard deviation of the annual salaries is different from \$10,300."
$H_0: \sigma = 10{,}300$; $H_a: \sigma \neq 10{,}300$ (claim)

(b) $\chi_L^2 = 5.629$, $\chi_R^2 = 26.119$ Rejection regions: $\chi^2 < 5.629$, $\chi^2 > 26.119$

(c) $s \approx 14{,}530.3$

$$\chi^2 = \frac{(n-1)s^2}{\sigma^2} = \frac{(14)(14{,}530.3)^2}{(10{,}300)^2} \approx 27.86$$

(d) Because $\chi^2 > 26.119$, reject H_0.

(e) There is enough evidence at the 5% level of significance to support the claim that the standard deviation of the annual salaries of senior-level graphic design specialists is different from $10,300.

31. $\chi^2 = 25.48$, d.f. $= n - 1 = 27$
P-value = {Area left of $\chi^2 = 25.48$} $= 0.4524$
Fail to reject H_0 because P-value $= 0.4524 > 0.10$.

33. $\chi^2 = 47.04$, d.f. $= n - 1 = 24$
P-value = {Area right of $\chi^2 = 47.04$} $= 0.0033$
Reject H_0 because P-value $= 0.0033 < 0.10$.

CHAPTER 7 REVIEW EXERCISE SOLUTIONS

1. $H_0: \mu \le 375$ (claim); $H_a: \mu > 375$

3. $H_0: p \ge 0.205$; $H_a: p < 0.205$ (claim)

5. $H_0: \sigma \le 1.9$; $H_a: \sigma > 1.9$ (claim)

7. (a) $H_0: p = 0.65$ (claim); $H_a: p \ne 0.65$

(b) A type I error will occur when the actual proportion of U.S. adults who have volunteered their time or donated money to help clean up the environment is 65%, but you reject $H_0: p = 0.65$.
A type II error will occur when the actual proportion is not 65%, but you fail to reject $H_0: p = 0.65$.

(c) Two-tailed because the alternative hypothesis contains $\ne$.

(d) There is enough evidence to reject the polling organization's claim that the proportion of U.S. adults who have volunteered their time or donated money to help clean up the environment is 65%.

(e) There is not enough evidence to reject the polling organization's claim that the proportion of U.S. adults who have volunteered their time or donated money to help clean up the environment is 65%.

9. (a) $H_0: \sigma \le 9.5$ (claim); $H_a: \sigma > 9.5$

(b) A type I error will occur when the actual standard deviation of the fuel economies is no more than 9.5 miles per gallon, but you reject $H_0: \sigma \le 9.5$.

A type II error will occur when the actual standard deviation of the fuel economies is more than 9.5 miles per gallon, but you fail to reject H_0: $\sigma \le 9.5$.

(c) Right-tailed because the alternative hypothesis contains >.

(d) There is enough evidence to reject the nonprofit consumer organization's claim that the standard deviation of the fuel economies of its top-rated vehicles for a recent year is no more than 9.5 miles per gallon.

(e) There is not enough evidence to reject the nonprofit consumer organization's claim that the standard deviation of the fuel economies of its top-rated vehicles for a recent year is no more than 9.5 miles per gallon.

11. P-value = {Area to left of $z = -0.94$} = 0.1736
Fail to reject H_0.

13. Critical value: $z_0 \approx -2.05$; Rejection region: $z < -2.05$

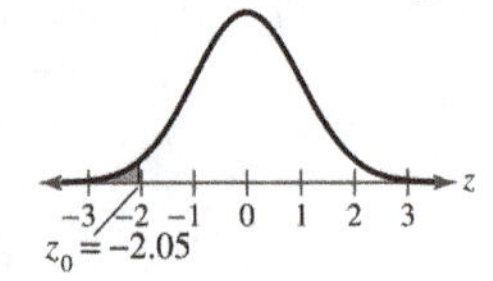

15. Critical value: $z_0 = 1.96$; Rejection region: $z > 1.96$

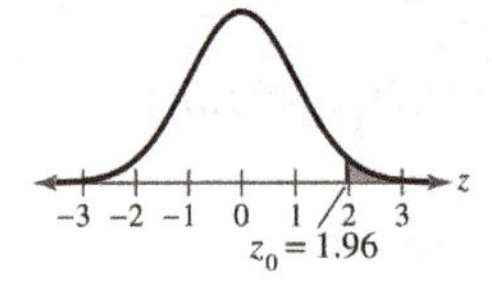

17. Fail to reject H_0 because $-1.645 < z < 1.645$.

19. Fail to reject H_0 because $-1.645 < z < 1.645$.

21. $H_0: \mu \le 45$ (claim); $H_a: \mu > 45$
$z_0 = 1.645$; Rejection region: $z > 1.645$

$$z = \frac{\bar{x} - \mu}{\frac{\sigma}{\sqrt{n}}} = \frac{47.2 - 45}{\frac{6.7}{\sqrt{22}}} \approx 1.54$$

Because $z < 1.645$, fail to reject H_0. There is not enough evidence at the 5% level of significance to reject the claim.

23. $H_0: \mu \ge 5.500$; $H_a: \mu < 5.500$ (claim)
$z_0 = -2.33$; Rejection region: $z < -2.33$

$$z = \frac{\bar{x} - \mu}{\frac{\sigma}{\sqrt{n}}} = \frac{5.497 - 5.500}{\frac{0.011}{\sqrt{36}}} \approx -1.636$$

Because $z > -2.33$, fail to reject H_0. There is not enough evidence at the 1% level of significance to support the claim.

25. (a) The claim is "the mean annual production of cotton is 3.5 million bales per country."
$H_0: \mu = 3.5$ (claim); $H_a: \mu \neq 3.5$

(b) $z = \dfrac{\bar{x} - \mu}{\frac{\sigma}{\sqrt{n}}} = \dfrac{2.1 - 3.5}{\frac{4.5}{\sqrt{44}}} \approx -2.06$

(c) P-value = 2{Area to left of $z = -2.06$} $= 2(0.0197) = 0.0394$

(d) Because P-value $< 0.05 = \alpha$, reject H_0.

(e) There is enough evidence at the 5% level of significance to reject the researcher's claim that the mean annual production of cotton is 3.5 million bales per country.

27. (a) The claim is "the mean amount of sulfur dioxide in the air in U.S. cities is 1.15 parts per billion."
$H_0: \mu = 1.15$ (claim); $H_a: \mu \neq 1.15$

(b) $-z_0 = -2.575,\ z_0 = 2.575$; Rejection regions: $z < -2.575,\ z > 2.575$

(c) $z = \dfrac{\bar{x} - \mu}{\frac{\sigma}{\sqrt{n}}} = \dfrac{0.93 - 1.15}{\frac{2.62}{\sqrt{134}}} \approx -0.97$

(d) Because $-2.575 < z < 2.575$, fail to reject H_0.

(e) There is not enough evidence at the 1% level of significance to reject the environmental researcher's claim that the mean amount of sulfur dioxide in the air in U.S. cities is 1.15 parts per billion.

29. Critical values: $-t_0 = -2.093,\ t_0 = 2.093$; Rejection region: $-t < -2.093,\ t > 2.093$

31. Critical value: $t_0 = 2.098$; Rejection region: $t > 2.098$

33. Critical value: $t_0 = -2.977$; Rejection region: $t < -2.977$

35. $H_0: \mu \leq 12{,}700$; $H_a: \mu > 12{,}700$ (claim)
$t_0 = 2.845$; Rejection region: $t > 2.845$

$t = \dfrac{\bar{x} - \mu}{\frac{s}{\sqrt{n}}} = \dfrac{12{,}855 - 12{,}700}{\frac{248}{\sqrt{21}}} \approx 2.864$

Because $t > 2.845$, reject H_0. There is enough evidence at the 0.5% level of significance to support the claim.

37. $H_0: \mu \leq 51$ (claim); $H_a: \mu > 51$
$t_0 = 2.426$; Rejection region: $t > 2.426$

$$t = \frac{\overline{x} - \mu}{\frac{s}{\sqrt{n}}} = \frac{52 - 51}{\frac{2.5}{\sqrt{40}}} \approx 2.530$$

Because $t > 2.426$, reject H_0. There is enough evidence at the 1% level of significance to reject the claim.

39. $H_0: \mu = 195$ (claim); $H_a: \mu \neq 195$

$-t_0 = -1.660; t_0 = 1.660$ Rejection regions: $t < -1.660; t > 1.660$

$$t = \frac{\overline{x} - \mu}{\frac{s}{\sqrt{n}}} = \frac{190 - 195}{\frac{36}{\sqrt{101}}} \approx -1.396$$

Because $-1.660 < t < 1.660$, fail to reject H_0. There is not enough evidence at the 10% level of significance to reject the claim.

41. (a) The claim is "the mean monthly cost of joining a health club is \$25."
$H_0: \mu = 25$ (claim); $H_a: \mu \neq 25$

(b) $-t_0 = -1.740$, $t_0 = 1.740$; Rejection regions: $t < -1.740$, $t > 1.740$

(c) $$t = \frac{\overline{x} - \mu}{\frac{s}{\sqrt{n}}} = \frac{26.25 - 25}{\frac{3.23}{\sqrt{18}}} \approx 1.642$$

(d) Because $-1.740 < t < 1.740$, fail to reject H_0.

(e) There is not enough evidence at the 10% level of significance to reject the claim that the mean monthly cost of joining a health club is \$25.

43. (a) The claim is "the mean score for grade 12 students on a science achievement test is more than 145."
$H_0: \mu \leq 145$; $H_a: \mu > 145$ (claim)

(b) $\overline{x} \approx 155.39$ $\qquad s \approx 43.93$

$$t = \frac{\overline{x} - \mu}{\frac{s}{\sqrt{n}}} = \frac{155.39 - 145}{\frac{43.93}{\sqrt{36}}} \approx 1.419$$

P-value = {Area to right of $t = 1.419$} ≈ 0.0824

(c) Because $P\text{-value} < 0.10 = \alpha$, reject H_0.

(d) There is enough evidence at the 10% level of significance to support the education publication's claim that the mean score for grade 12 students on a science achievement test is more than 145.

45. $np = (40)(0.15) = 6 > 5$

$nq = (40)(0.85) = 34 > 5 \rightarrow$ can use normal distribution

$H_0: p = 0.15$ (claim); $H_a: p \neq 0.15$

$-z_0 = -1.96,\ z_0 = 1.96$; Rejection regions: $z < -1.96,\ z > 1.96$

$$z = \frac{\hat{p} - p}{\sqrt{\frac{pq}{n}}} = \frac{0.09 - 0.15}{\sqrt{\frac{(0.15)(0.85)}{40}}} \approx -1.063$$

Because $-1.96 < z < 1.96$, fail to reject H_0. There is not enough evidence at the 5% level of significance to reject the claim.

47. $np = (68)(0.70) = 47.6 > 5$

$nq = (68)(0.30) = 20.4 > 5 \rightarrow$ can use normal distribution

$H_0: p \geq 0.70;\ H_a: p < 0.70$ (claim)

$z_0 = -2.33$; Rejection region: $z < -2.33$

$$z = \frac{\hat{p} - p}{\sqrt{\frac{pq}{n}}} = \frac{0.50 - 0.70}{\sqrt{\frac{(0.70)(0.30)}{68}}} \approx -3.599$$

Because $z < -2.33$, reject H_0. There is enough evidence at the 1% level of significance to support the claim.

49. (a) The claim is "over 40% of U.S. adults say they are less likely to travel to Europe in the next six months for fear of terrorist attacks."

$H_0: p \leq 0.40;\ H_a: p > 0.40$ (claim)

(b) $z_0 = 2.33$; Rejection region: $z > 2.33$

(c) $$z = \frac{\hat{p} - p}{\sqrt{\frac{pq}{n}}} \approx \frac{0.42 - 0.40}{\sqrt{\frac{(0.40)(0.60)}{1000}}} \approx 1.29$$

(d) Because $z < 2.33$, fail to reject H_0.

(e) There is not enough evidence at the 1% level of significance to support the polling agency's claim that over 40% of U.S. adults say they are less likely to travel to Europe in the next six months for fear of terrorist attacks.

51. Critical value: $\chi_0^2 = 30.144$; Rejection region: $\chi^2 > 30.144$

53. Critical values: $\chi_L^2 = 26.509,\ \chi_R^2 = 55.758$; Rejection regions: $\chi^2 < 26.509,\ \chi^2 > 55.758$

55. $H_0: \sigma^2 \leq 2;\ H_a: \sigma^2 > 2$ (claim)

$\chi_0^2 = 24.769$; Rejection region: $\chi^2 > 24.769$

$$\chi^2 = \frac{(n-1)s^2}{\sigma^2} = \frac{(17)(2.95)}{(2)} = 25.075$$

Because $\chi^2 > 24.769$, reject H_0. There is enough evidence at the 10% level of significance to support the claim.

57. $H_0: \sigma = 1.25$ (claim); $H_a: \sigma \neq 1.25$

$\chi_L^2 = 0.831,\ \chi_R^2 = 12.833$; Rejection regions: $\chi^2 < 0.831,\ \chi^2 > 12.833$

$$\chi^2 = \frac{(n-1)s^2}{\sigma^2} = \frac{(5)(1.03)^2}{(1.25)^2} \approx 3.395$$

Because $0.831 < \chi^2 < 12.833$, fail to reject H_0. There is not enough evidence at the 5% level of significance to reject the claim.

59. (a) The claim is "the variance of the bolt widths is at most 0.01."
$H_0: \sigma^2 \leq 0.01$ (claim); $H_a: \sigma^2 > 0.01$

(b) $\chi_0^2 = 49.645$; Rejection region: $\chi^2 > 49.645$

(c) $\chi^2 = \dfrac{(n-1)s^2}{\sigma^2} = \dfrac{(27)(0.064)}{(0.01)} = 172.8$

(d) Because $\chi^2 > 49.645$, reject H_0.

(e) There is enough evidence at the 0.5% level of significance to reject the claim that the variance is at most 0.01.

61. $\chi_0^2 = 46.963$; Rejection region: $\chi^2 > 46.963$

From Exercise 59, $\chi^2 = 172.8$.

You can reject H_0 at the 1% level of significance because $\chi^2 = 172.8 > 46.963$.

CHAPTER 7 QUIZ SOLUTIONS

1. (a) The claim is "the mean hat size for a male is at least 7.25."
$H_0: \mu \geq 7.25$ (claim); $H_a: \mu < 7.25$

(b) Left-tailed because the alternative hypothesis contains $<$; z-test because σ is known and the population is normally distributed.

(c) *Sample answer*: $z_0 = -2.33$; Rejection region: $z < -2.33$

$$z = \frac{\bar{x} - \mu}{\frac{\sigma}{\sqrt{n}}} = \frac{7.15 - 7.25}{\frac{0.27}{\sqrt{12}}} \approx -1.283$$

(d) Because $z > -2.33$, fail to reject H_0.

(e) There is not enough evidence at the 1% level of significance to reject the company's claim that the mean hat size for a male is at least 7.25.

2. (a) The claim is "the mean daily base price for renting a full-size or less expensive vehicle in Vancouver, Washington, is more than $36."
H_0: $\mu \leq 36$ H_a: $\mu > 36$ (claim)

(b) Right-tailed because the alternative hypothesis contains >; z-test because σ is known and $n \geq 30$.

(c) *Sample answer:* $z_0 = 1.28$; Rejection region: $z > 1.28$

$$z = \frac{\bar{x} - \mu}{\frac{\sigma}{\sqrt{n}}} = \frac{42 - 36}{\frac{19}{\sqrt{40}}} \approx 1.997$$

(d) Because $z > 1.28$, reject H_0.

(e) There is enough evidence at the 10% level of significance to support the travel analyst's claim that the mean daily base price for renting a full-size or less expensive vehicle in Vancouver, Washington, is more than $36.

3. (a) The claim is "the mean amount of earnings for full-time workers ages 18 to 24 with a bachelor's degree in a recent year is $47,254."
H_0: $\mu = 47{,}254$ (claim); H_a: $\mu \neq 47{,}254$

(b) Two-tailed because the alternative hypothesis contains ≠; t-test because σ is unknown and the population is normally distributed.

(c) *Sample answer:* $-t_0 = -2.145; t_0 = 2.145$ Rejection regions: $t < -2.145; t > 2.145$

$$t = \frac{\bar{x} - \mu}{\frac{s}{\sqrt{n}}} = \frac{50{,}781 - 47{,}254}{\frac{5290}{\sqrt{15}}} \approx 2.58$$

(d) Because $t > 2.145$, reject H_0.

(e) There is not enough evidence at the 5% level of significance to support the government agency's claim that the mean amount of earnings for full-time workers ages 18 to 24 with a bachelor's degree is a recent year is $47,254.

4. (a) The claim is "program participants have a mean weight loss of at least 10.5 pounds after 1 month."
H_0: $\mu \geq 10.5$ (claim); H_a: $\mu < 10.5$

(b) Left-tailed because the alternative hypothesis contains<; t-test because σ is unknown and $n \geq 30$.

(c) *Sample answer:* $t_0 = -2.426$; Rejection region: $t < -2.426$

$$t = \frac{\bar{x} - \mu}{\frac{s}{\sqrt{n}}} = \frac{9.233 - 10.5}{\frac{2.593}{\sqrt{40}}} \approx -3.09$$

(d) Because $t < -2.426$, reject H_0.

(e) There is enough evidence at the 1% level of significance to reject the weight loss program's claim that program participants have a mean weight loss of at least 10.5 pounds after 1 month.

5. (a) The claim is "less than 18% of the vehicles a nonprofit consumer organization rated in a recent year have an overall score of 78 or more."
H_0: $p \geq 0.18$; H_a: $p < 0.18$ (claim)

(b) Left-tailed because the alternative hypothesis contains <; *z*-test because $np \geq 5$ and $nq \geq 5$.

(c) *Sample answer:* $z_0 = -1.645$; Rejection region: $z < -1.645$

$$z = \frac{\hat{p} - p}{\sqrt{\frac{pq}{n}}} = \frac{0.20 - 0.18}{\sqrt{\frac{(0.18)(0.82)}{90}}} \approx 0.49$$

(d) Because $z > -1.645$, fail to reject H_0..

(e) There is not enough evidence at the 5% level of significance to support the nonprofit consumer organization's claim that less than 18% of the vehicles a nonprofit consumer organization rated in a recent year have an overall score of 78 or more.

6. (a) The claim is "the standard deviation of vehicle rating scores is 11.90."
H_0: $\sigma = 11.90$ (claim).18; H_a: $\sigma \neq 11.90$.

(b) Two-tailed because the alternative hypothesis contains ≠; chi-square test because the test is for a standard deviation and the population is normally distributed.

(c) *Sample answer:* $\chi_L^2 = 68.249$, $\chi_R^2 = 112.022$; Rejection regions: $\chi^2 < 68.249$, $\chi^2 > 112.022$

$$\chi^2 = \frac{(n-1)s^2}{\sigma^2} = \frac{(89)(11.96)^2}{(11.90)^2} \approx 89.90$$

(d) Because $68.249 < \chi^2 < 112.022$, fail to reject H_0.

(e) There is not enough evidence at the 10% level of significance to reject the nonprofit consumer organization's claim that the standard deviation of vehicle rating scores is 11.90.

CHAPTER 8

Hypothesis Testing with Two Samples

8.1 TESTING THE DIFFERENCE BETWEEN MEANS (LARGE INDEPENDENT SAMPLES)

8.1 TRY IT YOURSELF SOLUTIONS

Note: Answers may differ due to rounding.

1. (1) Independent
Because each sample represents blood pressures of different individuals, and it is not possible to form a pairing between the members of the samples.

(2) Dependent
Because the samples represent exam scores of the same students, the samples can be paired with respect to each student.

2. The claim is "there is a difference in the mean annual wages for forensic science technicians working for local and state governments."
$H_0: \mu_1 = \mu_2$; $H_a: \mu_1 \neq \mu_2$ (claim)
$\alpha = 0.10$
$-z_0 = -1.645$, $z_0 = 1.645$; Rejection regions: $z < -1.645$, $z > 1.645$

$$z = \frac{(\bar{x}_1 - \bar{x}_2) - (\mu_1 - \mu_2)}{\sqrt{\frac{\sigma_1^2}{n_1} + \frac{\sigma_2^2}{n_2}}} = \frac{(60{,}680 - 59{,}430) - (0)}{\sqrt{\frac{(6200)^2}{100} + \frac{(5575)^2}{100}}} \approx 1.499$$

Since $-1.645 < z < 1.645$, fail to reject H_0.

There is not enough evidence at the 10% level of significance to support the claim that there is a difference in the mean annual wages for forensic science technicians working for local and state governments.

3.
$$z = \frac{(\bar{x}_1 - \bar{x}_2) - (\mu_1 - \mu_2)}{\sqrt{\frac{\sigma_1^2}{n_1} + \frac{\sigma_2^2}{n_2}}} = \frac{(310 - 306) - (0)}{\sqrt{\frac{(25)^2}{15} + \frac{(20)^2}{20}}} \approx 0.509$$

$\rightarrow$ P-value = {area right of $z = 0.509$} ≈ 0.305
Because $P\text{-value} > 0.05 = \alpha$, fail to reject H_0.
There is not enough evidence at the 5% level of significance to support the travel agency's claim that the average daily cost of meals and lodging for vacationing in Alaska is greater than the average daily cost for vacationing in Colorado.

8.1 EXERCISE SOLUTIONS

1. Two samples are dependent if each member of one sample corresponds to a member of the other sample. Example: The weights of 22 people before starting an exercise program and the weights of the same 22 people 6 weeks after starting the exercise program.

Two samples are independent if the sample selected from one population is not related to the sample from the other population. Example: The weights of 25 cats and the weights of 20 dogs.

3. Use P-values.

5. Dependent because the same football players were sampled.

7. Independent because different boats were sampled.

9. Because $z \approx 2.96 > 1.96$ and $P \approx 0.0031 < 0.05$, reject H_0.

11. $H_0: \mu_1 = \mu_2$ (claim); $H_a: \mu_1 \neq \mu_2$

Rejection region: $z < -2.575$, $z > 2.575$

$$z = \frac{(\bar{x}_1 - \bar{x}_2) - (\mu_1 - \mu_2)}{\sqrt{\frac{\sigma_1^2}{n_1} + \frac{\sigma_2^2}{n_2}}} = \frac{(16-14)-(0)}{\sqrt{\frac{(3.4)^2}{29} + \frac{(1.5)^2}{28}}} \approx 2.89$$

Because $z > 2.575$, reject H_0. There is enough evidence at the 1% level of significance to reject the claim.

13. $H_0: \mu_1 \geq \mu_2$; $H_a: \mu_1 < \mu_2$ (claim)

Rejection region: $z < -1.645$

$$z = \frac{(\bar{x}_1 - \bar{x}_2) - (\mu_1 - \mu_2)}{\sqrt{\frac{\sigma_1^2}{n_1} + \frac{\sigma_2^2}{n_2}}} = \frac{(2435-2432)-(0)}{\sqrt{\frac{(75)^2}{35} + \frac{(105)^2}{90}}} \approx 0.18$$

Because $z > -1.645$, fail to reject H_0. There is not enough evidence at the 5% level of significance to support the claim.

15. (a) The claim is "the mean braking distances are different for the two makes of automobiles."
$H_0: \mu_1 = \mu_2$; $H_a: \mu_1 \neq \mu_2$ (claim)

(b) $-z_0 = -1.645$, $z_0 = 1.645$; Rejection regions: $z < -1.645$, $z > 1.645$

(c) $$z = \frac{(\bar{x}_1 - \bar{x}_2) - (\mu_1 - \mu_2)}{\sqrt{\frac{\sigma_1^2}{n_1} + \frac{\sigma_2^2}{n_2}}} = \frac{(137-132)-(0)}{\sqrt{\frac{(5.5)^2}{23} + \frac{(6.7)^2}{24}}} \approx 2.80$$

(d) Because $z > 1.645$, reject H_0.

(e) There is enough evidence at the 10% level of significance to support the safety engineer's claim that the mean braking distances are different for the two makes of automobiles.

17. (a) The claim is "the wind speed in Region A is less than the wind speed in Region B."
$H_0: \mu_1 \geq \mu_2;\ H_a: \mu_1 < \mu_2$ (claim)

(b) $z_0 = -1.645$; Rejection region: $z < -1.645$

(c) $$z = \frac{(\bar{x}_1 - \bar{x}_2) - (\mu_1 - \mu_2)}{\sqrt{\frac{\sigma_1^2}{n_1} + \frac{\sigma_2^2}{n_2}}} = \frac{(14 - 15.1) - (0)}{\sqrt{\frac{(2.9)^2}{60} + \frac{(3.3)^2}{60}}} \approx -1.94$$

(d) Because, $z < -1.645$ reject H_0.

(e) There is enough evidence at the 5% level of significance to support the claim that the wind speed in Region A is less than the wind speed in Region B.

19. (a) The claim is "ACT mathematics and science scores are equal."
$H_0: \mu_1 = \mu_2$ (claim); $H_a: \mu_1 \neq \mu_2$

(b) $-z_0 = -2.575,\ z_0 = 2.575$; Rejection regions: $z < -2.575,\ z > 2.575$

(c) $$z = \frac{(\bar{x}_1 - \bar{x}_2) - (\mu_1 - \mu_2)}{\sqrt{\frac{\sigma_1^2}{n_1} + \frac{\sigma_2^2}{n_2}}} = \frac{(20.6 - 20.8) - (0)}{\sqrt{\frac{(5.4)^2}{60} + \frac{(5.6)^2}{75}}} \approx -0.21$$

(d) Because $-2.575 < z < 2.575$, fail to reject H_0.

(e) There is not enough evidence at the 1% level of significance to reject the claim that ACT mathematics and science scores are equal.

21. The claim is "the mean home sales price in Casper, Wyoming, is the same as in Cheyenne, Wyoming."

(a) $H_0: \mu_1 = \mu_2$ (claim); $H_a: \mu_1 \neq \mu_2$

(b) $-z_0 = -2.575,\ z_0 = 2.575$; Rejection regions: $z < -2.575,\ z > 2.575$

(c) $$z = \frac{(\bar{x}_1 - \bar{x}_2) - (\mu_1 - \mu_2)}{\sqrt{\frac{\sigma_1^2}{n_1} + \frac{\sigma_2^2}{n_2}}} = \frac{(294{,}220 - 287{,}984) - (0)}{\sqrt{\frac{(135{,}387)^2}{25} + \frac{(151{,}996)^2}{25}}} \approx 0.15$$

(d) Because $-2.575 < z < 2.575$, fail to reject H_0.

(e) There is not enough evidence at the 1% level of significance to reject the real estate agency's claim that the mean home sales price in Casper, Wyoming, is the same as in Cheyenne, Wyoming.

23. (a) The claim is "the precipitation is Seattle, Washington, was greater than in Birmingham, Alabama."

$H_0: \mu_1 \le \mu_2$; $H_a: \mu_1 > \mu_2$ (claim)

(b) $z_0 = 1.645$; Rejection region: $z > 1.645$

(c) $\bar{x}_1 = 0.157,\ n_1 = 30$

$\bar{x}_2 \approx 0.081,\ n_2 = 30$

$$z = \frac{(\bar{x}_1 - \bar{x}_2) - (\mu_1 - \mu_2)}{\sqrt{\dfrac{\sigma_1^2}{n_1} + \dfrac{\sigma_2^2}{n_2}}} = \frac{(0.157 - 0.081) - (0)}{\sqrt{\dfrac{(0.24)^2}{30} + \dfrac{(0.33)^2}{30}}} \approx 1.02$$

(d) Because $z < 1.645$, fail to reject H_0.

(e) There is not enough evidence at the 5% level of significance to support the climatologist's claim that the precipitation is Seattle, Washington, was greater than in Birmingham, Alabama.

25. They are equivalent through algebraic manipulation of the equation.

$\mu_1 = \mu_2 \rightarrow \mu_1 - \mu_2 = 0$

27. $H_0: \mu_1 - \mu_2 \le 2000$; $H_a: \mu_1 - \mu_2 > 2000$ (claim)

$z_0 = 1.645$; Rejection region: $z > 1.645$

$$z = \frac{(\bar{x}_1 - \bar{x}_2) - (\mu_1 - \mu_2)}{\sqrt{\dfrac{\sigma_1^2}{n_1} + \dfrac{\sigma_2^2}{n_2}}} = \frac{(64{,}270 - 62{,}610) - (2000)}{\sqrt{\dfrac{(10{,}850)^2}{42} + \dfrac{(10{,}970)^2}{38}}} \approx -0.14$$

Because $z < 1.645$, fail to reject H_0. There is not enough evidence at the 5% level of significance to support the claim that the difference between the mean annual salaries of entry-level software engineers in Raleigh, North Carolina, and Wichita, Kansas, is more than \$2000.

29.

$$(\bar{x}_1 - \bar{x}_2) - z_c\sqrt{\frac{\sigma_1^2}{n_1} + \frac{\sigma_2^2}{n_2}} < \mu_1 - \mu_2 < (\bar{x}_1 - \bar{x}_2) + z_c\sqrt{\frac{\sigma_1^2}{n_1} + \frac{\sigma_2^2}{n_2}}$$

$$(64{,}270 - 62{,}610) - 1.96\sqrt{\frac{(10{,}850)^2}{42} + \frac{(10{,}970)^2}{38}} < \mu_1 - \mu_2 < (64{,}270 - 62{,}610) + 1.96\sqrt{\frac{(10{,}850)^2}{42} + \frac{(10{,}970)^2}{38}}$$

$$1660 - 1.96\sqrt{5{,}969{,}782.456} < \mu_1 - \mu_2 < 1660 + 1.96\sqrt{5{,}969{,}782.456}$$

$$-\$3129 < \mu_1 - \mu_2 < \$6449$$

8.2 TESTING THE DIFFERENCE BETWEEN MEANS (SMALL INDEPENDENT SAMPLES)

8.2 TRY IT YOURSELF SOLUTIONS

1. The claim is "there is a difference in the mean annual earnings based on level of education."
$H_0: \mu_1 = \mu_2$; $H_a: \mu_1 \neq \mu_2$ (claim)
$\alpha = 0.05$; $\text{d.f.} = \min\{n_1 - 1,\ n_2 - 1\} = \min\{25-1,\ 16-1\} = 15$
$-t_0 = -2.131,\ t_0 = 2.131$; Rejection regions: $t < -2.131,\ t > 2.131$

$$t = \frac{(\bar{x}_1 - \bar{x}_2) - (\mu_1 - \mu_2)}{\sqrt{\frac{s_1^2}{n_1} + \frac{s_2^2}{n_2}}} = \frac{(36{,}875 - 44{,}900) - (0)}{\sqrt{\frac{(5475)^2}{25} + \frac{(8580)^2}{16}}} \approx -3.33$$

Because $t < -2.131$, reject H_0.
There is enough evidence at the 5% level of significance to support the claim that there is a difference in the mean annual earnings based on level of education.

2. The claim is "the mean driving cost per mile of the manufacturer's minivans is less than that of its leading competitor."
$H_0: \mu_1 \geq \mu_2$; $H_a: \mu_1 < \mu_2$ (claim)
$\alpha = 0.10$; $\text{d.f.} = n_1 + n_2 - 2 = 34 + 38 - 2 = 70$
$t_0 = -1.294$; Rejection regions: $t < -1.294$

$$t = \frac{(\bar{x}_1 - \bar{x}_2) - (\mu_1 - \mu_2)}{\sqrt{\frac{(n_1 - 1)s_1^2 + (n_2 - 1)s_2^2}{n_1 + n_2 - 2}}\sqrt{\frac{1}{n_1} + \frac{1}{n_2}}} = \frac{(0.52 - 0.54) - (0)}{\sqrt{\frac{(34-1)(0.08)^2 + (38-1)(0.07)^2}{34 + 38 - 2}}\sqrt{\frac{1}{34} + \frac{1}{38}}} \approx -1.13$$

Because $t > -1.294$, fail to reject H_0.
There is not enough evidence at the 10% level of significance to support the manufacturer's claim that the mean driving cost per mile of its minivans is less than that of its leading competitor.

8.2 EXERCISE SOLUTIONS

1. (1) The population standard deviations are unknown.

(2) The samples are randomly selected.

(3) The samples are independent

(4) The populations are normally distributed or each sample size is at least 30.

3. (a) $\text{d.f.} = n_1 + n_2 - 2 = 23$
$-t_0 = -1.714,\ t_0 = 1.714$

5. (a) $\text{d.f.} = n_1 + n_2 - 2 = 16$
$t_0 = -1.746$

(b) $\text{d.f.} = \min\{n_1 - 1, n_2 - 1\} = 10$
$-t_0 = -1.812, t_0 = 1.812$

(b) $\text{d.f.} = \min\{n_1 - 1, n_2 - 1\} = 6$
$t_0 = -1.943$

7. (a) $\text{d.f.} = n_1 + n_2 - 2 = 19$
$t_0 = 1.729$

(b) $\text{d.f.} = \min\{n_1 - 1, n_2 - 1\} = 7$
$t_0 = 1.895$

9. $H_0: \mu_1 = \mu_2$ (claim); $H_a: \mu_1 \neq \mu_2$
$\text{d.f.} = n_1 + n_2 - 2 = 27$
Rejection regions: $t < -2.771, t > 2.771$

$$t = \frac{(\bar{x}_1 - \bar{x}_2) - (\mu_1 - \mu_2)}{\sqrt{\frac{(n_1 - 1)s_1^2 + (n_2 - 1)s_2^2}{n_1 + n_2 - 2}}\sqrt{\frac{1}{n_1} + \frac{1}{n_2}}} = \frac{(33.7 - 35.5) - (0)}{\sqrt{\frac{(12-1)(3.5)^2 + (17-1)(2.2)^2}{12 + 17 - 2}}\sqrt{\frac{1}{12} + \frac{1}{17}}} \approx -1.70$$

Because $-2.771 < t < 2.771$, fail to reject H_0. There is not enough evidence at the 1% level of significance to reject the claim.

11. $H_0: \mu_1 \leq \mu_2$ (claim); $H_a: \mu_1 > \mu_2$
$\text{d.f.} = \min\{n_1 - 1, n_2 - 1\} = 9$
Rejection region: $t > 1.833$

$$t = \frac{(\bar{x}_1 - \bar{x}_2) - (\mu_1 - \mu_2)}{\sqrt{\frac{s_1^2}{n_1} + \frac{s_2^2}{n_2}}} = \frac{(2410 - 2305) - (0)}{\sqrt{\frac{(175)^2}{13} + \frac{(52)^2}{10}}} \approx 2.05$$

Because $t > 1.833$, reject H_0. There is enough evidence at the 5% level of significance to reject the claim.

13. (a)The claim is "the mean annual costs of food for dogs and cats are the same."
$H_0: \mu_1 = \mu_2$ (claim); $H_a: \mu_1 \neq \mu_2$

(b) $\text{d.f.} = n_1 + n_2 - 2 = 32$
$-t_0 \approx -1.694, t_0 \approx 1.694$; Rejection regions: $t < -1.694, t > 1.694$

(c) $$t = \frac{(\bar{x}_1 - \bar{x}_2) - (\mu_1 - \mu_2)}{\sqrt{\frac{(n_1 - 1)s_1^2 + (n_2 - 1)s_2^2}{n_1 + n_2 - 2}}\sqrt{\frac{1}{n_1} + \frac{1}{n_2}}} = \frac{(263 - 183) - (0)}{\sqrt{\frac{(16-1)(30)^2 + (18-1)(27)^2}{16 + 18 - 2}}\sqrt{\frac{1}{16} + \frac{1}{18}}} \approx 8.19$$

(d) Because $t > 1.694$, reject H_0.

(e) There is enough evidence at the 10% level of significance to reject the pet association's claim that the mean annual costs of food for dogs and cats are the same.

15. (a) The claim is "the stomachs of blue crabs from one location contain more fish than the stomachs of blue crabs from another location."

$H_0: \mu_1 \le \mu_2;\ H_a: \mu_1 > \mu_2$ (claim)

(b) $\text{d.f.} = n_1 + n_2 - 2 = 38$
$t_0 = 2.429$; Rejection region: $t > 2.429$

(c) $$t = \frac{(\bar{x}_1 - \bar{x}_2) - (\mu_1 - \mu_2)}{\sqrt{\frac{(n_1-1)s_1^2 + (n_2-1)s_2^2}{n_1+n_2-2}}\sqrt{\frac{1}{n_1}+\frac{1}{n_2}}} = \frac{(320-280)-(0)}{\sqrt{\frac{(25-1)(60)^2 + (15-1)(80)^2}{25+15-2}}\sqrt{\frac{1}{25}+\frac{1}{15}}} \approx 1.80$$

(d) Because $t < 2.429$, fail to reject H_0.

(e) There is not enough evidence to support the claim at the 1% level of significance that the stomachs of blue crabs from one location contain more fish than the stomachs of blue crabs from another location.

17. (a) The claim is "the mean household income in a recent year is greater in Cuyahoga County, Ohio, than it is in Wayne County, Michigan."
$H_0: \mu_1 \le \mu_2;\ H_a: \mu_1 > \mu_2$ (claim)

(b) $\text{d.f.} = \min\{n_1 - 1,\ n_2 - 1\} = 14$
$t_0 = 1.761$; Rejection region: $t > 1.761$

(c) $$t = \frac{(\bar{x}_1 - \bar{x}_2) - (0)}{\sqrt{\frac{s_1^2}{n_1}+\frac{s_2^2}{n_2}}} = \frac{(45{,}600 - 41{,}500) - (0)}{\sqrt{\frac{(2800)^2}{19}+\frac{(1310)^2}{15}}} \approx 5.65$$

(d) Because $t > 1.761$, reject H_0.

(e) There is enough evidence at the 5% level of significance to support the demographics researcher's claim that the mean household income in a recent year is greater in Cuyahoga County, Ohio, than it is in Wayne County, Michigan.

19. (a) The claim is "an experimental method makes a difference in the tensile strength of steel bars."
$H_0: \mu_1 = \mu_2;\ H_a: \mu_1 \ne \mu_2$ (claim)

(b) $\text{d.f.} = n_1 + n_2 - 2 = 22$
$-t_0 = -2.819,\ t_0 = 2.819$; Rejection regions: $t < -2.819,\ t > 2.819$

(c) $\bar{x}_1 = 368.3,\ s_1 \approx 22.301,\ n_1 = 10$
$\bar{x}_2 \approx 388.214,\ s_2 \approx 14.797,\ n_2 = 14$
$$t = \frac{(\bar{x}_1 - \bar{x}_2) - (\mu_1 - \mu_2)}{\sqrt{\frac{(n_1-1)s_1^2 + (n_2-1)s_2^2}{n_1+n_2-2}}\sqrt{\frac{1}{n_1}+\frac{1}{n_2}}} \approx \frac{(368.3-388.214)-(0)}{\sqrt{\frac{(10-1)(22.301)^2 + (14-1)(14.797)^2}{10+14-2}}\sqrt{\frac{1}{10}+\frac{1}{14}}} \approx -2.64$$

(d) Because $-2.831 < t < 2.831$, fail to reject H_0.

(e) There is not enough evidence at the 1% level of significance to support the claim that an experimental method makes a difference in the tensile strength of steel bars.

21. (a) The claim is "the new method of teaching reading produces higher reading test scores than the old method"
$H_0: \mu_1 \geq \mu_2$; $H_a: \mu_1 < \mu_2$ (claim)

(b) $\text{d.f.} = n_1 + n_2 - 2 = 42$
$t_0 \approx -1.303$; Rejection region: $t < -1.303$

(c) $\bar{x}_1 \approx 56.684,\ s_1 \approx 6.961,\ n_1 = 19$
$\bar{x}_2 \approx 67.4,\ s_2 \approx 9.014,\ n_2 = 25$

$$t = \frac{(\bar{x}_1 - \bar{x}_2) - (\mu_1 - \mu_2)}{\sqrt{\frac{(n_1 - 1)s_1^2 + (n_2 - 1)s_2^2}{n_1 + n_2 - 2}}\sqrt{\frac{1}{n_1} + \frac{1}{n_2}}} \approx \frac{(56.684 - 67.4) - (0)}{\sqrt{\frac{(19-1)(6.961)^2 + (25-1)(9.014)^2}{19 + 25 - 2}}\sqrt{\frac{1}{19} + \frac{1}{25}}} \approx -4.295$$

(d) Because $t < -1.303$, reject H_0.

(e) There is enough evidence at the 10% level of significance to support the claim that the new method of teaching reading produces higher reading test scores than the old method.

23. $$(\bar{x}_1 - \bar{x}_2) \pm t_c\sqrt{\frac{s_1^2}{n_1} + \frac{s_2^2}{n_2}} \to (0.73 - 0.75) \pm 1.363\sqrt{\frac{(0.17)^2}{20} + \frac{(0.04)^2}{12}}$$
$$\to -0.02 \pm 0.054 \to -0.074 < \mu_1 - \mu_2 < 0.034 \to -0.07 < \mu_1 - \mu_2 < 0.03$$

25. $$\hat{\sigma} = \sqrt{\frac{(n_1 - 1)s_1^2 + (n_2 - 1)s_2^2}{n_1 + n_2 - 2}} = \sqrt{\frac{(20-1)(39.7)^2 + (17-1)(42.4)^2}{20 + 17 - 2}} \approx 40.956$$
$$(\bar{x}_1 - \bar{x}_2) \pm t_c\hat{\sigma}\sqrt{\frac{1}{n_1} + \frac{1}{n_2}} \to (28 - 26) \pm 1.69 \cdot 40.956\sqrt{\frac{1}{20} + \frac{1}{17}}$$
$$\to 2 - 22.833 \to < \mu_1 - \mu_2 < 2 + 22.833 \to -20.8 < \mu_1 - \mu_2 < 24.8$$

8.3 TESTING THE DIFFERENCE BETWEEN MEANS (DEPENDENT SAMPLES)

8.3 TRY IT YOURSELF SOLUTIONS

1. The claim is "athletes can decrease their times in the 40-yard dash."
$H_0: \mu_d \leq 0$; $H_a: \mu_d > 0$ (claim)
$\alpha = 0.05$, $\text{d.f.} = n - 1 = 11$

$t_0 = 1.796$; Rejection region: $t > 1.796$

Before	After	d	d^2
4.85	4.78	0.07	0.0049
4.90	4.90	0.00	0.0000
5.08	5.05	0.03	0.0009
4.72	4.65	0.07	0.0049
4.62	4.64	−0.02	0.0004
4.54	4.50	0.04	0.0016
5.25	5.24	0.01	0.0001
5.18	5.27	−0.09	0.0081
4.81	4.75	0.06	0.0036
4.57	4.43	0.14	0.0196
4.63	4.61	0.02	0.0004
4.77	4.82	−0.05	0.0025
		$\sum d = 0.28$	$\sum d^2 = 0.047$

$$\bar{d} = \frac{\sum d}{n} = \frac{0.28}{12} \approx 0.0233$$

$$s_d = \sqrt{\frac{\left(\sum d^2\right) - \left[\frac{\left(\sum d\right)^2}{n}\right]}{n-1}} = \sqrt{\frac{0.047 - \frac{(0.28)^2}{12}}{11}} \approx 0.0607$$

$$t = \frac{\bar{d} - \mu_d}{\frac{s_d}{\sqrt{n}}} \approx \frac{0.0233 - 0}{\frac{0.0607}{\sqrt{12}}} \approx 1.330$$

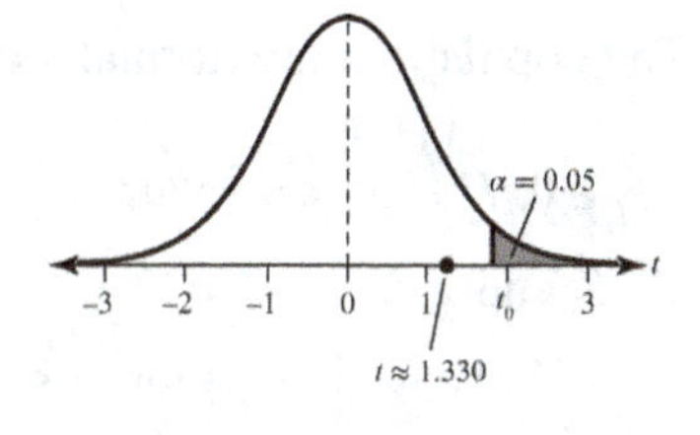

Because $t < 1.796$, fail to reject H_0.

There is not enough evidence at the 5% level of significance to support the claim that athletes can decrease their times in the 40-yard dash using new strength shoes.

2. The claim is "the drug changes the body's temperature."

$H_0: \mu_d = 0$; $H_a: \mu_d \neq 0$ (claim)

$\alpha = 0.05$, d.f. $= n - 1 = 6$

$-t_0 = -2.447$, $t_0 = 2.447$; Rejection regions: $t < -2.447$, $t > 2.447$

Before	After	d	d^2
101.8	99.2	2.6	6.76
98.5	98.4	0.1	0.01
98.1	98.2	−0.1	0.01
99.4	99	0.4	0.16
98.9	98.6	0.3	0.09
100.2	99.7	0.5	0.25
97.9	97.8	0.1	0.01
		$\sum d = 3.9$	$\sum d^2 = 7.29$

$$\bar{d} = \frac{\sum d}{n} = \frac{3.9}{7} \approx 0.5771$$

$$s_d = \sqrt{\frac{\sum d^2 - \left[\frac{\left(\sum d\right)^2}{n}\right]}{n-1}} = \sqrt{\frac{7.29 - \frac{(3.9)^2}{7}}{6}} \approx 0.9235$$

$$t = \frac{\bar{d} - \mu_d}{\frac{s_d}{\sqrt{n}}} \approx \frac{0.5571 - 0}{\frac{0.9235}{\sqrt{7}}} \approx 1.596$$

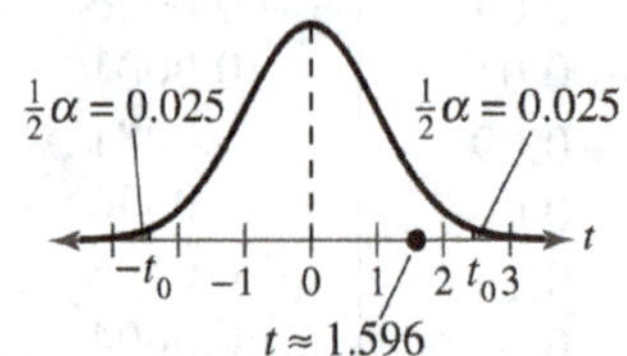

Because $-2.447 < t < 2.447$, fail to reject H_0.
There is not enough evidence at the 5% level of significance to conclude that the drug changes the body's temperature.

8.3 EXERCISE SOLUTIONS

1. (1) The samples are randomly selected.

(2) The samples are dependent.

(3) The populations are normally distributed or the number n of pairs of data is at least 30.

3. $H_0: \mu_d \geq 0$; $H_a: \mu_d < 0$ (claim)

$\alpha = 0.05$ and d.f. $= n - 1 = 13$

$t_0 = -1.771$; Rejection region: $t < -1.771$

$$t = \frac{\bar{d} - \mu_d}{\frac{s_d}{\sqrt{n}}} = \frac{1.5 - 0}{\frac{3.2}{\sqrt{14}}} \approx 1.754$$

Because $t > -1.771$, fail to reject H_0. There is not enough evidence at the 5% level of significance to support the claim.

5. $H_0: \mu_d \leq 0$ (claim); $H_a: \mu_d > 0$

$\alpha = 0.10$ and d.f. $= n - 1 = 15$

$t_0 = 1.341$; Rejection region: $t > 1.341$

$$t = \frac{\bar{d} - \mu_d}{\frac{s_d}{\sqrt{n}}} = \frac{6.5 - 0}{\frac{9.54}{\sqrt{16}}} \approx 2.725$$

Because $t > 1.341$, reject H_0. There is enough evidence at the 10% level of significance to reject the claim.

7. $H_0: \mu_d \geq 0$ (claim); $H_a: \mu_d < 0$

$\alpha = 0.01$ and $\text{d.f.} = n - 1 = 14$

$t_0 = -2.624$; Rejection region: $t < -2.624$

$$t = \frac{\bar{d} - \mu_d}{\frac{s_d}{\sqrt{n}}} = \frac{-2.3 - 0}{\frac{1.2}{\sqrt{15}}} \approx -7.423$$

Because $t < -2.624$, reject H_0. There is enough evidence at the 1% level of significance to reject the claim.

9. (a) The claim is "seven of the stocks that make up the Dow Jones Industrial Average lost value from one hour to the next on one business day."
$H_0: \mu_d \leq 0$; $H_a: \mu_d > 0$ (claim)

(b) $t_0 = 3.143$; Rejection region: $t > 3.143$

(c) $\bar{d} \approx 0.087$ and $s_d \approx 0.405$

(d) $$t = \frac{\bar{d} - \mu_d}{\frac{s_d}{\sqrt{n}}} = \frac{0.0871 - 0}{\frac{0.4053}{\sqrt{7}}} \approx 0.569$$

(e) Because $t < 3.143$, fail to reject H_0.

(f) There is not enough evidence at the 1% level of significance to support the stock market analyst's claim that seven of the stocks that make up the Dow Jones Industrial Average lost value from one hour to the next on one business day.

11. (a) The claim is "caffeine ingestion improves repeated freestyle sprints in trained male swimmers."
$H_0: \mu_d \leq 0$; $H_a: \mu_d > 0$ (claim)

(b) $t_0 = 3.365$; Rejection region: $t > 3.365$

(c) $\bar{d} \approx 0.533$ and $s_d \approx 0.350$

(d) $$t = \frac{\bar{d} - \mu_d}{\frac{s_d}{\sqrt{n}}} = \frac{0.533 - 0}{\frac{0.350}{\sqrt{6}}} \approx 3.730$$

(e) Because $t > 3.365$, reject H_0.

(f) There is enough evidence at the 1% level of significance to support the researcher's claim that caffeine ingestion improves repeated freestyle sprints in trained male swimmers.

13. (a) The claim is "soft tissue therapy helps to reduce the numbers of days per week patients suffer from headaches."
$H_0: \mu_d \leq 0$; $H_a: \mu_d > 0$ (claim)

(b) $t_0 = 2.567$; Rejection region; $t > 2.567$

(c) $\bar{d} = 1.5$ and $s_d \approx 2.249$

(d) $t = \dfrac{\bar{d} - \mu_d}{\frac{s_d}{\sqrt{n}}} = \dfrac{1.5 - 0}{\frac{1.249}{\sqrt{18}}} \approx 5.095$

(e) Because $t > 2.567$, reject H_0.

(f) There is enough evidence at the 1% level of significance to support the claim that soft tissue therapy helps to reduce the numbers of days per week patients suffer from headaches.

15. (a) The claim is "student housing rates have increased from one academic year to the next."
$H_0: \mu_d \geq 0$; $H_a: \mu_d < 0$ (claim)

(b) $t_0 = -1.796$; Rejection region: $t < -1.796$

(c) $\bar{d} = -254.5$ and $s_d \approx 291.767$

(d) $t = \dfrac{\bar{d} - \mu_d}{\frac{s_d}{\sqrt{n}}} = \dfrac{-254.5 - 0}{\frac{291.767}{\sqrt{12}}} \approx -3.022$

(e) Because $t < -1.796$, reject H_0.

(f) There is enough evidence at the 5% level of significance to support the college administrator's claim that student housing rates have increased from one academic year to the next.

17. (a) The claim is "the product ratings have changed from last year to this year."
$H_0: \mu_d = 0$; $H_a: \mu_d \neq 0$ (claim)

(b) $-t_0 = -2.365$, $t_0 = 2.365$; Rejection regions: $t < -2.365$, $t > 2.365$

(c) $\bar{d} = -1$ and $s_d \approx 1.309$

(d) $t = \dfrac{\bar{d} - \mu_d}{\frac{s_d}{\sqrt{n}}} = \dfrac{-1 - 0}{\frac{1.309}{\sqrt{8}}} \approx -2.161$

(e) Because $-2.365 < t < 2.365$, fail to reject H_0.

(f) There is not enough evidence at the 5% level of significance to support the claim that the product ratings have changed from last year to this year.

19. (a) The claim is "eating new cereal as part of a daily diet lowers total blood cholesterol levels."
$H_0: \mu_d \le 0$; $H_a: \mu_d > 0$ (claim)

(b) $t_0 = 1.943$; Rejection region: $t > 1.943$

(c) $\bar{d} \approx 2.857$ and $s_d \approx 4.451$

(d) $t = \dfrac{\bar{d} - \mu_d}{\frac{s_d}{\sqrt{n}}} = \dfrac{2.857 - 0}{\frac{4.451}{\sqrt{7}}} \approx 1.698$

(e) Because $t < 1.943$, fail to reject H_0.

(f) There is not enough evidence at the 5% level of significance to support the claim that eating new cereal as part of a daily diet lowers total blood cholesterol levels.

21. Yes; $P \approx 0.0058 < 0.05$, so you reject H_0.

23. $\bar{d} = -1.525$ and $s_d \approx 0.542$

$$\bar{d} - t_{\alpha/2}\frac{s_d}{\sqrt{n}} < \mu_d < \bar{d} - t_{\alpha/2}\frac{s_d}{\sqrt{n}}$$

$$-1.525 - 1.753\left(\frac{0.542}{\sqrt{16}}\right) < \mu_d < -1.525 + 1.753\left(\frac{0.542}{\sqrt{16}}\right)$$

$$-1.525 - 0.238 < \mu_d < -1.525 + 0.238$$

$$-1.76 < \mu_d < -1.29$$

8.4 TESTING THE DIFFERENCE BETWEEN PROPORTIONS

8.4 TRY IT YOURSELF SOLUTIONS

1. $\bar{p} = \dfrac{x_1 + x_2}{n_1 + n_2} = \dfrac{367 + 6290}{1593 + 29948} \approx 0.2111$, $\bar{q} = 1 - \bar{p} \approx 0.7889$

$n_1\bar{p} = 1593(0.2111) \approx 336.3 > 5$, $n_1\bar{q} = 1593(0.7889) \approx 1256.7 > 5$

$n_2\bar{p} = 29{,}948(0.2111) \approx 6322.0 > 5$, $n_2\bar{q} = 29{,}948(0.7889) \approx 23{,}626.0 > 5$

The claim is "there is a difference between the proportion 40- to 49-year olds who are yoga users and the proportion of 40- to 49-year-olds who are non-yoga users."

$H_0: p_1 = p_2$; $H_a: p_1 \ne p_2$ (claim)

$\alpha = 0.05$

$-z_0 = -1.96$, $z_0 = 1.96$; Rejection regions: $z < -1.96$, $z > 1.96$

$$z=\frac{(\hat{p}_1-\hat{p}_2)-(p_1-p_2)}{\sqrt{\bar{p}\bar{q}\left(\frac{1}{n_1}+\frac{1}{n_2}\right)}}=\frac{(0.230-0.210)-(0)}{\sqrt{0.211(0.789)\left(\frac{1}{1593}+\frac{1}{29,948}\right)}}\approx 1.91$$

Because $-1.96<z<1.96$, fail to reject H_0.

There is not enough evidence at the 5% level of significance to support the claim that there is a difference between the proportion 40- to 49-year olds who are yoga users and the proportion of 40- to 49-year-olds who are non-yoga users.

2. $\bar{p}=\frac{x_1+x_2}{n_1+n_2}=\frac{239+5990}{1593+29948}\approx 0.1975,\ \bar{q}=1-\bar{p}\approx 0.8025$

$n_1\bar{p}=1593(0.1975)\approx 314.6>5,\ n_1\bar{q}=1593(0.8025)\approx 1278.4>5$

$n_2\bar{p}=29,948(0.1975)\approx 5914.7>5,\ n_2\bar{q}=29,948(0.8025)\approx 24,033.3>5$

The claim is "the proportion of yoga users with incomes of \$20,000 to \$34,499 is less than the proportion of non-yoga users with incomes of \$20,000 to \$34,499."

$H_0: p_1\geq p_2;\ H_a: p_1<p_2$ (claim)

$\alpha=0.05$

$z_0=-1.645$; Rejection region: $z<-1.645$

$$z=\frac{(\hat{p}_1-\hat{p}_2)-(p_1-p_2)}{\sqrt{\bar{p}\bar{q}\left(\frac{1}{n_1}+\frac{1}{n_2}\right)}}=\frac{(0.15-0.20)-(0)}{\sqrt{0.1975\cdot 0.8025\left(\frac{1}{1593}+\frac{1}{29,948}\right)}}\approx -4.88$$

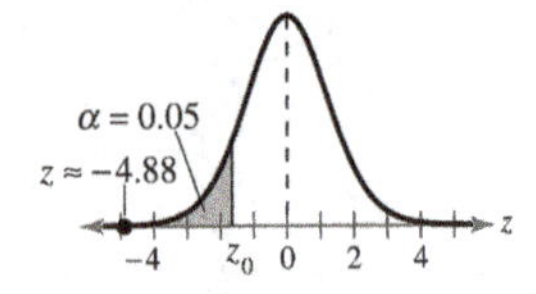

Because $z<-1.645$, reject H_0.

There is enough evidence at the 5% level of significance to support the claim that the proportion of yoga users with incomes of \$20,000 to \$34,499 is less than the proportion of non-yoga users with incomes of \$20,000 to \$34,499.

8.4 EXERCISE SOLUTIONS

1. (1) The samples are randomly selected.

(2) The samples are independent.

(3) $n_1\bar{p}\geq 5,\ n_1\bar{q}\geq 5,\ n_2\bar{p}\geq 5,\ n_2\bar{q}\geq 5$

3. $\bar{p}=\dfrac{x_1+x_2}{n_1+n_2}=\dfrac{35+36}{70+60}\approx 0.5462;\ \bar{q}=1-\bar{p}\approx 0.4538$

$n_1\bar{p}=70(0.5462)\approx 38.2>5,\ n_1\bar{q}=70(0.4538)\approx 31.8>5$

$n_2\bar{p}=60(0.5462)\approx 32.8>5,\ n_2\bar{q}=60(0.4538)\approx 27.2>5$

Because all conditions are met above, the normal sampling distribution can be used.

$H_0: p_1=p_2;\ H_a: p_1\neq p_2$ (claim)

$-z_0=-2.575,\ z_0=2.575;$ Rejection regions: $z<-2.575,\ z>2.575$

$$z=\frac{(\hat{p}_1-\hat{p}_2)-(p_1-p_2)}{\sqrt{\bar{p}\bar{q}\left(\frac{1}{n_1}+\frac{1}{n_2}\right)}}=\frac{(0.5-0.6)-(0)}{\sqrt{0.5462\cdot 0.4538\left(\frac{1}{70}+\frac{1}{60}\right)}}\approx -1.142$$

Because $-2.575<z<2.575$, fail to reject H_0.

There is not enough evidence at the 1% level of significance to support the claim.

5. $\bar{p}=\dfrac{x_1+x_2}{n_1+n_2}=\dfrac{42+76}{150+200}\approx 0.337,\ \bar{q}=1-\bar{p}\approx 0.663$

$n_1\bar{p}\approx 150(0.337)\approx 50.57>5,\ n_1\bar{q}\approx 150(0.663)\approx 99.44>5$.

$n_2\bar{p}\approx 200(0.337)\approx 67.42>5,$ and $n_2\bar{q}\approx 200(0.663)\approx 132.58>5$

Because all conditions are met above, the normal sampling distribution can be used.

$H_0: p_1=p_2$ (claim); $H_a: p_1\neq p_2$

$-z_0=-1.645,\ z_0=1.645;$ Rejection regions: $z<-1.645,\ z>1.645$

$$z=\frac{(\hat{p}_1-\hat{p}_2)-(p_1-p_2)}{\sqrt{\bar{p}\bar{q}\left(\frac{1}{n_1}+\frac{1}{n_2}\right)}}=\frac{(0.28-0.38)-(0)}{\sqrt{(0.337)(0.663)\left(\frac{1}{150}+\frac{1}{200}\right)}}\approx -1.96$$

Because $z<-1.645$, reject H_0.

There is enough evidence at the 10% level of significance to reject the claim.

7. (a) The claim is "there is a difference in the proportion of subjects who had no 12-week confirmed disability progression."

$H_0: p_1=p_2;\ H_a: p_1\neq p_2$ (claim)

(b) $-z_0=-2.575,\ z_0=2.575;$ Rejection regions: $z<-2.575,\ z>2.575$

(c) $\bar{p}=\dfrac{x_1+x_2}{n_1+n_2}=\dfrac{327+148}{488+244}\approx 0.649,\ \bar{q}=1-\bar{p}\approx 0.351$

$$z=\frac{(\hat{p}_1-\hat{p}_2)-(p_1-p_2)}{\sqrt{\bar{p}\bar{q}\left(\frac{1}{n_1}+\frac{1}{n_2}\right)}}=\frac{(0.670-0.607)-(0)}{\sqrt{(0.649)(0.341)\left(\frac{1}{488}+\frac{1}{244}\right)}}\approx 1.7$$

(d) Because $-2.575<z<2.575$, fail to reject H_0.

(e) There is not enough evidence at the 1% level of significance to support the claim that there is a difference in the proportion of subjects who had no 12-week confirmed disability progression.

9. (a) The claim is "there is a difference in the proportion of those employed between females ages 20 to 24 and males ages 20 to 24."
$H_0: p_1 = p_2$; $H_a: p_1 \neq p_2$ (claim)

(b) $-z_0 = -2.575$, $z_0 = 2.575$; Rejection regions: $z < -2.575$, $z > 2.575$

(c) $$\bar{p} = \frac{x_1 + x_2}{n_1 + n_2} = \frac{1127 + 1464}{1750 + 2000} \approx 0.691, \ \bar{q} = 1 - \bar{p} \approx 0.309$$
$$z = \frac{(\hat{p}_1 - \hat{p}_2) - (p_1 - p_2)}{\sqrt{\bar{p}\bar{q}\left(\frac{1}{n_1} + \frac{1}{n_2}\right)}} = \frac{(0.644 - 0.732) - (0)}{\sqrt{(0.691)(0.309)\left(\frac{1}{1750} + \frac{1}{2000}\right)}} \approx -5.82$$

(d) Because $z < -2.575$, reject H_0.

(e) There is enough evidence at the 1% level of significance to support the claim that there is a difference in the proportion of those employed between females ages 20 to 24 and males ages 20 to 24.

11. (a) The claim is "the proportion of drivers who wear seat belts is greater in the West than in the Northeast."
$H_0: p_1 \leq p_2$; $H_a: p_1 > p_2$ (claim)

(b) $z_0 = 1.645$; Rejection region: $z > 1.645$

(c) $$\bar{p} = \frac{x_1 + x_2}{n_1 + n_2} = \frac{934 + 909}{1000 + 1000} = 0.9215, \ \bar{q} = 1 - \bar{p} = 0.0785$$
$$z = \frac{(\hat{p}_1 - \hat{p}_2) - (p_1 - p_2)}{\sqrt{\bar{p}\bar{q}\left(\frac{1}{n_1} + \frac{1}{n_2}\right)}} = \frac{(0.934 - 0.909) - (0)}{\sqrt{(0.9215)(0.0785)\left(\frac{1}{1000} + \frac{1}{1000}\right)}} \approx 2.08$$

(d) Because $z > 1.645$, reject H_0.

(e) There is enough evidence at the 5% level of significance to support the claim that the proportion of drivers who wear seat belts is greater in the West than in the Northeast.

13. The claim is "the proportion of newlywed Asians who have a spouse of a different race or ethnicity is the same as the proportion of newlywed Hispanics who have a spouse of a different race or ethnicity."
$H_0: p_1 = p_2$ (claim); $H_a: p_1 \neq p_2$
$-z_0 = -1.96$, $z_0 = 1.96$; Rejection regions: $z < -1.96$, $z > 1.96$
$$\bar{p} = \frac{x_1 + x_2}{n_1 + n_2} = \frac{290 + 270}{1000 + 1000} = 0.28, \ \bar{q} = 1 - \bar{p} = 0.72$$

$$z=\frac{(\hat{p}_1-\hat{p}_2)-(p_1-p_2)}{\sqrt{\overline{p}\overline{q}\left(\frac{1}{n_1}+\frac{1}{n_2}\right)}}=\frac{(0.29-0.27)-(0)}{\sqrt{(0.28)(0.72)\left(\frac{1}{1000}+\frac{1}{1000}\right)}}\approx 0.996$$

Because $-1.96<z<1.96$, fail to reject H_0.

No, there is not enough evidence at the 5% level of significance to reject the claim that the proportion of newlywed Asians who have a spouse of a different race or ethnicity is the same as the proportion of newlywed Hispanics who have a spouse of a different race or ethnicity.

15. The claim is "the proportion of newlywed Asians who have a spouse of a different race or ethnicity is greater than the proportion of newlywed whites who have a spouse of a different race or ethnicity."

$H_0: p_1 \le p_2$; $H_a: p_1 > p_2$ (claim)

$z_0 = 2.33$; Rejection region: $z > 2.33$

$$\overline{p}=\frac{x_1+x_2}{n_1+n_2}=\frac{290+110}{1000+1000}=0.2,\ \overline{q}=1-\overline{p}=0.8$$

$$z=\frac{(\hat{p}_1-\hat{p}_2)-(p_1-p_2)}{\sqrt{\overline{p}\overline{q}\left(\frac{1}{n_1}+\frac{1}{n_2}\right)}}=\frac{(0.29-0.11)-(0)}{\sqrt{(0.2)(0.8)\left(\frac{1}{1000}+\frac{1}{1000}\right)}}\approx 10.1$$

Because $z > 2.33$, reject H_0.

Yes, there is enough evidence at the 1% level of significance to support the claim that the proportion of newlywed Asians who have a spouse of a different race or ethnicity is greater than the proportion of newlywed whites who have a spouse of a different race or ethnicity.

17. The claim is "the proportion of newlywed whites who have a spouse of a different race or ethnicity is less than the proportion of newlywed blacks who have a spouse of a different race or ethnicity."

$H_0: p_1 \ge p_2$; $H_a: p_1 < p_2$ (claim)

$z_0 = -2.33$; Rejection region: $z < -2.33$

$$\overline{p}=\frac{x_1+x_2}{n_1+n_2}=\frac{110+180}{1000+1000}=0.145,\ \overline{q}=1-\overline{p}=0.855$$

$$z=\frac{(\hat{p}_1-\hat{p}_2)-(p_1-p_2)}{\sqrt{\overline{p}\overline{q}\left(\frac{1}{n_1}+\frac{1}{n_2}\right)}}=\frac{(0.11-0.18)-(0)}{\sqrt{(0.145)(0.855)\left(\frac{1}{1000}+\frac{1}{1000}\right)}}\approx -4.45$$

Because $z < -2.33$, reject H_0.

Yes, there is enough evidence at the 1% level of significance to support the claim that the proportion of newlywed whites who have a spouse of a different race or ethnicity is less than the proportion of newlywed blacks who have a spouse of a different race or ethnicity.

19. The claim is "the proportion of men who work 40 hours per week is the same as the proportion of men who work more than 40 hours per week."

$H_0: p_1 = p_2$ (claim); $H_a: p_1 \ne p_2$

$-z_0 = -2.576$; $z_0 = 2.576$ Rejection regions: $z < -2.576$; $z > 2.576$

$$\overline{p}=\frac{x_1+x_2}{n_1+n_2}=\frac{141+141}{300+300}=0.47,\ \overline{q}=1-\overline{p}=0.53$$

$$z=\frac{(\hat{p}_1-\hat{p}_2)-(p_1-p_2)}{\sqrt{\overline{p}\overline{q}\left(\frac{1}{n_1}+\frac{1}{n_2}\right)}}=\frac{(0.47-0.47)-(0)}{\sqrt{(0.47)(0.53)\left(\frac{1}{300}+\frac{1}{300}\right)}}=0$$

Because $-2.576 < z < 2.576$, fail to reject H_0.

No, there is not enough evidence at the 1% level of significance to reject claim that the proportion of men who work 40 hours per week is the same as the proportion of men who work more than 40 hours per week.

21. The claim is "the proportion of the U.S. workforce that works 40 hours per week is greater for women than for men."

$H_0: p_1 \le p_2$; $H_a: p_1 > p_2$ (claim)

$z_0 = 1.645$; Rejection region: $z > 1.645$

$$\overline{p}=\frac{x_1+x_2}{n_1+n_2}=\frac{140+141}{250+300}\approx 0.511,\ \overline{q}=1-\overline{p}\approx 0.489$$

$$z=\frac{(\hat{p}_1-\hat{p}_2)-(p_1-p_2)}{\sqrt{\overline{p}\overline{q}\left(\frac{1}{n_1}+\frac{1}{n_2}\right)}}=\frac{(0.56-0.47)-(0)}{\sqrt{(0.511)(0.489)\left(\frac{1}{250}+\frac{1}{300}\right)}}\approx 2.10$$

Because $z > 1.645$, reject H_0.

Yes, there is enough evidence at the 5% level of significance to support the claim that the proportion of the U.S. workforce that works 40 hours per week is greater for women than for men.

23.

$$(\hat{p}_1-\hat{p}_2)-z_c\sqrt{\frac{\hat{p}_1\hat{q}_1}{n_1}+\frac{\hat{p}_2\hat{q}_2}{n_2}} < p_1-p_2 < (\hat{p}_1-\hat{p}_2)+z_c\sqrt{\frac{\hat{p}_1\hat{q}_1}{n_1}+\frac{\hat{p}_2\hat{q}_2}{n_2}}$$

$$(0.07-0.09)-1.96\sqrt{\frac{(0.07)(0.93)}{10{,}000}+\frac{(0.09)(0.91)}{8000}} < p_1-p_2 < (0.07-0.09)+1.96\sqrt{\frac{(0.07)(0.93)}{10{,}000}+\frac{(0.09)(0.91)}{8000}}$$

$$-0.02-0.008 < p_1-p_2 < -0.02+0.008$$

$$-0.028 < p_1-p_2 < -0.012$$

25.

$$(\hat{p}_1-\hat{p}_2)-z_c\sqrt{\frac{\hat{p}_1\hat{q}_1}{n_1}+\frac{\hat{p}_2\hat{q}_2}{n_2}} < p_1-p_2 < (\hat{p}_1-\hat{p}_2)+z_c\sqrt{\frac{\hat{p}_1\hat{q}_1}{n_1}+\frac{\hat{p}_2\hat{q}_2}{n_2}}$$

$$(0.69-0.65)-1.96\sqrt{\frac{(0.69)(0.31)}{2000}+\frac{(0.65)(0.35)}{2000}} < p_1-p_2 < (0.69-0.65)+1.96\sqrt{\frac{(0.69)(0.31)}{2000}+\frac{(0.65)(0.35)}{2000}}$$

$$0.04-0.029 < p_1-p_2 < 0.04+0.029$$

$$0.011 < p_1-p_2 < 0.069$$

Answers will vary.

CHAPTER 8 REVIEW EXERCISE SOLUTIONS

1. Dependent because the same adults were sampled.

3. Independent because different vehicles were sampled.

5. $H_0: \mu_1 \geq \mu_2$ (claim); $H_a: \mu_1 < \mu_2$

$z_0 = -1.645$; Rejection region: $z < -1.645$

$$z = \frac{(\bar{x}_1 - \bar{x}_2) - (\mu_1 - \mu_2)}{\sqrt{\frac{\sigma_1^2}{n_1} + \frac{\sigma_2^2}{n_2}}} = \frac{(1.28 - 1.34) - (0)}{\sqrt{\frac{(0.30)^2}{96} + \frac{(0.23)^2}{85}}} \approx -1.519$$

Because $z > -1.645$, fail to reject H_0. There is not enough evidence at the 5% level of significance to reject the claim.

7. $H_0: \mu_1 \geq \mu_2$; $H_a: \mu_1 < \mu_2$ (claim)

$z_0 = -1.28$; Rejection regions: $z < -1.28$

$$z = \frac{(\bar{x}_1 - \bar{x}_2) - (\mu_1 - \mu_2)}{\sqrt{\frac{\sigma_1^2}{n_1} + \frac{\sigma_2^2}{n_2}}} = \frac{(0.28 - 0.33) - (0)}{\sqrt{\frac{(0.11)^2}{41} + \frac{(0.10)^2}{34}}} \approx -2.060$$

Because $z < -1.28$, reject H_0. There is enough evidence at the 10% level of significance to support the claim.

9. (a) The claim is "the mean sodium content of sandwiches at Restaurant A is less than the mean sodium content of sandwiches at Restaurant B."
$H_0: \mu_1 \geq \mu_2$; $H_a: \mu_1 < \mu_2$ (claim)

(b) $z_0 = -1.645$; Rejection region: $z < -1.645$

(c) $$z = \frac{(\bar{x}_1 - \bar{x}_2) - (\mu_1 - \mu_2)}{\sqrt{\frac{\sigma_1^2}{n_1} + \frac{\sigma_2^2}{n_2}}} = \frac{(670 - 690) - (0)}{\sqrt{\frac{(20)^2}{22} + \frac{(30)^2}{28}}} \approx -2.82$$

(d) Because $z < -1.645$, reject H_0.

(e) There is enough evidence at the 5% level of significance to support the claim that the mean sodium content of sandwiches at Restaurant A is less than the mean sodium content of sandwiches at Restaurant B.

11. $H_0: \mu_1 = \mu_2$ (claim); $H_a: \mu_1 \neq \mu_2$

$\text{d.f.} = n_1 + n_2 - 2 = 31$

$-t_0 = -2.040$, $t_0 = 2.040$; Rejection regions: $t < -2.040$, $t > 2.040$

$$t = \frac{(\bar{x}_1 - \bar{x}_2) - (\mu_1 - \mu_2)}{\sqrt{\frac{(n_1 - 1)s_1^2 + (n_2 - 1)s_2^2}{n_1 + n_2 - 2}}\sqrt{\frac{1}{n_1} + \frac{1}{n_2}}} = \frac{(228 - 207) - (0)}{\sqrt{\frac{(20-1)(27)^2 + (13-1)(25)^2}{20 + 13 - 2}}\sqrt{\frac{1}{20} + \frac{1}{13}}} \approx 2.25$$

Because $t > 2.040$, reject H_0. There is enough evidence at the 5% level of significance to reject the claim.

13. $H_0: \mu_1 \le \mu_2$ (claim); $H_a: \mu_1 > \mu_2$

d.f. $= \min\{n_1 - 1,\ n_2 - 1\} = 39$

$t_0 = 1.304$; Rejection region: $t > 1.304$

$$t = \frac{(\bar{x}_1 - \bar{x}_2) - (\mu_1 - \mu_2)}{\sqrt{\frac{s_1^2}{n_1} + \frac{s_2^2}{n_2}}} = \frac{(664.5 - 665.5) - (0)}{\sqrt{\frac{(2.4)^2}{40} + \frac{(4.1)^2}{40}}} \approx -1.33$$

Because $t < 1.304$, fail to reject H_0. There is not enough evidence at the 10% level of significance to reject the claim.

15. $H_0: \mu_1 = \mu_2$; $H_a: \mu_1 \ne \mu_2$ (claim)

d.f. $= n_1 + n_2 - 2 = 10$

$-t_0 = -3.169,\ t_0 = 3.169$; Rejection regions: $t < -3.169,\ t > 3.169$

$$t = \frac{(\bar{x}_1 - \bar{x}_2) - (\mu_1 - \mu_2)}{\sqrt{\frac{(n_1 - 1)s_1^2 + (n_2 - 1)s_2^2}{n_1 + n_2 - 2}} \cdot \sqrt{\frac{1}{n_1} + \frac{1}{n_2}}} = \frac{(61 - 55) - (0)}{\sqrt{\frac{(5-1)(3.3)^2 + (7-1)(1.2)^2}{5 + 7 - 2}} \cdot \sqrt{\frac{1}{5} + \frac{1}{7}}} \approx 4.484$$

Because $t > 3.169$, reject H_0. There is enough evidence at the 1% level of significance to support the claim.

17. (a) The claim is "the new method of teaching mathematics produces higher mathematics test scores than the old method does."

Note: μ_1 represents the new curriculum and μ_2 represents the old curriculum.

$H_0: \mu_1 \le \mu_2$; $H_a: \mu_1 > \mu_2$ (claim)

(b) d.f. $= n_1 + n_2 - 2 = 70$

$t_0 = 1.667$; Rejection region: $t > 1.667$

(c) $\bar{x}_1 \approx 62.806,\ s_1 \approx 22.845,\ n_1 = 36$

$\bar{x}_2 \approx 48.694,\ s_2 \approx 28.592,\ n_2 = 36$

$$t = \frac{(\bar{x}_1 - \bar{x}_2) - (\mu_1 - \mu_2)}{\sqrt{\frac{(n_1 - 1)s_1^2 + (n_2 - 1)s_2^2}{n_1 + n_2 - 2}} \sqrt{\frac{1}{n_1} + \frac{1}{n_2}}} \approx \frac{(62.806 - 48.694) - (0)}{\sqrt{\frac{(36-1)(22.845)^2 + (36-1)(28.592)^2}{36 + 36 - 2}} \sqrt{\frac{1}{36} + \frac{1}{36}}} \approx 2.31$$

(d) Because $t > 1.667$, reject H_0.

(e) There is enough evidence at the 5% level of significance to support the claim the new method of teaching mathematics produces higher mathematics test scores than the old method does.

19. $H_0: \mu_d = 0$ (claim); $H_a: \mu_d \ne 0$

$\alpha = 0.01$ and d.f. $= n - 1 = 15$

$-t_0 = -2.947,\ t_0 = 2.947$; Rejection regions: $t < -2.947,\ t > 2.947$

$$t = \frac{\bar{d} - \mu_d}{\frac{s_d}{\sqrt{n}}} = \frac{8.5 - 0}{\frac{10.7}{\sqrt{16}}} \approx 3.178$$

Because $t > 2.947$, reject H_0. There is enough evidence at the 1% level of significance to reject the claim.

21. $H_0 : \mu_d \le 0$ (claim); $H_a : \mu_d > 0$

$\alpha = 0.10$ and $\text{d.f.} = n - 1 = 33$

$t_0 = 1.308$; Rejection region: $t > 1.308$

$$t = \frac{\bar{d} - \mu_d}{\frac{s_d}{\sqrt{n}}} = \frac{10.3 - 0}{\frac{18.19}{\sqrt{33}}} \approx 3.253$$

Because $t > 1.308$, reject H_0. There is enough evidence at the 10% level of significance to reject the claim.

23. (a) The claim is "the numbers of passing yards for college football quarterbacks change from their junior to their senior years."

$H_0 : \mu_d = 0$; $H_a : \mu_d \ne 0$ (claim)

(b) $-t_0 = -2.262,\ t_0 = 2.262$; ; Rejection regions: $t < -2.262,\ t > 2.262$

(c) $\bar{d} = -12.3$ and $s_d \approx 553.0877$

(d) $$t = \frac{\bar{d} - \mu_d}{\frac{s_d}{\sqrt{n}}} = \frac{-12.3 - 0}{\frac{553.0877}{\sqrt{10}}} \approx -0.070$$

(e) Because $-2.262 < t < 2.262$, fail to reject H_0.

(f) There is not enough evidence at the 5% level of significance to support the sports statistician's claim the numbers of passing yards for college football quarterbacks change from their junior to their senior years.

25. $$\bar{p} = \frac{x_1 + x_2}{n_1 + n_2} = \frac{425 + 410}{840 + 760} \approx 0.522,\ \bar{q} = 1 - \bar{p} \approx 0.478$$

$n_1\bar{p} \approx 840(0.522) \approx 438.48 > 5,\ n_1\bar{q} \approx 840(0.478) \approx 401.52 > 5,$

$n_2\bar{p} \approx 760(0.522) \approx 396.72 > 5,$ and $n_2\bar{q} \approx 760(0.478) \approx 363.28 > 5$.

Can use normal sampling distribution.

$H_0 : p_1 = p_2$ (claim); $H_a : p_1 \ne p_2$

$-z_0 = -1.96,\ z_0 = 1.96$; Rejection regions: $z < -1.96,\ z > 1.96$

$$z=\frac{(\hat{p}_1-\hat{p}_2)-(p_1-p_2)}{\sqrt{\bar{p}\bar{q}\left(\frac{1}{n_1}+\frac{1}{n_2}\right)}}=\frac{(0.506-0.539)-(0)}{\sqrt{0.522(0.478)\left(\frac{1}{840}+\frac{1}{760}\right)}}\approx -1.320$$

Because $-1.96<z<1.96$, fail to reject H_0. There is not enough evidence at the 5% level of significance to reject the claim.

27. $\bar{p}=\dfrac{x_1+x_2}{n_1+n_2}=\dfrac{261+207}{556+483}\approx 0.450, \bar{q}=1-\bar{p}\approx 0.550$

$n_1\bar{p}\approx 556(0.450)\approx 250.2>5,\ n_1\bar{q}\approx 556(0.550)\approx 305.8>5,$
$n_2\bar{p}\approx 483(0.450)\approx 217.35>5,$ and $n_2\bar{q}\approx 483(0.550)\approx 265.65>5$.

Can use normal sampling distribution.

$H_0: p_1\le p_2;\ H_a: p_1>p_2$ (claim)

$z_0=1.28;$ Rejection region: $z>1.28$

$$z=\frac{(\hat{p}_1-\hat{p}_2)-(p_1-p_2)}{\sqrt{\bar{p}\bar{q}\left(\frac{1}{n_1}+\frac{1}{n_2}\right)}}=\frac{(0.469-0.429)-(0)}{\sqrt{0.450(0.550)\left(\frac{1}{556}+\frac{1}{483}\right)}}\approx 1.293$$

Because $z>1.28$, reject H_0. There is enough evidence at the 10% level of significance to support the claim.

29. (a) The claim is "the proportion of subjects who had at least 24 weeks of accrued remission is the same for the two groups."
$H_0: p_1=p_2$ (claim); $H_a: p_1\ne p_2$

(b) $-z_0=-1.96,\ z_0=1.96;$ Rejection regions: $z<-1.96,\ z>1.96$.

(c) $\bar{p}=\dfrac{x_1+x_2}{n_1+n_2}=\dfrac{19+2}{68+68}\approx 0.1544, \bar{q}=1-\bar{p}\approx 0.8456$

$$z=\frac{(\hat{p}_1-\hat{p}_2)-(p_1-p_2)}{\sqrt{\bar{p}\bar{q}\left(\frac{1}{n_1}+\frac{1}{n_2}\right)}}=\frac{(0.2794-0.0294)-(0)}{\sqrt{0.1544(0.8456)\left(\frac{1}{68}+\frac{1}{68}\right)}}\approx 4.03$$

(d) Because $z>1.96$, reject H_0.

(e) There is enough evidence at the 5% level of significance to reject the medical research team's claim that the proportion of subjects who had at least 24 weeks of accrued remission is the same for the two groups.

CHAPTER 8 QUIZ SOLUTIONS

1. (a) The claim is "the mean score on the reading assessment test for male high school students is greater than the mean score for female high school students."
$H_0: \mu_1 \le \mu_2$; $H_a: \mu_1 > \mu_2$(claim)

(b) Right-tailed because H_a contains >; z-test because σ_1 and σ_2 are known, the samples are random samples, the samples are independent, and $n_1 \ge 30$ and $n_2 \ge 30$.

(c) $z_0 = 1.645$; Rejection region: $z > 1.645$

(d) $$z = \frac{(\bar{x}_1 - \bar{x}_2) - (\mu_1 - \mu_2)}{\sqrt{\frac{\sigma_1^2}{n_1} + \frac{\sigma_2^2}{n_2}}} = \frac{(279 - 278) - (0)}{\sqrt{\frac{(41)^2}{49} + \frac{(39)^2}{50}}} \approx 0.12$$

(e) Because $z < 1.645$, fail to reject H_0.

(f) There is not enough evidence at the 5% level of significance to support the claim that the mean score on the reading assessment test for male high school students is greater than for the female high school students.

2. (a) The claim is "the mean scores on a music assessment test for eighth grade boys and girls are equal."
$H_0: \mu_1 = \mu_2$ (claim); $H_a: \mu_1 \ne \mu_2$

(b) Two-tailed because H_a contains ≠; t-test because σ_1 and σ_2 are unknown, the samples are random samples, the samples are independent, and the populations are normally distributed.

(c) d.f. $= n_1 + n_2 - 2 = 26$
$-t_0 = -1.706$, $t_0 = 1.706$; Rejection regions: $t < -1.706$, $t > 1.706$

(d) $$t = \frac{(\bar{x}_1 - \bar{x}_2) - (\mu_1 - \mu_2)}{\sqrt{\frac{(n_1 - 1)s_1^2 + (n_2 - 1)s_2^2}{n_1 + n_2 - 2}}\sqrt{\frac{1}{n_1} + \frac{1}{n_2}}} = \frac{(142 - 156) - (0)}{\sqrt{\frac{(13-1)(49)^2 + (15-1)(42)^2}{13 + 15 - 2}}\sqrt{\frac{1}{13} + \frac{1}{15}}} \approx -0.814$$

(e) Because $-1.706 < t < 1.706$, fail to reject H_0.

(f) There is not enough evidence at the 10% level of significance to reject the teacher's claim that the mean scores on the music assessment test are the same for eighth grade boys and girls.

3. (a) The claim is "the seminar helps adults increase their credit scores."
$H_0: \mu_d \ge 0$; $H_a: \mu_d < 0$ (claim)

(b) Left-tailed because H_a contains <; t-test because both populations are normally distributed and the samples are dependent.

(c) $\text{d.f.} = n - 1 = 11$
$t_0 = -2.718$; Rejection region: $t < -2.718$

(d) $t = \dfrac{\bar{d} - \mu_d}{\frac{s_d}{\sqrt{n}}} = \dfrac{-51.167 - 0}{\frac{34.938}{\sqrt{12}}} \approx -5.073$

(e) Because $t < -2.718$, reject H_0.

(f) There is enough evidence at the 1% level of significance to support the claim that the seminar helps adults increase their credit scores.

4. (a) The claim is "the proportion of U.S. adults who approve of the job the Supreme Court is doing is less than it was 3 years prior."
$H_0: p_1 \geq p_2$; $H_a: p_1 < p_2$ (claim)

(b) Left-tailed because H_a contains $<$; z-test because you are testing proportions, the samples are random samples, the samples are independent and the quantities $n_1\bar{p}$, $n_1\bar{q}$, $n_2\bar{p}$, and $n_2\bar{q}$ are at least 5.

(c) $z_0 = -1.645$; Rejection region: $z < -1.645$

(d) $\bar{p} = \dfrac{x_1 + x_2}{n_1 + n_2} = \dfrac{459 + 694}{1020 + 1510} \approx 0.4557$, $\bar{q} = 1 - \bar{p} \approx 0.5443$

$$z = \frac{(\hat{p}_1 - \hat{p}_2) - (p_1 - p_2)}{\sqrt{\bar{p}\bar{q}\left(\frac{1}{n_1} + \frac{1}{n_2}\right)}} = \frac{(0.4500 - 0.4596) - (0)}{\sqrt{0.4557(0.5443)\left(\frac{1}{1020} + \frac{1}{1510}\right)}} \approx -0.48$$

(e) Because $z > -1.645$, fail to reject H_0.

(f) There is not enough evidence at the 5% level of significance to support the claim that the proportion of U.S. adults who approve of the job the Supreme Court is doing is less than it was 3 years prior.

CUMULATIVE REVIEW, CHAPTERS 6–8

1. (a) $\hat{p} = 0.80$, $\hat{q} = 0.20$

$$\hat{p} \pm z_c\sqrt{\frac{\hat{p}\hat{q}}{n}} = 0.80 \pm 1.96\sqrt{\frac{(0.80)(0.20)}{3015}} \approx 0.80 \pm 0.014 = (0.786,\ 0.814)$$

(b) $H_0: p \le 0.75$; $H_a: p > 0.75$ (claim)

$z_0 = 1.645$; Rejection region: $z > 1.645$

$$z = \frac{\hat{p} - p}{\sqrt{\frac{pq}{n}}} = \frac{0.80 - 0.75}{\sqrt{\frac{(0.75)(0.25)}{3015}}} \approx 6.34$$

Because $z > 1.645$, reject H_0.

There is enough evidence at the 5% level of significance to support the researcher's claim that more than 75% of U.S. adults say their household contains a desktop or a laptop computer.

3. $\bar{x} \pm z_c \frac{\sigma}{\sqrt{n}} = 26.97 \pm 1.96 \frac{3.4}{\sqrt{42}} \approx 26.97 \pm 1.03 = (25.94,\ 28.00)$; z-distribution

5. $\bar{x} \pm t_c \frac{s}{\sqrt{n}} = 12.1 \pm 2.787 \frac{2.64}{\sqrt{26}} \approx 12.1 \pm 1.4 = (10.7,\ 13.5)$; t-distribution

7. $H_0: \mu \ge 33$

$H_a: \mu < 33$ (claim)

9. $H_0: \sigma = 0.63$ (claim)

$H_a: \sigma \ne 0.63$

11. $H_0: \mu_1 \le \mu_2$; $H_a: \mu_1 > \mu_2$ (claim)

$z_0 = 1.28$; Rejection region: $z > 1.28$

$$z = \frac{(\bar{x}_1 - \bar{x}_2) - 0}{\sqrt{\frac{\sigma_1^2}{n_1} + \frac{\sigma_2^2}{n_2}}} = \frac{(3086 - 2263) - (0)}{\sqrt{\frac{(563)^2}{85} + \frac{(624)^2}{68}}} \approx 8.464$$

Because $z > 1.28$, reject H_0. There is enough evidence at the 10% level of significance to support the claim that the mean birth weight of a single-birth baby is greater than the mean birth weight of a baby that has a twin.

13. $H_0: \mu_1 = \mu_2$; $H_a: \mu_1 \ne \mu_2$ (claim)

$\text{d.f.} = n_1 + n_2 - 2 = 26 + 18 - 2 = 42$

$-t_0 = -2.018$ $t_0 = 2.018$; Rejection regions: $t < -2.018$, $t > 2.018$

$$t = \frac{(\bar{x}_1 - \bar{x}_2) - (0)}{\sqrt{\frac{(n_1 - 1)s_1^2 + (n_2 - 1)s_2^2}{n_1 + n_2 - 2}}\sqrt{\frac{1}{n_1} + \frac{1}{n_2}}} = \frac{1189 - 1376}{\sqrt{\frac{(26-1)(218)^2 + (18-1)(186)^2}{26 + 18 - 2}}\sqrt{\frac{1}{26} + \frac{1}{18}}} \approx -2.97$$

Because $t < -2.018$, reject H_0. There is enough evidence at the 5% level of significance to support the organization's claim that the mean SAT scores for male athletes and male non-athletes at a college are different.

15. $H_0: p_1 \le p_2$; $H_a: p_1 > p_2$ (claim)

$$\bar{p} = \frac{x_1 + x_2}{n_1 + n_2} = \frac{6 + 1}{52 + 56} \approx 0.065,\ \bar{q} = 1 - \bar{p} \approx 0.935$$

$z_0 = 1.28$; Rejection regions: $z > 1.28$;

$$z_0 = \frac{(\hat{p}_1 - \hat{p}_2) - 0}{\sqrt{\bar{p}\bar{q}\left(\frac{1}{n_1} + \frac{1}{n_2}\right)}} = \frac{(0.1154 - 0.0179)}{\sqrt{(0.065)(0.935)\left(\frac{1}{52} + \frac{1}{56}\right)}} \approx 2.05$$

Because $z > 1.28$, reject H_0. There is enough evidence at the 10% level of significance to support the medical research team's claim that the proportion of monthly convulsive seizure reduction is greater for the group that received the extract than for the group that received the placebo.

17. A type I error will occur when the actual proportion of people who purchase their eyeglasses online is 0.05, but you reject H_0. A type II error will occur when the actual proportion of people who purchase their eyeglasses online is different from 0.05, but you fail to reject H_0.

CHAPTER 9

Correlation and Regression

9.1 CORRELATION

9.1 TRY IT YOURSELF SOLUTIONS

1.

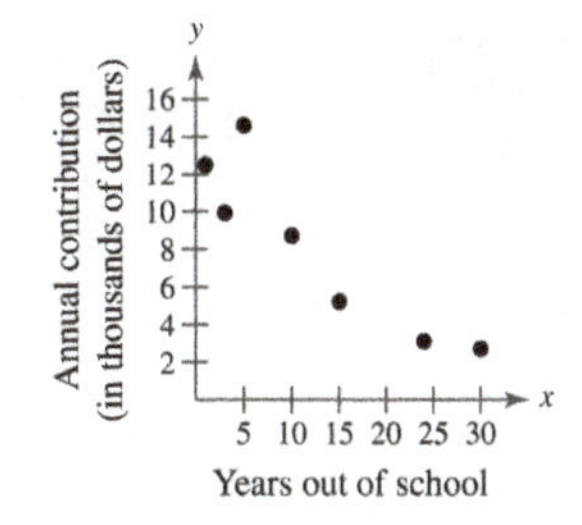

It appears that there is a negative linear correlation. As the number of years out of school increases, the annual contribution tends to decreases.

2.

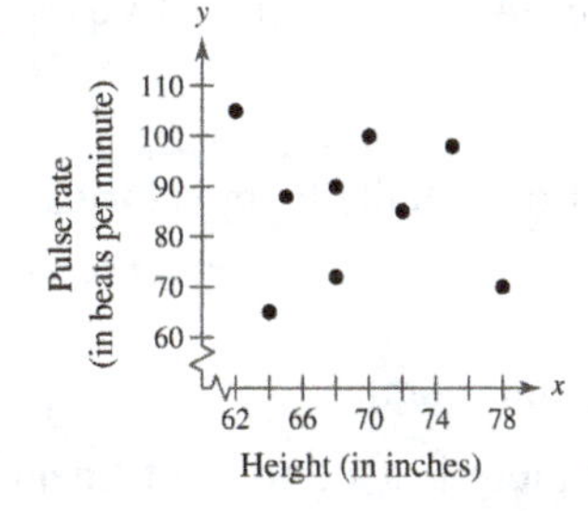

It appears that there is no linear correlation between height and pulse rate.

3.

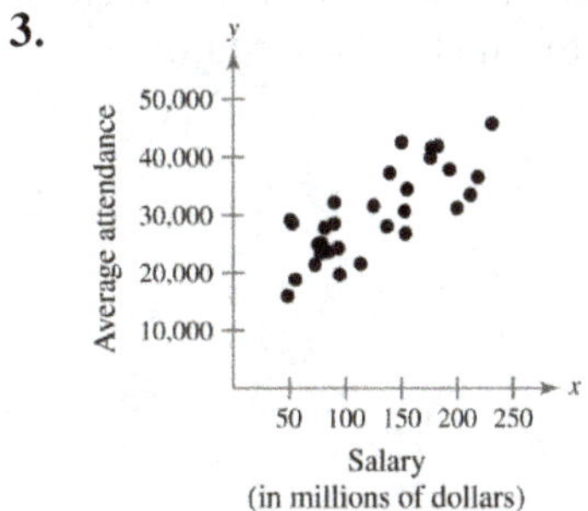

It appears that there is a positive linear correlation. As the team salary increases, the average attendance per home game tends to increase.

4. $n = 7$

x	y	xy	x^2	y^2
1	12.5	12.5	1	156.25
10	8.7	87.0	100	75.69
5	14.6	73.0	25	213.16
15	5.2	78.0	225	27.04
3	9.9	29.7	9	98.01
24	3.1	74.4	576	9.61
30	2.7	81.0	900	7.29
$\sum x = 88$	$\sum y = 56.7$	$\sum xy = 435.6$	$\sum x^2 = 1836$	$\sum y^2 = 587.05$

$$r = \frac{n\sum xy - (\sum x)(\sum y)}{\sqrt{n\sum x^2 - (\sum x)^2}\sqrt{n\sum y^2 - (\sum y)^2}}$$
$$= \frac{(7)(435.6) - (88)(56.7)}{\sqrt{(7)(1836) - (88)^2}\sqrt{(7)(587.05) - (56.7)^2}}$$
$$= \frac{-1940.4}{\sqrt{5108}\sqrt{894.46}} \approx -0.908$$

Because r is close to -1, this suggests a strong negative linear correlation. As the number of years out of school increases, the annual contribution tends to decrease.

5. 0.775; Because r is close to 1, this suggest a strong positive linear correlation. As the team salaries increase, the average attendance per home game tends to increase.

6. $n = 7$, $\alpha = 0.01$, 0.875, $|r| \approx |-0.908| > 0.875$; The correlation is significant.
There is enough evidence at the 1% level of significance to conclude that there is a significant linear correlation between the number of years out of school and the annual contribution.

7. There is enough evidence at the 1% level of significance to conclude that there is a significant linear correlation between the salaries and average attendances per home game for the teams in Major League Baseball.

9.1 EXERCISE SOLUTIONS

1. Increase; Decrease

3. The sample correlation coefficient r measures the strength and direction of a linear relationship between two variables; $r = -0.932$ indicates a stronger correlation because $|-0.932| = 0.932$ is closer to 1 than $|0.918| = 0.918$.

5. A table can be used to compare r with a critical value, or a hypothesis test can be performed using a t-test.

7. Since the t-test is non-directional, the null hypothesis, or hypothesis of no difference, will include =, instead of > or <, and may be interpreted as saying there is no significant linear correlation between the variables. The alternate hypothesis is the opposite of the null hypothesis and states that there is a significant linear correlation between the variables.
The two hypotheses are written as:
$H_0 : \rho = 0$ (no significant linear correlation)
$H_a : \rho \neq 0$ (significant linear correlation)
The null hypothesis is rejected when the calculated t is greater than the critical t-value. In this case, t is in the rejection area of the standard normal curve.

9. Strong negative linear correlation

11. No linear correlation

13. Explanatory variable: Amount of water consumed
Response variable: Weight loss

15. (c), You would expect a positive linear correlation between age and income.

17. (b), You would expect a negative linear correlation between age and balance on student loans.

19. Answers will vary. Sample answer: People who can afford an expensive home likely have enough money to seek medical attention and purchase necessary medication.

21. Answers will vary. Sample answer: Ice cream sales increase in the warm months of the year, when more people spend time outdoors. Homicide rates increase when more people are outside.

23. (a)

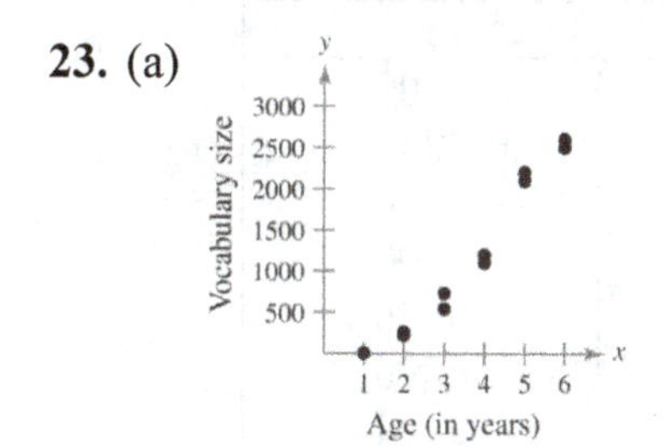

(b)

x	y	xy	x^2	y^2
1	3	3	1	9
2	220	440	4	48,400
3	540	1,620	9	291,600
4	1,100	4,400	16	1,210,000
5	2,100	10,500	25	4,410,000
6	2,600	15,600	36	6,760,000
3	730	2,190	9	532,900
5	2,200	11,000	25	4,840,000
2	260	520	4	67,600
4	1,200	4,800	16	1,440,000
6	2,500	15,000	36	6,250,000
$\Sigma x=41$	$\Sigma y=13,453$	$\Sigma xy=66,073$	$\Sigma x^2=181$	$\Sigma y^2=25,850,509$

$$r = \frac{n\sum xy - (\sum x)(\sum y)}{\sqrt{n\sum x^2 - (\sum x)^2}\sqrt{n\sum y^2 - (\sum y)^2}}$$

$$r = \frac{11(66{,}073) - (41)(13{,}453)}{\sqrt{11(181) - (41)^2}\sqrt{11(25{,}850{,}509) - (13{,}453)^2}}$$

$$= \frac{175{,}230}{\sqrt{310}\sqrt{103{,}372{,}390}} \approx 0.979$$

(c) Strong positive linear correlation. As age increases, the number of words in the children's vocabulary increases, as well.

(d) There is enough evidence at the 1% level of significance to conclude that there is a significant linear correlation between children's ages and number of words in their vocabulary.

25. (a)

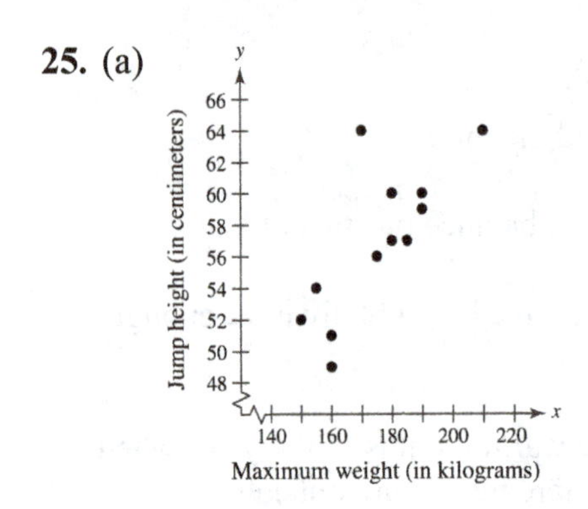

(b)

x	y	xy	x^2	y^2
190	60	11,400	36,100	3600
185	57	10,545	34,225	3249
155	54	8370	24,025	2916
180	60	10,800	32,400	3600
175	56	9800	30,625	3136
170	64	10,880	28,900	4096
150	52	7800	22,500	2704
160	51	8160	25,600	2601
160	49	7840	25,600	2401
180	57	10,260	32,400	3249
190	59	11,210	36,100	3481
210	64	13,440	44,100	4096
$\sum x = 2105$	$\sum y = 683$	$\sum xy = 120{,}505$	$\sum x^2 = 372{,}575$	$\sum y^2 = 39{,}129$

$$r = \frac{n\sum xy - (\sum x)(\sum y)}{\sqrt{n\sum x^2 - (\sum x)^2}\sqrt{n\sum y^2 - (\sum y)^2}}$$

$$= \frac{12(120{,}505) - (2105)(683)}{\sqrt{12(372{,}575) - (2105)^2}\sqrt{12(39{,}129) - (683)^2}}$$

$$= \frac{8345}{\sqrt{39{,}875}\sqrt{3059}} \approx 0.756$$

(c) Strong positive linear correlation; As the maximum weight for one repetition of a half squat increases, the jump height tends to increase.

(d) There is enough evidence at the 1% level of significance to conclude that there is a significant linear correlation between maximum weight for one repetition of a half squat and jump height.

27. (a)

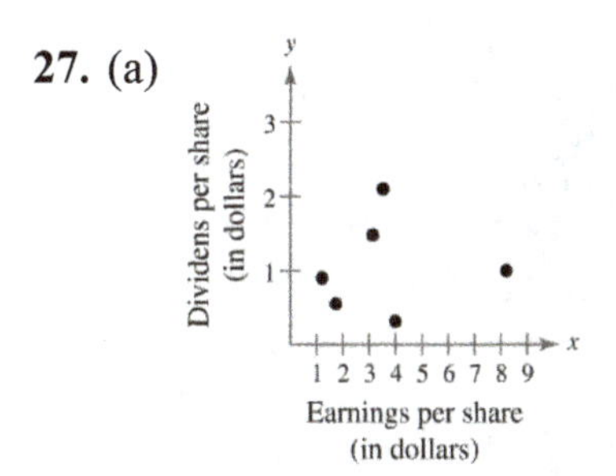

(b)

x	y	xy	x^2	y^2
1.22	0.9	1.098	1.4884	0.81
4	0.31	1.24	16	0.0961
3.53	2.1	7.413	12.4609	4.41
8.21	1	8.21	67.4041	1
1.74	0.55	0.957	3.0276	0.3025
3.14	1.48	4.6472	9.8596	2.1904
$\sum x=21.84$	$\sum y=6.34$	$\sum xy=23.5652$	$\sum x^2=110.2406$	$\sum y^2=8.809$

$$r=\frac{n\sum xy-(\sum x)(\sum y)}{\sqrt{n\sum x^2-(\sum x)^2}\sqrt{n\sum y^2-(\sum y)^2}}$$

$$r=\frac{6(23.5652)-(21.84)(6.34)}{\sqrt{6(110.2406)-(21.84)^2}\sqrt{6(8.809)-(6.34)^2}}$$

$$=\frac{2.9256}{\sqrt{184.458}\sqrt{12.6584}}\approx 0.061$$

(c) No linear correlation; The earnings per share for the companies do not appear to be related to their dividends per share.

(d) There is not enough evidence at the 1% level of significance to conclude that there is a significant linear correlation between earnings per share for the companies and their dividends per share.

29. The new correlation coefficient is approximately 0.863, showing a decrease in r by approximately 0.116. Thus, the strength of the linear relationship becomes weaker with the addition of the one data point.

31. The correlation coefficient gets stronger, going from $r\approx 0.756$ to $r\approx 0.908$.

33. $r \approx 0.623$

$n = 8$ and $\alpha = 0.01$

critical value $= 0.834$

$|r| \approx 0.623 < 0.834 \Rightarrow$ The correlation is not significant.

OR

$H_0: \rho = 0;\ H_a: \rho \neq 0$

$\alpha = 0.01$

$\text{d.f.} = n - 2 = 6$

$-t_0 = -3.707, t_0 = 3.707$; Rejection regions: $t < -3.707$ or $t > 3.707$.

$$t = \frac{r}{\sqrt{\frac{1-r^2}{n-2}}} = \frac{0.623}{\sqrt{\frac{1-(0.623)^2}{8-2}}} \approx 1.951$$

Because $-3.707 < t < 3.707$, fail to reject H_0. There is not enough evidence at the 1% level of significance to conclude that there is a significant linear correlation between vehicle weight and the variability in braking distance on a dry surface.

35.

x	y	xy	x^2	y^2
1.80	175	315	3.24	30,625
1.77	180	318.6	3.1329	32,400
2.05	155	317.75	4.2025	24,025
1.42	210	298.2	2.0164	44,100
2.04	150	306	4.1616	22,500
1.61	190	305.9	2.5921	36,100
1.70	185	314.5	2.89	34,225
1.91	160	305.6	3.6481	25,600
1.60	190	304	2.56	36,100
1.63	180	293.4	2.6569	32,400
1.98	160	316.8	3.9204	25,600
1.90	170	323	3.61	28,900
$\sum x = 21.41$	$\sum y = 2105$	$\sum xy = 3718.75$	$\sum x^2 = 38.6309$	$\sum y^2 = 372,575$

$$r = \frac{n\sum xy - (\sum x)(\sum y)}{\sqrt{n\sum x^2 - (\sum x)^2}\sqrt{n\sum y^2 - (\sum y)^2}}$$

$$= \frac{12(3718.75) - (21.41)(2105)}{\sqrt{12(38.6309) - (21.41)^2}\sqrt{12(372,575) - (2105)^2}}$$

$$= \frac{-443.05}{\sqrt{5.1827}\sqrt{39,875}} \approx -0.975$$

The correlation coefficient remains unchanged when the x-values and the y-values are switched.

37.

x	y	xy	x^2	y^2
1.80	175	315	3.24	30,625
1.77	180	318.6	3.1329	32,400
2.05	155	317.75	4.2025	24,025
1.42	210	298.2	2.0164	44,100
2.04	150	306	4.1616	22,500
1.61	190	305.9	2.5921	36,100
1.70	185	314.5	2.89	34,225
1.91	160	305.6	3.6481	25,600
1.60	190	304	2.56	36,100
1.63	180	293.4	2.6569	32,400
1.98	160	316.8	3.9204	25,600
1.90	170	323	3.61	28,900
$\sum x=21.41$	$\sum y=2105$	$\sum xy=3718.75$	$\sum x^2=38.6309$	$\sum y^2=372{,}575$

$$r=\frac{n\sum xy-(\sum x)(\sum y)}{\sqrt{n\sum x^2-(\sum x)^2}\sqrt{n\sum y^2-(\sum y)^2}}$$
$$=\frac{12(3718.75)-(21.41)(2105)}{\sqrt{12(38.6309)-(21.41)^2}\sqrt{12(372{,}575)-(2105)^2}}$$
$$=\frac{-443.05}{\sqrt{5.1827}\sqrt{39{,}875}}\approx-0.975$$

The correlation coefficient remains unchanged when the x-values and the y-values are switched.

9.2 LINEAR REGRESSION

9.2 TRY IT YOURSELF SOLUTIONS

1. $n=7,\ \sum x=88,\ \sum y=56.7,\ \sum xy=435.6,\ \sum x^2=1836$

$$m=\frac{n\sum xy-(\sum x)(\sum y)}{n\sum x^2-(\sum x)^2}=\frac{(7)(435.6)-(88)(56.7)}{(7)(1836)-(88)^2}=\frac{-1940.4}{5108}\approx-0.379875$$

$$b=\bar{y}-m\bar{x}=\frac{\sum y}{n}-m\frac{\sum x}{n}\approx\frac{(56.7)}{7}-(-0.379875)\frac{(88)}{7}\approx12.8756$$

$\hat{y}=-0.380x+12.876$

2. $\hat{y}=108.022x+16{,}586.282$

3 (1) $\hat{y}=12.481(2)+33.683$

$\hat{y}=58.645$

58.645 minutes

(2) $\hat{y}=12.481(3.32)+33.683$

$\hat{y}=75.120$

75.120 minutes

9.2 EXERCISE SOLUTIONS

1. A residual is the difference between the observed y-value of a data point and the predicted y-value on the regression line for the x-coordinate of the data point. A residual is positive when the data point is above the line, negative when the point is below the line, and zero when the observed y-value equals the predicted y-value.

3. Substitute a value of x into the equation of a regression line and solve for $\hat{y}$.

5. The correlation between variables must be significant.

7. b **9.** e **11.** f

13. b

15. d

17.

x	y	xy	x^2
1,002	75	75,150	1,004,004
992	71	70,432	984,064
901	64	57,664	811,801
780	56	43,680	608,400
762	53	40,386	580,644
756	55	41,580	571,536
752	48	36,096	565,504
741	47	34,827	549,081
732	53	38,796	535,824
$\Sigma x=7,418$	$\Sigma y=522$	$\Sigma xy=438,611$	$\Sigma x^2=6,210,858$

$$m=\frac{n\Sigma xy-(\Sigma x)(\Sigma y)}{n\Sigma x^2-(\Sigma x)^2}$$

$$m=\frac{9(438,611)-(7,418)(522)}{9(6,210,858)-(7,418)^2}=\frac{75,303}{870,998}\approx 0.08646;$$

$$b=\bar{y}-m\bar{x}=\frac{\Sigma y}{n}-m\frac{\Sigma x}{n}$$

$$b=\frac{522}{9}-\left(\frac{75,303}{870,998}\right)\left(\frac{7,418}{9}\right)\approx -13.259$$

$$\hat{y}=0.086x-13.259$$

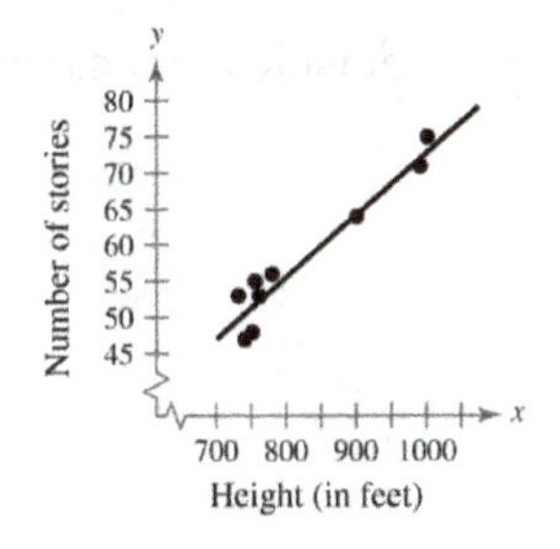

(a) $\hat{y} = 0.086(950) - 13.259 \approx 68$ stories

(b) $\hat{y} = 0.086(850) - 13.259 \approx 60$ stories

(c) $\hat{y} = 0.086(800) - 13.259 \approx 56$ stories

(d) $\hat{y} = 0.086(700) - 13.259 \approx 47$ stories

19.

x	y	xy	x^2
0	40	0	0
2	51	102	4
4	64	256	16
5	69	345	25
5	73	365	25
5	75	375	25
6	93	558	36
7	90	630	49
8	95	760	64
$\sum x = 42$	$\sum y = 650$	$\sum xy = 3391$	$\sum x^2 = 244$

$$m = \frac{n\sum xy - (\sum x)(\sum y)}{n\sum x^2 - (\sum x)^2} = \frac{(9)(3391) - (42)(650)}{(9)(244) - (42)^2} = \frac{3219}{432} \approx 7.451389$$

$$b = \bar{y} - m\bar{x} \approx \left(\frac{650}{9}\right) - (7.451389)\left(\frac{42}{9}\right) \approx 37.4491$$

$$\hat{y} = 7.451x + 37.449$$

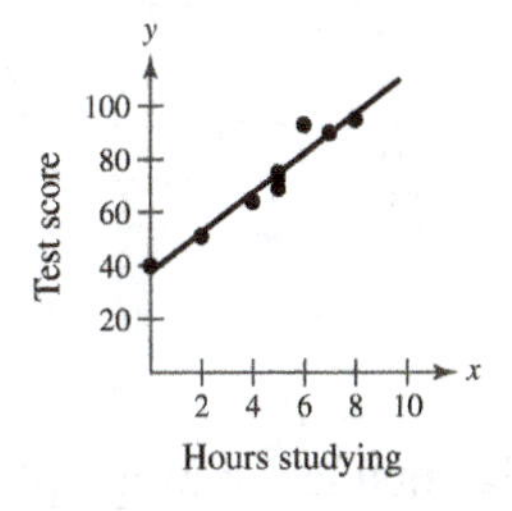

(a) $\hat{y} = 7.451(3) + 37.449 \approx 60$

(b) $\hat{y} = 7.451(6.5) + 37.449 \approx 86$

(c) It is not meaningful to predict the value of y for $x = 13$ because $x = 13$ is outside the range of the original data.

(d) $\hat{y} = 7.451(4.5) + 37.449 \approx 71$

21.

x	y	xy	x^2
60	403	24,180	3600
75	363	27,225	5625
62	381	23,622	3844
68	367	24,956	4624
84	341	28,644	7056
97	317	30,749	9409
66	401	26,466	4356
65	384	24,960	4225
86	342	29,412	7396
78	377	29,406	6084
93	329	30,597	8649
75	377	28,275	5625
88	349	30,712	7744
$\sum x = 997$	$\sum y = 4731$	$\sum xy = 359,204$	$\sum x^2 = 78,237$

$$m = \frac{x\sum xy - (\sum x)(\sum y)}{n\sum x^2 - (\sum x)^2}$$

$$= \frac{(13)(359,204) - (997)(4731)}{(13)(78,237) - (997)^2}$$

$$= \frac{-47,155}{23,072} \approx -2.0438$$

$$b = \bar{y} - m\bar{x} \approx \left(\frac{4731}{13}\right) - (-2.04382)\left(\frac{997}{13}\right) \approx 520.6683$$

$$\hat{y} = -2.044x + 520.668$$

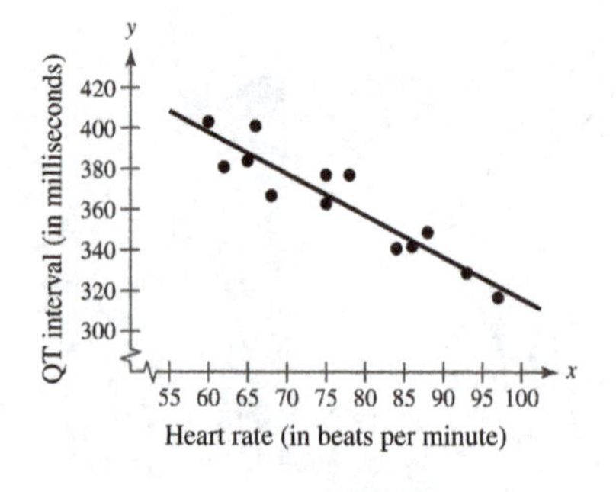

(a) It is not meaningful to predict the value of y for $x = 120$ because $x = 120$ is outside the range of the original data.

(b) $\hat{y} = -2.044(67) + 520.668 = 384$ milliseconds

(c) $\hat{y} = -2.044(90) + 520.668 = 337$ milliseconds

(d) $\hat{y} = -2.044(83) + 520.668 = 351$ milliseconds

23.

x	y	xy	x^2
180	510	91,800	32,400
220	740	162,800	48,400
230	740	170,200	52,900
90	280	25,200	8,100
160	530	84,800	25,600
190	580	110,200	36,100
150	490	73,500	22,500
110	480	52,800	12,100
110	330	36,300	12,100
160	640	102,400	25,600
140	480	67,200	19,600
150	460	69,000	22,500
$\sum x = 1,890$	$\sum y = 6,260$	$\sum xy = 1,046,200$	$\sum x^2 = 317,900$

$$m = \frac{n\sum xy - (\sum x)(\sum y)}{n\sum x^2 - (\sum x)^2}$$

$$m = \frac{12(1,046,200) - (1,890)(6,260)}{12(317,900) - (1,890)^2} = \frac{723,000}{242,700} \approx 2.979;$$

$$b = \bar{y} - m\bar{x} = \frac{\sum y}{n} - m\frac{\sum x}{n}$$

$$b = \frac{6,260}{12} - \left(\frac{723,000}{242,700}\right)\left(\frac{1,890}{12}\right) \approx 52.476$$

$$\hat{y} = 2.979x + 52.476$$

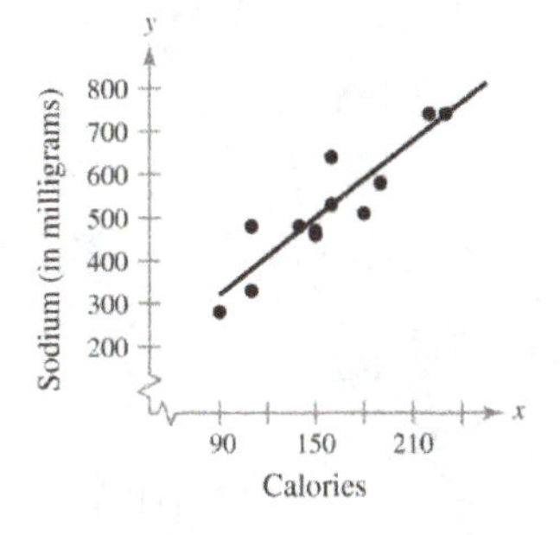

(a) $\hat{y} = 2.979(170) + 52.476 \approx 559$ milligrams

(b) $\hat{y} = 2.979(100) + 52.476 \approx 350$ milligrams

(c) It is not meaningful to predict the value of y for $x = 260$ because $x = 260$ is outside the range of the observed data.

(d) $\hat{y} = 2.979(210) + 52.476 \approx 678$ milligrams

25.

x	y	xy	x^2
8.5	66.0	561.00	72.25
9.0	68.5	616.50	81.00
9.0	67.5	607.50	81.00
9.5	70.0	665.00	90.25
10.0	70.0	700.00	100.00
10.0	72.0	720.00	100.00
10.5	71.5	750.75	110.25
10.5	69.5	729.75	110.25
11.0	71.5	786.50	121.00
11.0	72.0	792.00	121.00
11.0	73.0	803.00	121.00
12.0	73.5	882.00	144.00
12.0	74.0	888.00	144.00
12.5	74.0	925.00	156.25
$\sum x = 146.5$	$\sum y = 993$	$\sum xy = 10,427$	$\sum x^2 = 1552.25$

$$m = \frac{x\sum xy - (\sum x)(\sum y)}{n\sum x^2 - (\sum x)^2} = \frac{(14)(10,427) - (146.5)(993)}{(14)(1552.25) - (146.5)^2} = \frac{503.5}{269.25} \approx 1.870009$$

$$b = \bar{y} - m\bar{x} \approx \left(\frac{993.0}{14}\right) - (1.870)\left(\frac{146.5}{14}\right) \approx 51.3603$$

$$\hat{y} = 1.870x + 51.360$$

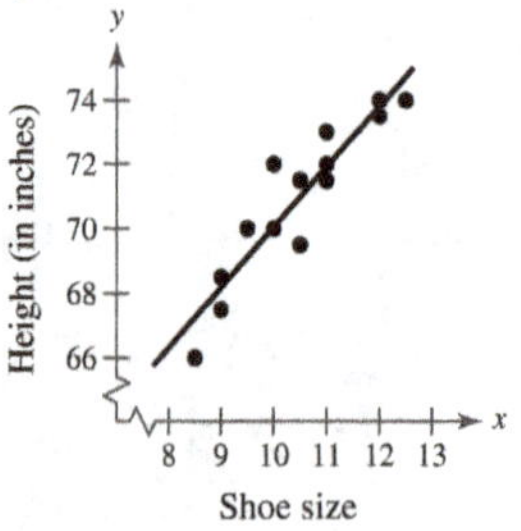

(a) $\hat{y} = 1.870(11.5) + 51.360 = 72.865$ inches

(b) $\hat{y} = 1.870(8.0) + 51.360 = 66.32$ inches

(c) It is not meaningful to predict the value of y for $x = 15.5$ because $x = 15.5$ is outside the range of the original data.

(d) $\hat{y} = 1.870(10.0) + 51.360 = 70.06$ inches

27. Strong positive linear correlation; As the years of experience of the registered nurses increase, their salaries tend to increase.

29. No, it is not meaningful to predict a salary for a registered nurse with 28 years of experience because $x = 28$ is outside the range of the original data.

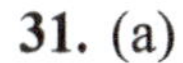

31. (a)

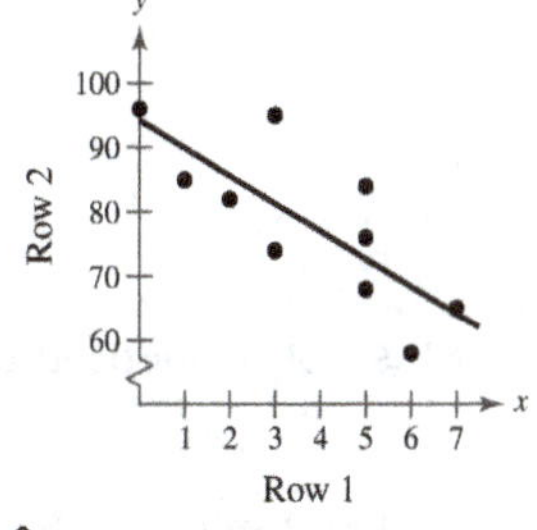

$\hat{y} = -4.297x + 94.200$

(b)

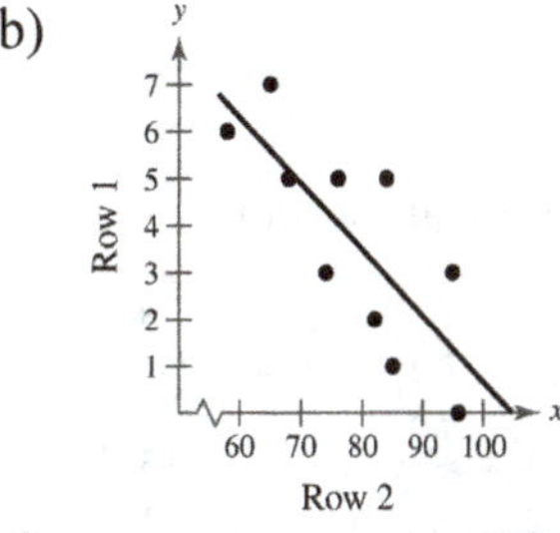

$\hat{y} = -0.141x + 14.763$

(c) The slope of the line keeps the same sign, but the values of m and b change.

33. (a) $m \approx 0.139$

$b \approx 21.024$

$\hat{y} = 0.139x + 21.024$

(b)

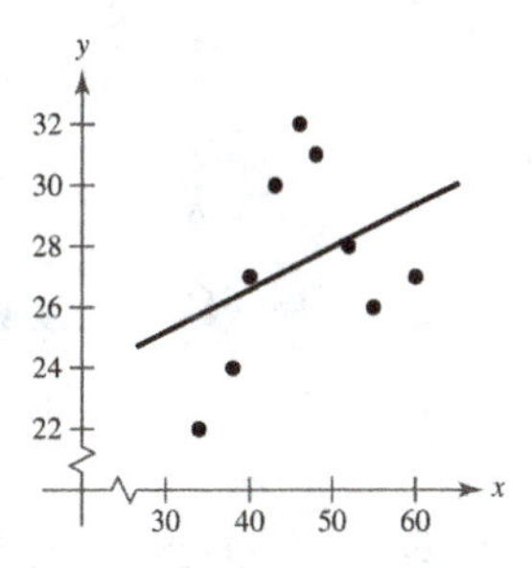

(c)

x	y	$\hat{y} = 0.139x + 21.024$	$y - \hat{y}$
38	24	26.306	−2.306
34	22	22.750	−3.750
40	27	26.584	0.416
46	32	27.418	4.582
43	30	27.001	2.999
48	31	27.696	3.304
60	27	29.364	−2.364
55	26	28.669	−2.669
52	28	28.252	−0.252

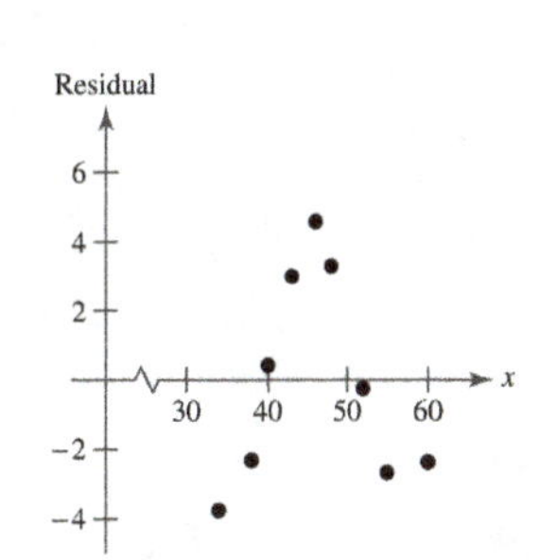

(d) The residual plot shows a pattern because the residuals do not fluctuate about 0. This implies that the regression line is not a good representation of the relationship between the variables.

35. (a)

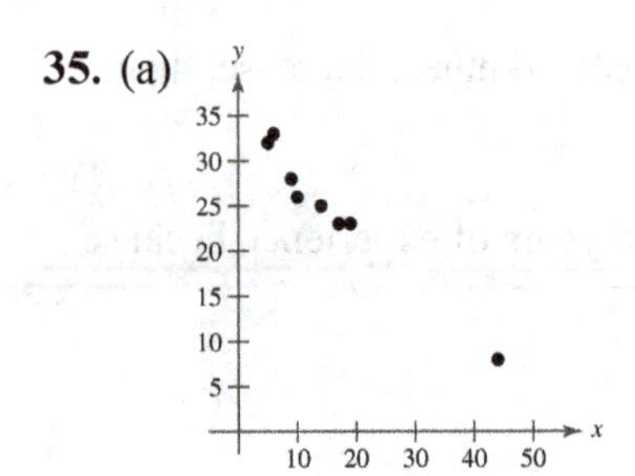

(b) The point (44, 8) may be an outlier.

(c) Excluding the point (44, 8) $\Rightarrow \hat{y} = -0.711x + 35.263$. The point (44, 8) is not influential because using all 8 points $\Rightarrow \hat{y} = -0.607x + 34.160$.
The slopes and y-intercepts of the regression lines with the point included and without the point are not significantly different.

37. $m \approx 654.536$

$b \approx -1214.857$

$\hat{y} = 654.536x - 1214.857$

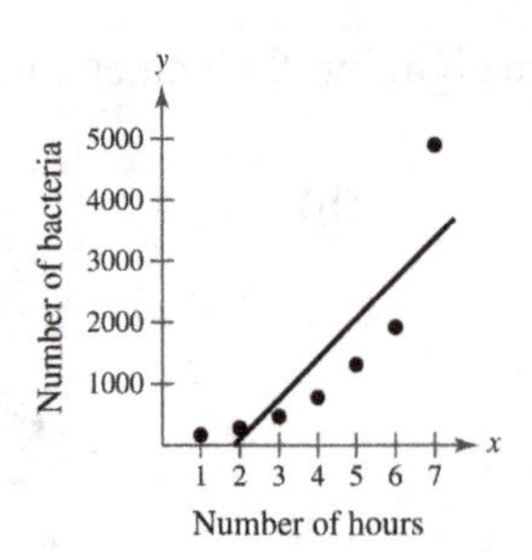

39. Using a technology tool $\Rightarrow y = 93.028(1.712)^x$.

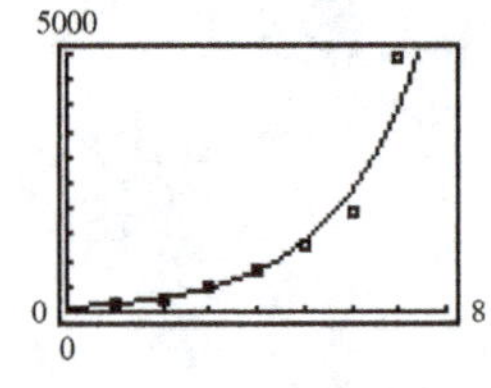

41. $m = -78.929$

$b = 576.179$

$\hat{y} = -78.929x + 576.179$

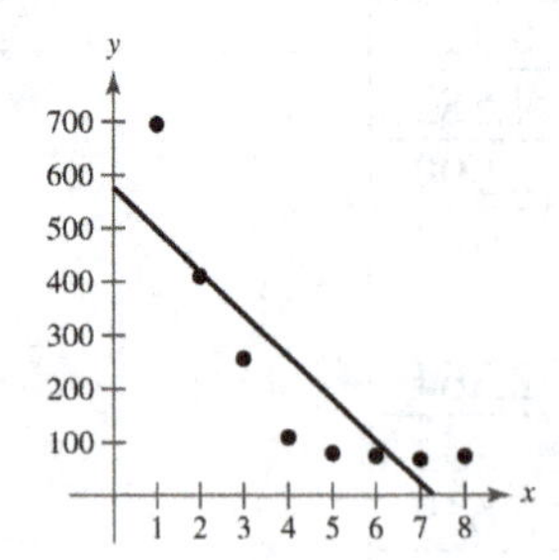

43. Using a technology tool $\Rightarrow y = 782.300x^{-1.251}$.

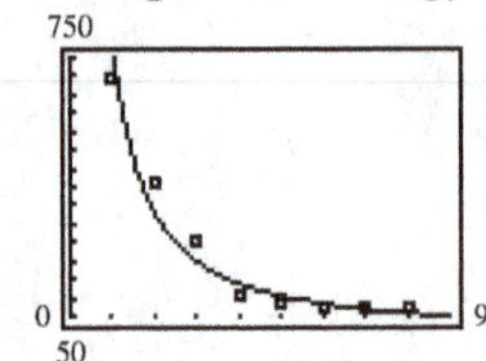

45. $y = a + b\ln x = 25.035 + 19.599\ln x$

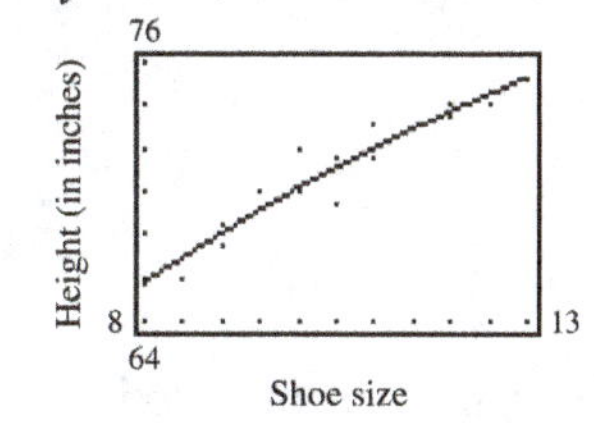

47. The logarithmic equation is a better model for the data. The graph of the logarithmic equation fits the data better than the regression line.

9.3 MEASURES OF REGRESSION AND PREDICTION INTERVALS

9.3 TRY IT YOURSELF SOLUTIONS

1. $r \approx 0.979$ $\quad r^2 \approx (0.979)^2 \approx 0.958$

About 95.8% of the variation in the times is explained.
About 4.2% of the variation is unexplained.

2.

x_i	y_i	$\hat{y}_i$	$y_i - \hat{y}_i$	$(y_i - \hat{y}_i)^2$
15	26	28.386	–2.386	5.692996
20	32	35.411	–3.411	11.634921
20	38	35.411	2.589	6.702921
30	56	49.461	6.539	42.758521
40	54	63.511	–9.511	90.459121
45	78	70.536	7.464	55.711296
50	80	77.561	2.439	5.948721
60	88	91.611	–3.611	13.039321
				$\Sigma = 231.947818$

$n = 8$

$$s_e = \sqrt{\frac{\Sigma(y_i - \hat{y}_i)^2}{n-2}} = \sqrt{\frac{231.947818}{6}} \approx 6.218$$

The standard error of estimate of the weekly sales for a specific radio ad time is about $621.80.

3. $n = 10$, d.f. $= 8$, $t_c = 2.306$, $s_e \approx 141.935$

$\hat{y} = 199.535(4) + 56.036 = 854.176$

$$E = t_c s_e \sqrt{1 + \frac{1}{n} + \frac{n(x - \bar{x})^2}{n(\Sigma x^2) - (\Sigma x)^2}}$$

$$E = (2.306)(141.935)\sqrt{1+\frac{1}{10}+\frac{10(4-2.29)^2}{10(65.49)-(22.9)^2}} \approx 376.624$$

$$\hat{y}-E = 854.176-376.624 = 477.552$$

$$\hat{y}+E = 854.176+376.624 = 1230.8$$

You can be 95% confident that when the gross domestic product is $4 trillion, the carbon dioxide emissions will be between 477.552 and 1230.8 million metric tons.

9.3 EXERCISE SOLUTIONS

1. The total variation is the sum of the squares of the differences between the y-values of each ordered pair and the mean of the y-values of the ordered pairs or $\Sigma\left(y_i-\bar{y}\right)^2$.

3. The unexplained variation is the sum of the squares of the differences between the observed y-values and the predicted y-values or $\Sigma\left(y_i-\hat{y}_i\right)^2$.

5. Two variables that have perfect positive or perfect negative linear correlation have a correlation coefficient of 1 or -1, respectively. In either case, the coefficient of determination is 1, which means that 100% of the variation in the response variable is explained by the variation in the explanatory variable.

7. $r^2 = (0.465)^2 \approx 0.216$; About 21.6% of the variation is explained. About 78.4% of the variation is unexplained.

9. $r^2 = (-0.957)^2 \approx 0.916$; About 91.6% of the variation is explained. About 8.4% of the variation is unexplained.

11. (a) $r^2 = \dfrac{\Sigma\left(\hat{y}_i-\bar{y}\right)^2}{\Sigma\left(y_i-\bar{y}\right)^2} \approx 0.798$

About 79.8% of the variation in proceeds can be explained by the relationship between the number of issues and proceeds, and about 20.2% of the variation is unexplained.

(b) $s_e = \sqrt{\dfrac{\Sigma\left(y_i-\hat{y}_i\right)^2}{n-2}} = \sqrt{\dfrac{648{,}721{,}935}{10}} \approx 8054.328$

The standard error of estimate of the proceeds for a specific number of issues is about $8,054,328,000.

13. (a) $r^2 = \dfrac{\Sigma(\hat{y}_i - \bar{y})^2}{\Sigma(y_i - \bar{y})^2} \approx .730$

About 73% of the variation in points earned can be explained by the relationship between the number of goals allowed and points earned, and about 27% of the variation is unexplained.

(b) $s_e = \sqrt{\dfrac{\Sigma(y_i - \hat{y}_i)^2}{n-2}} = \sqrt{\dfrac{1068.938017}{12}} \approx 9.438$

The standard error of estimate of the points earned for a specific number of goals allowed is about 9.438.

15. (a) $r^2 = \dfrac{\Sigma(\hat{y}_i - \bar{y})^2}{\Sigma(y_i - \bar{y})^2} \approx 0.651$

About 65.1% of the variation in mean annual wages can be explained by the relationship between percentages of employment in STEM occupations and mean annual wages, and about 34.9% of the variation is unexplained.

(b) $s_e = \sqrt{\dfrac{\Sigma(y_i - \hat{y}_i)^2}{n-2}} = \sqrt{\dfrac{927.7893389}{14}} \approx 8.141$.

The standard error of estimate of the mean annual wages for a specific percentage of employment in STEM occupations is about $8141.

17. (a) $r^2 = \dfrac{\Sigma(\hat{y}_i - \bar{y})^2}{\Sigma(y_i - \bar{y})^2} \approx 0.884$

About 88.4% of the variation in the amount of crude oil imported can be explained by the relationship between the amount of crude oil produced and the amount imported, and about 11.6% of the variation is unexplained.

(b) $s_e = \sqrt{\dfrac{\Sigma(y_i - \hat{y}_i)^2}{n-2}} = \sqrt{\dfrac{392,233.7292}{5}} \approx 280.083$.

The standard error of estimate of the amount of crude oil imported for a specific amount of crude oil produced is about 280,083 barrels per day.

19. (a) $r^2 = \dfrac{\Sigma(\hat{y}_i - \bar{y})^2}{\Sigma(y_i - \bar{y})^2} \approx 0.816$

About 81.6% of the variation in the new-vehicle sales of General Motors can be explained by the relationship between the new-vehicle sales of Ford and General Motors, and about 18.4% of the variation is unexplained.

(b) $s_e = \sqrt{\frac{\Sigma\left(y_i - \hat{y}_i\right)^2}{n-2}} = \sqrt{\frac{1,079,571.213}{9}} \approx 346.341$.

The standard error of estimate of the new-vehicle sales of General Motors for a specific amount of new-vehicle sales of Ford is about 346,341 new vehicles.

21. $n = 12$, d.f. $= 10$, $t_c = 2.228$, $s_e \approx 8054.328$

$\hat{y} = 104.965x + 14,093.666 = 104.965(450) + 14,093.666 = 61,327.916$

$$E = t_c s_e \sqrt{1 + \frac{1}{n} + \frac{n\left(x - \bar{x}\right)^2}{n\left(\Sigma x^2\right) - \left(\Sigma x\right)^2}}$$

$$\approx (2.228)(8054.328)\sqrt{1 + \frac{1}{12} + \frac{12\left(450 - 2138/12\right)^2}{12(613,126) - (2138)^2}}$$

$\approx 21,244.664$

$\hat{y} \pm E \rightarrow 61,327.916 \pm 21,244.664 \rightarrow (40,083.251,\ 82,572.581)$

You can be 95% confident that the proceeds will be between \$40,083,251,000 and \$82,572,581,000 when the number of initial offerings is 450.

23. $n = 14$, d.f. $= 12$, $t_c = 1.782$, $s_e \approx 9.438$

$\hat{y} = -0.573(250) + 220.087 = 76.837$

$$E = t_c s_e \sqrt{1 + \frac{1}{n} + \frac{n\left(x - \bar{x}\right)^2}{n\left(\Sigma x^2\right) - \left(\Sigma x\right)^2}}$$

$$E = (1.782)(9.438)\sqrt{1 + \frac{1}{14} + \frac{14(250 - 228.6428571)^2}{14(740,657) - (3,201)^2}} \approx 17.826$$

$\hat{y} - E = 76.837 - 17.826 = 59.011$; $\hat{y} + E = 76.837 + 17.826 = 94.663$

$59.011 < y < 94.663$

You can be 90% confident that the total points earned will be between 59.011 and 94.663 when the number of goals allowed by the team is 250.

25. $n = 16$, d.f. $= 14$, $t_c = 2.977$, $s_e \approx 8.141$

$\hat{y} = 1.153(13) + 46.374 = 61.363$

$$E = t_c s_e \sqrt{1 + \frac{1}{n} + \frac{n\left(x - \bar{x}\right)^2}{n\left(\Sigma x^2\right) - \left(\Sigma x\right)^2}}$$

$$E = (2.977)(8.141)\sqrt{1 + \frac{1}{16} + \frac{16(13 - 9.38125)^2}{16(2,712.05) - (150.1)^2}} \approx 25.1$$

$\hat{y} - E = 61.363 - 25.1 = 36.263$; $\hat{y} + E = 61.363 + 24.974 = 86.462$

$36.263 < y < 86.462$

You can be 99% confident that the mean annual wage is between \$36,263 and \$86,462 when the percentage of employment in STEM occupations is 13% in the industry.

27. $n=7,\ \text{d.f.}=5,\ t_c=2.571,\ s_e\approx 280.083$

$\hat{y}=-0.438(8000)+11{,}404.947=7900.947$.

$$E=t_c s_e\sqrt{1+\frac{1}{n}+\frac{n\left(x-\bar{x}\right)^2}{n\left(\sum x^2\right)-\left(\sum x\right)^2}}$$

$$E=(2.571)(280.083)\sqrt{1+\frac{1}{7}+\frac{7(8000-7447.142857)^2}{7(403{,}920{,}680)-(52{,}130)^2}}\approx 776.341.$$

$\hat{y}-E=7900.947-776.341=7124.606$; $\hat{y}+E=7900.947+776.341=8677.288$.

$7124.606<y<8677.288$

You can be 95% confident that the amount of crude oil imported by the United States is between 7,124,606 and 8,677,288 barrels when the amount of crude oil produced by the United States is 8 million barrels per day.

29. $n=11,\ \text{d.f.}=9,\ t_c=2.262,\ s_e\approx 346.341$

$\hat{y}=1.624(2628)-747.304=3520.568$.

$$E=t_c s_e\sqrt{1+\frac{1}{n}+\frac{n\left(x-\bar{x}\right)^2}{n\left(\sum x^2\right)-\left(\sum x\right)^2}}$$

$$E=(2.262)(346.341)\sqrt{1+\frac{1}{11}+\frac{11(2628-2334.454545)^2}{11(61{,}763{,}389)-(25{,}679)^2}}\approx 835.856.$$

$\hat{y}-E=3520.568-835.856=2684.712$; $\hat{y}+E=3520.568+835.856=4356.424$.

$2684.712<y<4356.424$

You can be 95% confident that the new-vehicle sales for General Motors is between 2,684,712 and 4,356,424 when the number of new vehicles sold by Ford is 2628 thousand.

31.

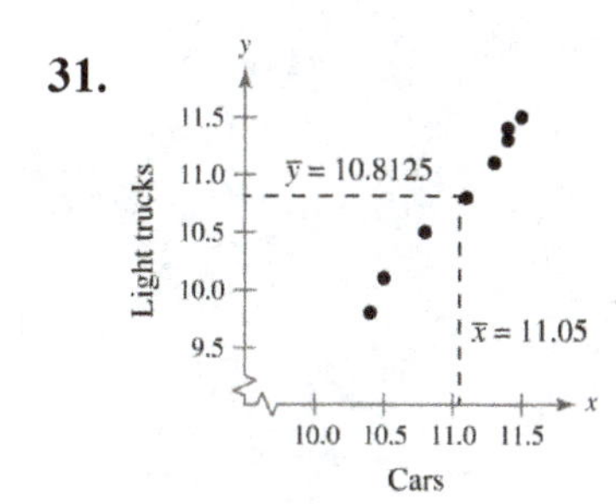

33. $r^2\approx 0.987$; ; About 98.7% of the variation in the median ages of light trucks can be explained by the relationship between the median ages of cars and light trucks, and about 1.3% of the variation is unexplained.

35. Critical value $= \pm 3.707$

$m \approx -0.205$

$s_e \approx 0.554$

$$t = \frac{m}{s_e}\sqrt{\sum x^2 - \frac{(\sum x)^2}{n}} \approx \frac{-0.205}{0.554}\sqrt{838.55 - \frac{(79.5)^2}{8}} \approx -2.578$$

Because $-3.707 < t < 3.707$, fail to reject $H_0: M = 0$. There is not enough evidence at the 1% level of significance to support the claim that there is a linear relationship between weight and number of hours slept.

37. $$E = t_c s_e \sqrt{\frac{1}{n} + \frac{\bar{x}^2}{\sum x^2 - \left[(\sum x)^2 / n\right]}} \approx (2.306)(141.935)\sqrt{\frac{1}{10} + \frac{(2.29)^2}{65.49 - \frac{(22.9)^2}{10}}} \approx 231.8716672.$$

$b + E = 56.036 + 231.8716672 \approx 287.908$; $b - E = 56.036 - 231.8716672 \approx -175.836$.

$$E = \frac{t_c s_e}{\sqrt{\sum x^2 - \frac{(\sum x)^2}{n}}} \approx \frac{(2.306)(141.935)}{\sqrt{65.49 - \frac{(22.9)^2}{10}}} = 90.60667419$$

$m + E = 199.535 + 90.60667419 \approx 290.142$; $m - E = 199.535 - 90.60667419 \approx 108.928$.

Confidence intervals for B is $-175.836 < B < 287.908$; Confidence intervals for M is $108.928 < M < 290.142$

9.4 MULTIPLE REGRESSION

9.4 TRY IT YOURSELF SOLUTIONS

1. Enter the data. $\hat{y} = 46.385 + 0.540x_1 - 4.897x_2$

2. (1) $\hat{y} = 46.385 + 0.540(89) - 4.897(1)$

$\hat{y} = 89.548 = 90$

(2) $\hat{y} = 46.385 + 0.540(78) - 4.897(3)$

$\hat{y} = 73.814 = 74$

(3) $\hat{y} = 46.385 + 0.540(83) - 4.897(2)$

$\hat{y} = 81.411 = 81$

9.4 EXERCISE SOLUTIONS

1. $\hat{y} = 24,791 + 4.508x_1 - 4.723x_2$

(a) $\hat{y} = 24,791 + 4.508(36,500) - 4.723(36,100) = 18,832.7$ pounds per acre

(b) $\hat{y} = 24,791 + 4.508(38,100) - 4.723(37,800) = 18,016.4$ pounds per acre

(c) $\hat{y} = 24,791 + 4.508(39,000) - 4.723(38,800) = 17,350.6$ pounds per acre

(d) $\hat{y} = 24,791 + 4.508(42,200) - 4.723(42,100) = 16,190.3$ pounds per acre

3. $\hat{y} = -52.2 + 0.3x_1 + 4.5x_2$

(a) $\hat{y} = -52.2 + 0.3(70) + 4.5(8.6) = 7.5$ cubic feet

(b) $\hat{y} = -52.2 + 0.3(65) + 4.5(11.0) = 16.8$ cubic feet

(c) $\hat{y} = -52.2 + 0.3(83) + 4.5(17.6) = 51.9$ cubic feet

(d) $\hat{y} = -52.2 + 0.3(87) + 4.5(19.6) = 62.1$ cubic feet

5. (a) $\hat{y} = 17,899 - 606.58x_1 - 52.9x_2$

(b) $s_e = 564.314$; ; The standard error of estimate of the predicted price given specific age and mileage of pre-owned Honda Civic Sedans is about \$564.31.

(c) $r^2 = 0.966$; ; The multiple regression model explains about 96.6% of the variation.

7. $n = 9$ $k = 2$, $r^2 = 0.966$

$$r^2_{adj} = 1 - \left[\frac{(1-0.966)(9-1)}{9-2-1}\right] \approx 0.955$$

About 95.5% of the variation in y can be explained by the relationship between variables; $r^2_{adj} < r^2$.

CHAPTER 9 REVIEW EXERCISE SOLUTIONS

1. (a)

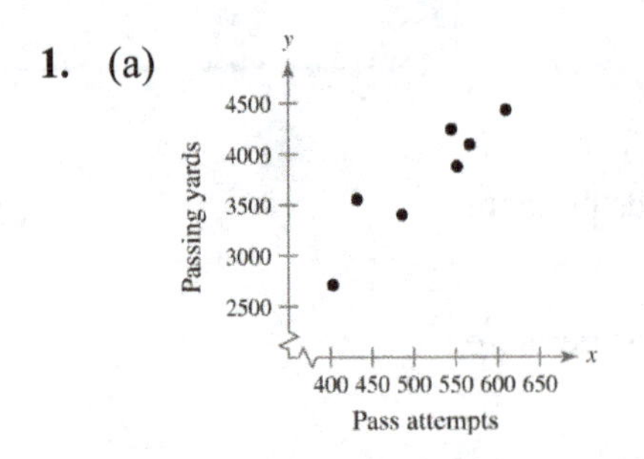

(b)

x	y	xy	x^2	y^2
610	4,428	2,701,080	372,100	19,607,184
545	4,240	2,310,800	297,025	17,977,600
567	4,090	2,319,030	321,489	16,728,100
552	3,877	2,140,104	304,704	15,031,129
432	3,554	1,535,328	186,624	12,630,916
486	3,401	1,652,886	236,196	11,566,801
403	2,710	1,092,130	162,409	7,344,100
$\sum x=3595$	$\sum y=26,300$	$\sum xy=13,751,358$	$\sum x^2=1,880,547$	$\sum y^2=100,885,830$

$$r=\frac{n\sum xy-(\sum x)(\sum y)}{\sqrt{n\sum x^2-(\sum x)^2}\sqrt{n\sum y^2-(\sum y)^2}}$$

$$=\frac{7(13,751,358)-(3595)(26,300)}{\sqrt{7(1,880,547)-(3595)^2}\sqrt{7(100,885,830)-(26,300)^2}}$$

$$=\frac{1,711,006}{\sqrt{239,804}\sqrt{14,510,810}}\approx 0.917$$

(c) Strong positive linear correlation; As the number of pass attempts increase, the number of passing yards tends to increase

3. (a)

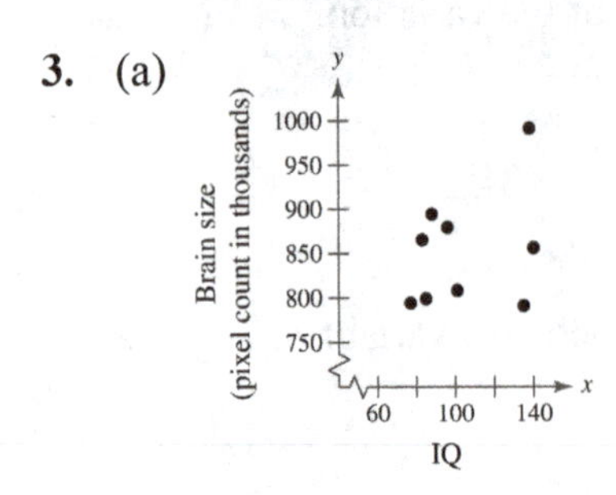

(b)

x	y	xy	x^2	y^2
138	991	136,758	19,044	982081
140	856	119,840	1960	732736
96	879	84,384	9216	772641
83	865	71,795	6889	748225
101	808	81,608	10,201	652864
135	791	106,785	18,225	625681
85	799	67,915	7225	638401
77	794	61,138	5929	630436
88	894	78,672	7744	799236
$\sum x = 943$	$\sum y = 7677$	$\sum xy = 808,895$	$\sum x^2 = 104,073$	$\sum y^2 = 6,582,301$

$$r = \frac{n\Sigma xy - (\Sigma x)(\Sigma y)}{\sqrt{n\Sigma x^2 - (\Sigma x)^2}\sqrt{n\Sigma y^2 - (\Sigma y)^2}}$$

$$= \frac{9(808,865) - (943)(7677)}{\sqrt{9(104,073) - (943)^2}\sqrt{9(6,582,301) - (7677)^2}}$$

$$= \frac{40,644}{\sqrt{47,408}\sqrt{304,380}} \approx 0.338$$

(c) Weak positive linear correlation; The IQ does not appear to be related to the brain size.

5. $H_0: \rho = 0;\ H_a: \rho \neq 0$

$\alpha = 0.05$, d.f. $= n - 2 = 5$

$-t_0 = -2.571,\ t_0 = 2.571$; Rejection regions: $t < -2.571$ or $t > 2.571$

$$t = \frac{r}{\sqrt{\frac{1-r^2}{n-2}}} = \frac{0.917}{\sqrt{\frac{1-(0.917)^2}{7-2}}} \approx 5.14$$

Because $t > 2.571$, reject H_0.There is enough evidence at the 5% level of significance to conclude that there is a significant linear correlation between a quarterback's pass attempts and passing yards.

7. $H_0: \rho = 0;\ H_a: \rho \neq 0$

$\alpha = 0.01$, d.f. $= n - 2 = 7$

$-t_0 = -3.499,\ t_0 = 3.499$; Rejection regions: $t < -3.499$ or $t > 3.499$

$$t = \frac{r}{\sqrt{\frac{1-r^2}{n-2}}} = \frac{0.338}{\sqrt{\frac{1-(0.338)^2}{9-2}}} \approx 0.950$$

Because $-3.499 < t < 3.499$, fail to reject H_0. There is not enough evidence at the 1% level of significance to conclude that a significant linear correlation between IQ and brain size.

9. $\hat{y} = 0.106x - 781.327$

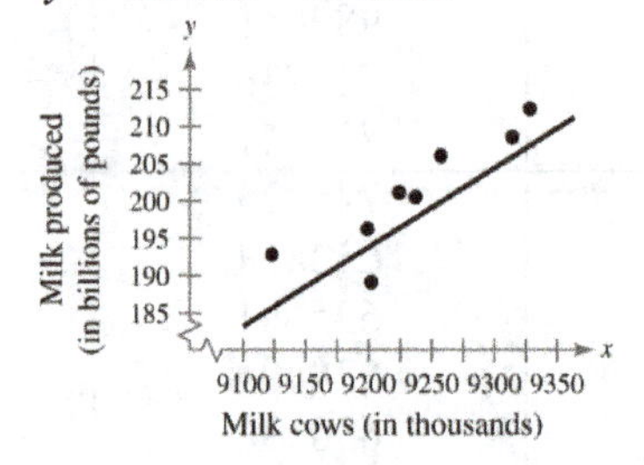

(a) It is not meaningful to predict the value of y for $x = 9080$ because $x = 9080$ is outside the range of the observed data.

(b) $\hat{y} = 0.106(9230) - 781.327 = 197.053$ billions of pounds

(c) $\hat{y} = 0.106(9250) - 781.327 = 199.173$ billions of pounds

(d) $\hat{y} = 0.106(9300) - 781.327 = 204.473$ billions of pounds

11. $\hat{y} = -0.086x + 10.450$

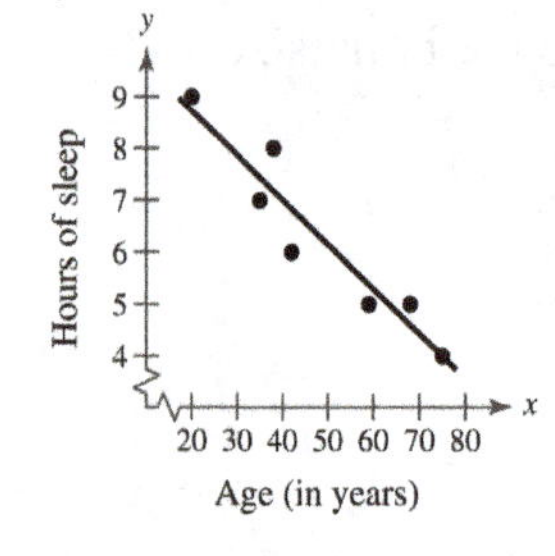

(a) It is not meaningful to predict the value of y for $x = 16$ because $x = 16$ is outside the range of the original data.

(b) $\hat{y} = -0.086(25) + 10.450 = 8.3$ hours

(c) It is not meaningful to predict the value of y for $x = 85$ because $x = 85$ is outside the range of the original data.

(d) $\hat{y} = -0.086(50) + 10.450 = 6.15$ hours

13. $r^2 = (-0.450)^2 \approx 0.203$

About 20.3% of the variation in y is explained.
About 79.7% of the variation in y is unexplained.

15. $r^2 = (0.642)^2 \approx 0.412$

About 41.2% of the variation in y is explained.
About 58.8% of the variation in y is unexplained.

17. (a) $r^2 = \dfrac{\Sigma(\hat{y}_i - \bar{y})^2}{\Sigma(y_i - \bar{y})^2} \approx 0.690$.

About 69.0% of the variation in top speed for hybrid and electric cars can be explained by the relationship between their fuel efficiencies and top speeds, and about 31.0% of the variation is unexplained.

(b) $s_e = \sqrt{\dfrac{\Sigma(y_i - \hat{y}_i)^2}{n-2}} = \sqrt{\dfrac{239.665646}{7}} \approx 5.851$.

The standard error of estimate of the top speed for hybrid and electric cars for a specific fuel efficiency is about 5.851 miles per hour.

19. $\hat{y} = 0.106(9275) - 781.327 = 201.823$

$$E = t_c s_e \sqrt{1 + \frac{1}{n} + \frac{n(x-\bar{x})^2}{n(\Sigma x^2) - (\Sigma x)^2}} = (1.943)(4.014)\sqrt{1 + \frac{1}{8} + \frac{8(9275 - 9235.5)^2}{8(682{,}386{,}108) - (73{,}884)^2}}$$

≈ 8.459

$\hat{y} + E = 201.823 + 8.459 = 210.282$; $\hat{y} - E = 201.823 - 8.459 = 193.364$

You can be 90% confident that the amount of milk produced will be between 193.364 billion pounds and 210.282 billion pounds when the average number of cows is 9275 thousand.

21. $\hat{y} = -0.086(45) + 10.450 = 6.580$

$$E = t_c s_e \sqrt{1 + \frac{1}{n} + \frac{n(x-\bar{x})^2}{n(\Sigma x^2) - (\Sigma x)^2}} \approx (2.571)(0.622)\sqrt{1 + \frac{1}{7} + \frac{7(45 - 337/7)^2}{7(18{,}563) - (337)^2}}$$

≈ 1.714

$\hat{y} \pm E \to 6.580 \pm 1.714 \to (4.866,\ 8.294)$

You can be 95% confident that the hours slept will be between 4.866 and 8.294 hours for a person who is 45 years old.

23. $\hat{y} = -0.465(90) + 139.433 = 97.583$

$$E = (3.499)(5.851)\sqrt{1 + \frac{1}{9} + \frac{9(90 - 103)^2}{9(97{,}955) - (927)^2}} \approx 22.234$$

$\hat{y} - E = 97.583 - 22.234 = 75.349$;

$\hat{y} + E = 97.583 + 22.234 = 119.817$.

$75.349 < y < 119.817$; You can be 99% confident that the top speed of a hybrid or electric car will be between 75.349 and 119.817 miles per hour when the combined city and highway fuel efficiency is 90 miles per gallon equivalent.

25. (a) $\hat{y} = 3.6738 + 1.2874x_1 - 7.531x_2$

(b) $s_e \approx 0.710$; The standard error of estimate of the carbon monoxide content given specific tar and nicotine contents is about 0.710 milligram.

(c) $r^2 \approx 0.943$; The multiple regression model explains about 94.3% of the variation in y.

27. (a) 21.705 miles per gallon

(b) 25.21 miles per gallon

(c) 30.1 miles per gallon

(d) 25.86 miles per gallon

CHAPTER 9 QUIZ SOLUTIONS

1.

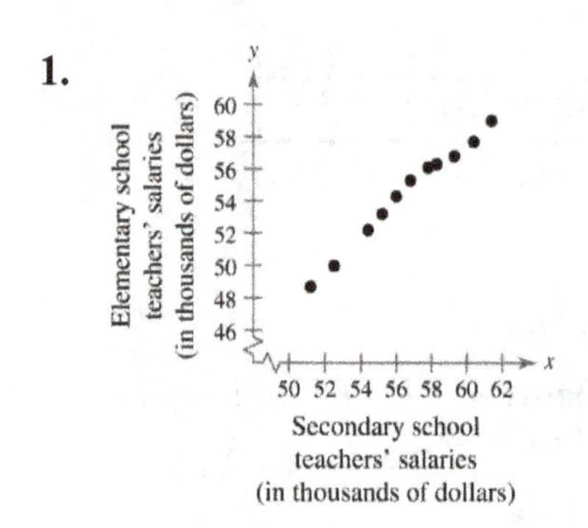

The data appear to have a positive linear correlation. As x increases, y tends to increase.

2. $r \approx 0.992$; ; Strong positive linear correlation; As the average annual salaries of secondary school teachers increase, the average annual salaries of elementary school teachers tend to increase.

3. $H_0: \rho = 0$; $H_a: \rho \neq 0$

$\alpha = 0.05$, d.f. $= n - 2 = 9$

$-t_0 = -2.262$, $t_0 = 2.262$; Rejection regions: $t < -2.262$ or $t > 2.262$

$$t = \frac{r}{\sqrt{\frac{1-r^2}{n-2}}} \approx \frac{0.992}{\sqrt{\frac{1-(0.992)^2}{11-2}}} \approx 23.57$$

Because $t > 2.262$, reject H_0.

There is enough evidence at the 5% level of significance to conclude that there is a significant linear correlation between the average annual salaries of secondary school teachers and the average annual salaries of elementary school teachers.

4.

x	y	xy	x^2
51.2	48.7	2493.44	2621.44
52.5	50	2625	2756.25
54.4	52.2	2839.68	2959.36
55.2	53.2	2936.64	3047.04
56	54.3	3040.8	3136
56.8	55.3	3141.04	3226.24
57.8	56.1	3242.58	3340.84
58.3	56.3	3282.29	3398.89
59.3	56.8	3368.24	3516.49
60.4	57.7	3485.08	3648.16
61.4	59	3622.6	3769.96
$\sum x = 623.3$	$\sum y = 599.6$	$\sum xy = 34077.39$	$\sum x^2 = 35420.67$

$$m = \frac{n\sum xy - (\sum x)(\sum y)}{n\sum x^2 - (\sum x)^2} = \frac{11(34{,}077.39) - (623.3)(599.6)}{11(35{,}420.67) - (623.3)^2} \approx 0.997$$

$$b = \bar{y} - m\bar{x} = \frac{\sum y}{n} - m\left(\frac{\sum x}{n}\right) \approx \frac{599.6}{11} - \left(\frac{1{,}120.61}{1{,}124.48}\right)\left(\frac{623.3}{11}\right) \approx -1.960$$

$$\hat{y} = 0.997x - 1.960$$

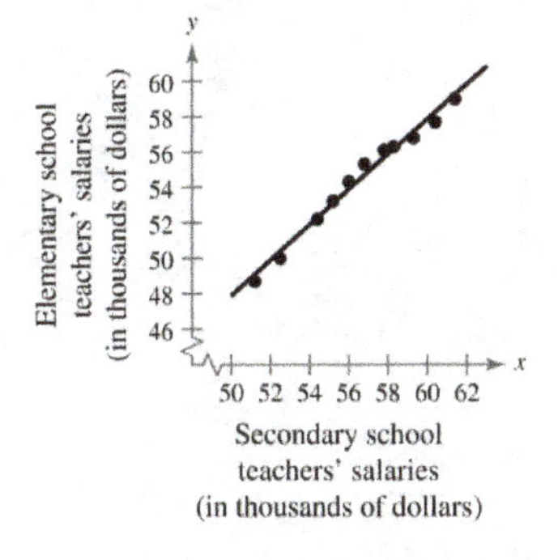

5. $\hat{y} = 0.997(52.5) - 1.960 = 50.3825$

6. $r^2 \approx (0.9921)^2 \approx 0.984$

About 98.4% of the variation in the average annual salaries of elementary school teachers can be explained by the relationship between the average annual salaries of secondary school teachers and elementary school teachers, and about 1.6% of the variation is unexplained.

7. $s_e \approx 0.422$;

The standard error of estimate of the average annual salaries of elementary school teachers for a specific average annual salary of secondary school teachers is about $422.

8. $\hat{y} = 0.997(52.5) - 1.960 = 50.3825$.

$s_e = \sqrt{\dfrac{1.61271403}{9}} \approx 0.423$.

$$E = (2.262)(0.423)\sqrt{1 + \frac{1}{11} + \frac{11(52.5 - 56.66363636)^2}{11(35{,}420.67) - (623.3)^2}} \approx 1.074$$

$\hat{y} - E = 50.3825 - 1.074 = 49.31$;

$\hat{y} + E = 50.3825 + 1.074 = 51.46$

$49.31 < y < 51.46$

You can be 95% confident that the average annual salary of elementary school teachers is between $49,310 and $51,460 when the average annual salary of secondary school teachers is $52,500.

9. (a) $\hat{y} = -86 + 7.46(27.6) - 1.61(15.3) = \$95.26$

(b) $\hat{y} = -86 + 7.46(24.1) - 1.61(14.6) = \$70.28$

(c) $\hat{y} = -86 + 7.46(23.5) - 1.61(13.4) = \$67.74$

(d) $\hat{y} = -86 + 7.46(22.8) - 1.61(15.3) = \$59.46$

CHAPTER 10

Chi-Square Tests and the *F*-Distribution

10.1 GOODNESS OF FIT TEST

10.1 TRY IT YOURSELF SOLUTIONS

1.

Tax Preparation Method	% of people	Expected frequency
Accountant	24%	500(0.24) = 120
By hand	20%	500(0.20) = 100
Computer software	35%	500(0.35) = 175
Friend/family	6%	500(0.06) = 30
Tax preparation service	15%	500(0.15) = 75

2. The expected frequencies are 64, 80, 32, 56, 60, 48, 40, and 20, all of which are at least 5.
Claimed distribution:

Ages	Distribution
0–9	16%
10–19	20%
20–29	8%
30–39	14%
40–49	15%
50–59	12%
60–69	10%
70+	5%

H_0: The distribution of the ages are 16% ages 0–9, 20% ages 10–19, 8% ages 20–29, 14% ages 30–39, 15% ages 40–49, 12% ages 50–59, 10% ages 60-69 and 5% ages 70+ (claim).
H_a: The distribution of ages differs from the claimed distribution.

$\alpha = 0.05$ **d.** $\text{d.f.} = n - 1 = 7$

$\chi_0^2 = 14.067$; Rejection region: $\chi^2 > 14.067$

Ages	Distribution	Observed	Expected	$\frac{(O-E)^2}{E}$
0–9	16%	76	64	2.250
10–19	20%	84	80	0.200
20–29	8%	30	32	0.125
30–39	14%	60	56	0.286
40–49	15%	54	60	0.600
50–59	12%	40	48	1.333
60–69	10%	42	40	0.100
70+	5%	14	20	1.800

$\chi^2 \approx 6.694$

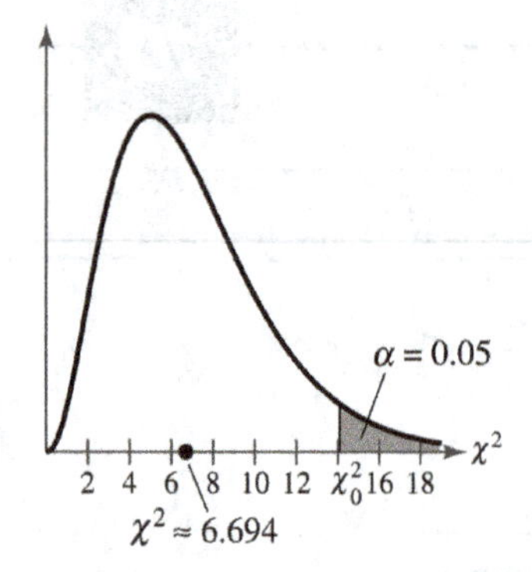

Because $\chi^2 < 14.067$, fail to reject H_0.
There is not enough evidence at the 5% level of significance to support the sociologist's claim that the distribution of ages differs from the age distribution 10 years ago.

3. The expected frequency for each category is 30 which is at least 5.
Claimed distribution:

Color	Distribution
Brown	$16.\overline{6}\%$
Yellow	$16.\overline{6}\%$
Red	$16.\overline{6}\%$
Blue	$16.\overline{6}\%$
Orange	$16.\overline{6}\%$
Green	$16.\overline{6}\%$

H_0: The distribution of colors is uniform. (claim)
H_a: The distribution of colors is not uniform.
$\alpha = 0.05$
$\text{d.f.} = n - 1 = 5$
$\chi_0^2 = 11.071$; Rejection region: $\chi^2 > 11.071$

Color	Distribution	Observed	Expected	$\dfrac{(O-E)^2}{E}$
Brown	16.6%	22	30	$2.1\overline{33}$
Yellow	16.6%	27	30	0.300
Red	16.6%	22	30	$2.1\overline{33}$
Blue	16.6%	41	30	$4.0\overline{33}$
Orange	16.6%	41	30	$4.0\overline{33}$
Green	16.6%	27	30	0.300
				12.933

$\chi^2 \approx 12.933$

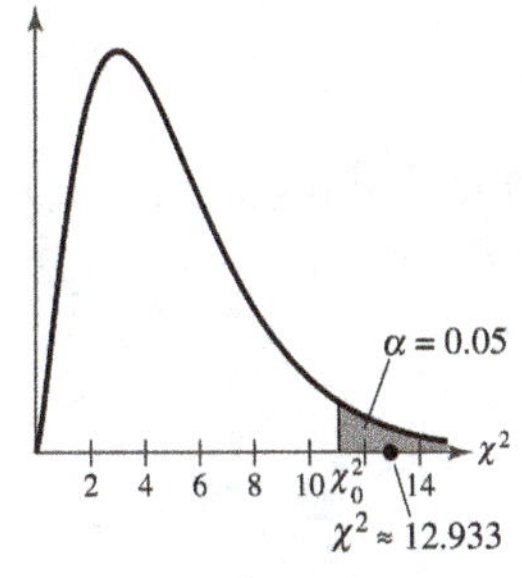

Because $\chi^2 > 11.071$, reject H_0.

There is enough evidence at the 5% level of significance to dispute the claim that the distribution of different colored candies in bags of peanut M&M's is uniform.

10.1 EXERCISE SOLUTIONS

1. A multinomial experiment is a probability experiment consisting of a fixed number of independent trials in which there are more than two possible outcomes for each trial. The probability of each outcome is fixed, and each outcome is classified into categories.

3. $E_i = np_i = (150)(0.3) = 45$

5. $E_i = np_i = (230)(0.25) = 57.5$

7. (a) Claimed Distribution:

Age	Distribution
2–17	23%
18–24	20%
25–39	22%
40–49	9%
50+	26%

H_0: The distribution of the ages of moviegoers is 23% ages 2–17, 20% ages 18–24, 22% ages 25–39, 9% ages 40–49, and 26% ages 50+ (claim).

H_a: The distribution of the ages differs from the claimed or expected distribution.

(b) $\chi_0^2 = 7.779$, Rejection region: $\chi^2 > 7.779$

(c)

Age	Distribution	Observed	Expected	$\frac{(O-E)^2}{E}$
2–17	23%	240	230	0.4348
18–24	20%	209	200	0.405
25–39	22%	203	220	1.3136
40–49	9%	106	90	2.8444
50+	26%	242	260	1.2462
				6.244

$\chi^2 \approx 6.244$

(d) Because $\chi^2 < 7.779$, fail to reject H_0.

(e) There is not enough evidence at the 10% level of significance to reject the claim that the distribution of the ages of moviegoers and the expected distribution are the same.

9. (a) Claimed distribution:

Day	Distribution
Sunday	7%
Monday	4%
Tuesday	6%
Wednesday	13%
Thursday	10%
Friday	36%
Saturday	24%

H_0: The distribution of the days people order food for delivery is 7% Sunday, 4% Monday, 6% Tuesday, 13% Wednesday, 10% Thursday, 36% Friday, and 24% Saturday.
H_a: The distribution of days differs from the claimed or expected distribution.

(b) $\chi_0^2 = 16.812$, Rejection region: $\chi^2 > 16.812$

(c)

	Distribution	Observed	Expected	$\frac{(O-E)^2}{E}$
Sunday	7%	43	35	1.8286
Monday	4%	16	20	0.8000
Tuesday	6%	25	30	0.8333
Wednesday	13%	49	65	3.9385
Thursday	10%	46	50	0.3200
Friday	36%	168	180	0.8000
Saturday	24%	153	120	9.0750
				17.5954

$\chi^2 \approx 17.5954$

(d) Reject H_0.

(e) There is enough evidence at the 1% level of significance to conclude that the distribution of delivery days has changed.

11. (a) Claimed Distribution:

County	Distribution
Alameda	6.25%
Contra Costa	6.25%
Fresno	6.25%
Kern	6.25%
Los Angeles	6.25%
Monterey	6.25%
Orange	6.25%
Riverside	6.25%
Sacramento	6.25%
San Bernardino	6.25%
San Diego	6.25%
San Francisco	6.25%
San Joaquin	6.25%
Santa Clara	6.25%
Stanislaus	6.25%
Tulare	6.25%

H_0: The distribution of the number of homicide crimes in California by county is uniform. (claim)
H_a: The distribution of homicides by county is not uniform.

(b) $\chi_0^2 = 30.578$, Rejection region: $\chi^2 > 30.578$

(c)

County	Distribution	Observed	Expected	$\frac{(O-E)^2}{E}$
Alameda	6.25%	116	62.5	45.796
Contra Costa	6.25%	55	62.5	0.9
Fresno	6.25%	57	62.5	0.484
Kern	6.25%	62	62.5	0.004
Los Angeles	6.25%	101	62.5	23.716
Monterey	6.25%	58	62.5	0.324
Orange	6.25%	30	62.5	16.9
Riverside	6.25%	65	62.5	0.1
Sacramento	6.25%	90	62.5	12.1
San Bernardino	6.25%	89	62.5	11.236
San Diego	6.25%	45	62.5	4.9
San Francisco	6.25%	51	62.5	2.116
San Joaquin	6.25%	62	62.5	0.004
Santa Clara	6.25%	39	62.5	8.836
Stanislaus	6.25%	37	62.5	10.404
Tulare	6.25%	43	62.5	6.084
				143.904

$\chi^2 \approx 143.904$

(d) Because $\chi^2 > 30.578$, reject H_0.

(e) There is enough evidence at the 1% level of significance to reject the claim that the distribution of the number of homicide crimes in California by county is uniform.

13. (a) Claimed distribution:

Month	Distribution
Strongly agree	55%
Somewhat agree	30%
Neither agree nor disagree	5%
Somewhat disagree	6%
Strongly disagree	4%

H_0: The distribution of the opinions of U.S. parents on whether a college education is worth the expense is 55% strongly agree, 30% somewhat agree, 5% neither agree nor disagree, 6% somewhat disagree, 4% strongly disagree.
H_a: The distribution of opinions differs from the expected distribution. (claim)

(b) $\chi_0^2 = 9.488$, Rejection region: $\chi^2 > 9.488$

(c)

Month	Distribution	Observed	Expected	$\frac{(O-E)^2}{E}$
Strongly agree	55%	86	110	5.2364
Somewhat agree	30%	62	60	0.0667
Neither agree nor disagree	5%	34	10	57.6000
Somewhat disagree	6%	14	12	0.3333
Strongly disagree	4%	4	8	2.0000
				65.2364

$\chi^2 \approx 65.236$

(d) Because $\chi^2 > 9.488$, reject H_0.

(e) There is enough evidence at the 5% level of significance to conclude that the distribution of opinions of U.S. parents on whether a college education is worth the expense differs from the claimed or expected distribution.

15. (a) Claimed distribution:

Response	Distribution
Larger	$33.\overline{3}\%$
Same size	$33.\overline{3}\%$
Smaller	$33.\overline{3}\%$

H_0: The distribution of prospective home buyers by the size they want their next house to be is uniform.

H_a: The distribution of prospective home buyers by the size they want their next house to be is not uniform. (claim)

(b) $\chi_0^2 = 5.991$, Rejection region: $\chi^2 > 5.991$

(c)

Response	Distribution	Observed	Expected	$\frac{(O-E)^2}{E}$
Larger	$33.\overline{3}\%$	285	$266.\overline{66}$	1.2604
Same size	$33.\overline{3}\%$	224	$266.\overline{66}$	6.8267
Smaller	$33.\overline{3}\%$	291	$266.\overline{66}$	2.2204
				10.3075

$\chi^2 \approx 10.308$

(d) Because $\chi^2 > 5.991$, reject H_0.

(e) There is enough evidence at the 5% level of significance to conclude that the distribution of prospective home buyers by the size they want their next house to be is not uniform.

17. (a) Frequency distribution: $\mu = 69.435$; $\sigma \approx 8.337$

Lower Boundary	Upper Boundary	Lower *z*-score	Upper *z*-score	Area
49.5	58.5	−2.39	−1.31	0.0867
58.5	67.5	−1.31	−0.23	0.3139
67.5	76.5	−0.23	0.85	0.3933
76.5	85.5	0.85	1.93	0.1709
85.5	94.5	1.93	3.01	0.0255

Class Boundaries	Distribution	Frequency	Expected	$\frac{(O-E)^2}{E}$
49.5–58.5	8.67%	19	17	0.2353
58.5–67.5	31.39%	61	63	0.0635
67.5–76.5	39.33%	82	79	0.1139
76.5–85.5	17.09%	34	34	0
85.5–94.5	2.55%	4	5	0.2000
		200		0.6127

H_0: Test scores have a normal distribution. (claim)

H_a: Test scores do not have a normal distribution.

(b) $\chi_0^2 = 13.277$; Rejection region . $\chi^2 > 13.277$

(c) $\chi^2 = 0.613$

(d) Because $\chi^2 < 13.277$, fail to reject H_0.

(e) There is not enough evidence at the 1% level of significance to reject the claim that the test scores are normally distributed.

10.2 INDEPENDENCE

10.2 TRY IT YOURSELF SOLUTIONS

1.

	Hotel	Leg Room	Rental Size	Other	Total
Business	36	108	14	22	180
Leisure	38	54	14	14	120
Total	74	162	28	36	300

$n = 300$

	Hotel	Leg Room	Rental Size	Other
Business	44.4	97.2	16.8	21.6
Leisure	29.6	64.8	11.2	14.4

2. The claim is that "the travel concerns depend on the purpose of travel."
H_0: Travel concern is independent of travel purpose.
H_a: Travel concern is dependent on travel purpose. (claim)
$\alpha = 0.01$ **c.** $(r-1)(c-1) = 3$
$\chi_0^2 = 11.345$; Rejection region: $\chi^2 > 11.345$

O	E	$O-E$	$(O-E)^2$	$\frac{(O-E)^2}{E}$
36	44.4	−8.4	70.56	1.5892
108	97.2	10.8	116.64	1.2000
14	16.8	−2.8	7.84	0.4667
22	21.6	0.4	0.16	0.0074
38	29.6	8.4	70.56	2.3838
54	64.8	−10.8	116.64	1.8000
14	11.2	2.8	7.84	0.7000
14	14.4	−0.4	0.16	0.0111
				8.1582

$\chi^2 \approx 8.158$

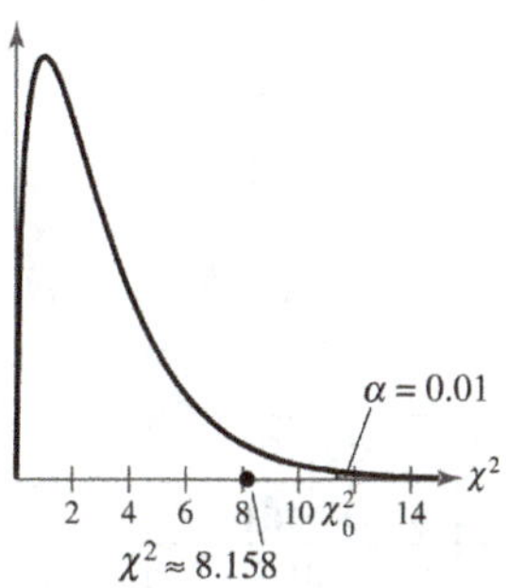

Because $\chi^2 < 11.345$, fail to reject H_0.

There is not enough evidence at the 1% level of significance for the consultant to conclude that travel concern is dependent on travel purpose.

3. H_0 : Whether or not a tax cut would influence an adult to purchase a hybrid vehicle is independent of age.

H_a : Whether or not a tax cut would influence an adult to purchase a hybrid vehicle is dependent on age. (claim)

Enter the data.

$\chi_0^2 = 9.210$; Rejection region: $\chi^2 > 9.210$

$\chi^2 \approx 15.306$

Because $\chi^2 > 9.210$, reject H_0.

There is enough evidence at the 1% level of significance to conclude that whether or not a tax cut would influence an adult to purchase a hybrid vehicle is dependent on age.

10.2 EXERCISE SOLUTIONS

1. Find the sum of the row and the sum of the column in which the cell is located. Find the product of these sums. Divide the product by the sample size.

3. Answer will vary. *Sample answer:* For both the chi-square test for independence and the chi-square goodness-of-fit test, you are testing a claim about data that are in categories. However, the chi-square goodness-of-fit test has only one data value per category, while the chi-square test for independence has multiple data values per category.

Both tests compare observed and expected frequencies. However, the chi-square goodness-of-fit test simply compares the distributions, whereas the chi-square test for independence compares them and then draws a conclusion about the dependence or independence of the variables.

5. False. If the two variables of a chi-square test for independence are dependent, then you can expect a large difference between the observed frequencies and the expected frequencies.

7. (a)

	Athlete has		
Result	**Stretched**	**Not stretched**	**Total**
Inquiry	18	22	40
No Inquiry	211	189	400
	229	211	440

(b)

Result	Athlete has Stretched	Not stretched
Inquiry	20.82	19.18
No Inquiry	208.18	191.82

9. (a)

Bank employee	Preference New procedure	Old Procedure	No preference	Total
Teller	92	351	50	493
Customer service	76	42	8	126
Total	168	393	58	619

(b)

Bank employee	Preference New procedure	Old Procedure	No preference
Teller	133.80	313.00	46.19
Customer service	34.20	80.00	11.81

11. (a)

Gender	Type of car Compact	Full-size	SUV	Truck/Van	Total
Male	28	39	21	22	110
Female	24	32	20	14	90
	52	71	41	36	200

(b)

Gender	Type of car Compact	Full-size	SUV	Truck/Van
Male	28.6	39.05	22.55	19.8
Female	23.4	31.95	18.45	16.2

13. (a) The claim is "an athlete's injury result is not related to whether or not the athlete has stretched."
H_0: An athlete's injury result is independent of whether or not the athlete has stretched. (claim)
H_a: An athlete's injury result is dependent on whether or not the athlete has stretched.

(b) d.f. $= (r-1)(c-1) = 1$
$\chi_0^2 = 6.635$; Rejection region: $\chi^2 > 6.635$

(c)

O	E	$O-E$	$(O-E)^2$	$\frac{(O-E)^2}{E}$
18	20.82	−2.82	7.9524	0.3820
22	19.18	2.82	7.9524	0.4146
211	208.18	2.82	7.9524	0.0382
189	191.82	−2.82	7.9524	0.0415
				0.8763

$\chi^2 \approx 0.8763$

(d) Because $\chi^2 < 6.635$, fail to reject H_0.

(e) There is not enough evidence at the 1% level of significance to reject the claim that an athlete's injury result is independent of whether or not the athlete has stretched.

15. (a) The claim is "result is related to type of treatment."
H_0: The result is independent of the type of treatment.
H_a: The result is dependent on the type of treatment. (claim)

(b) $\text{d.f.} = (r-1)(c-1) = 1$
$\chi_0^2 = 2.706$; Rejection region: $\chi^2 > 2.706$

(c)

O	E	$O-E$	$(O-E)^2$	$\frac{(O-E)^2}{E}$
58	69.5	−11.5	132.25	1.9029
81	69.5	11.5	132.25	1.9029
42	30.5	11.5	132.25	4.3361
19	30.5	−11.5	132.25	4.3361
				12.478

$\chi^2 \approx 12.478$

(d) Because $\chi^2 > 2.706$, reject H_0.

(e) There is enough evidence at the 10% level of significance to conclude that the result is dependent on the type of treatment.

17. (a) The claim is "the number of times former smokers tried to quit before they were habit-free is related to gender."
H_0: The number of times former smokers tried to quit is independent of gender.
H_a: The number of times former smokers tried to quit is dependent of gender. (claim)

(b) $\text{d.f.} = (r-1)(c-1) = 2$
$\chi_0^2 = 5.991$; Rejection region: $\chi^2 > 5.991$

(c)

O	E	$O-E$	$(O-E)^2$	$\frac{(O-E)^2}{E}$
271	270.930	0.070	0.004900	0
257	257.286	−0.286	0.081796	0.0003
149	148.784	0.216	0.046656	0.0003
146	146.070	−0.070	0.004900	0
139	138.714	0.286	0.081796	0.0006
80	80.216	−0.216	0.046656	0.0006
				0.0018

$\chi^2 \approx 0.002$

(d) Because $\chi^2 < 5.991$, fail to reject H_0.

(e) There is not enough evidence at the 5% level of significance to conclude that the number of times former smokers tried to quit is dependent on gender.

19. (a) The claim is "the reason and the type of worker are dependent."
H_0 : Reasons are independent of the type of worker.
H_a : Reasons are dependent on the type of worker. (claim)

(b) d.f. $= (r-1)(c-1) = 2$
$\chi_0^2 = 9.210$; Rejection region: $\chi^2 > 9.210$

(c)

O	E	$O-E$	$(O-E)^2$	$\frac{(O-E)^2}{E}$
30	39.421	–9.421	88.7552	2.2515
36	31.230	4.770	22.7529	0.7286
41	36.349	4.651	21.6318	0.5951
47	37.579	9.421	88.7552	2.3618
25	29.770	–4.770	22.7529	0.7643
30	34.651	–4.651	21.6318	0.6243
				7.3256

$\chi^2 \approx 7.326$

(d) Because $\chi^2 < 9.210$, fail to reject H_0.

(e) There is not enough evidence at the 1% level of significance to conclude that reasons for continuing education are dependent on the type or worker.

21. (a) The claim is "a family borrowing money for college is related to race."
H_0 : A family borrowing money for college is independent of race.
H_a : A family borrowing money for college is dependent on race. (claim)

(b) d.f. $= (r-1)(c-1) = 2$
$\chi_0^2 = 9.210$; Rejection region: $\chi^2 > 9.210$

(c)

O	E	$O-E$	$(O-E)^2$	$\frac{(O-E)^2}{E}$
49	42.23	6.77	45.8329	1.0853
64	70.77	–6.77	45.8329	0.6476
85	77.73	7.27	52.8529	0.6800
123	130.27	–7.27	52.8529	0.4057
85	99.04	–14.04	197.1216	1.9903
180	165.96	14.04	197.1216	1.1878
				5.9967

$\chi^2 \approx 5.9967$

(d) Because $\chi^2 < 9.210$, fail to reject H_0.

(e) There is not enough evidence at the 1% level of significance to conclude that a family borrowing money for college is dependent on race.

23. (a) The claim is "the type of crash is depends on the type of vehicle."
H_0: Type of crash is independent of the type of vehicle.
H_a: Type of crash is dependent on the type of vehicle. (claim)

(b) $\text{d.f.} = (r-1)(c-1) = 2$
$\chi_0^2 = 5.991$; Rejection region: $\chi^2 > 5.991$

(c)

O	E	$O-E$	$(O-E)^2$	$\frac{(O-E)^2}{E}$
1059	1221.19	−162.19	26,305.5961	21.5410
507	414.77	92.23	8,506.3729	20.5087
491	421.03	69.97	4,895.8009	11.6282
1476	1313.81	162.19	26,305.5961	20.0224
354	446.23	−92.23	8,506.3729	19.0628
383	452.97	−69.97	4,895.8009	10.8082
				103.5713

$\chi^2 \approx 103.5713$

(d) Because $\chi^2 > 5.991$, reject H_0.

(e) There is enough evidence at the 5% level of significance to conclude that the type of crash is dependent on the type of vehicle.

25. (a) The claim is "procedure preference is related to bank employee."
H_0: Procedure preference is independent of bank employee.
H_a: Procedure preference is dependent on bank employee. (claim)

(b) $\text{d.f.} = (r-1)(c-1) = 2$
$\chi_0^2 = 5.991$; Rejection region: $\chi^2 > 5.991$

(c)

O	E	$O-E$	$(O-E)^2$	$\frac{(O-E)^2}{E}$
92	133.80	−41.8	1747.24	13.0586
351	313.00	38	1444	4.6134
50	46.19	3.81	14.5161	0.3143
76	34.20	41.8	1747.24	51.0889
42	80.00	−38	1444	18.05
8	11.81	−3.81	14.5161	1.2291
				88.3543

$\chi^2 \approx 88.3543$

(d) Because $\chi^2 > 5.991$, reject H_0.

(e) There is enough evidence at the 5% level of significance to conclude that procedure preference is dependent on bank employee.

27. (a) The claim is "Type of car is not related to gender."
H_0: Type of car is independent of gender (claim)
H_a: Type of car is dependent on gender.

(b) $\text{d.f.} = (r-1)(c-1) = 3$
$\chi_0^2 = 6.251$; Rejection region: $\chi^2 > 6.251$

(c)

O	E	$O-E$	$(O-E)^2$	$\frac{(O-E)^2}{E}$
28	28.60	–0.60	0.36	0.0126
39	39.05	–0.05	0.0025	0.0001
21	22.55	–1.55	2.4025	0.1065
22	19.80	2.2	4.84	0.2444
24	23.40	0.6	0.36	0.0154
32	31.95	0.05	0.0025	0.0001
20	18.45	1.55	2.4025	0.1302
14	16.20	–2.2	4.84	0.2988
				0.8081

$\chi^2 \approx 0.8081$

(d) Because $\chi^2 < 6.251$, fail to reject H_0.

(e) There is not enough evidence at the 10% level of significance to conclude that type of car is dependent on gender.

29. The claim is "the proportions of motor vehicle crash deaths involving males and females are the same for each group."
H_0: The proportions are equal. (claim)
H_a: At least one of proportion is different from the others.
$\text{d.f.} = (r-1)(c-1) = 7$
$\chi_0^2 = 14.067$; Rejection region: $\chi^2 > 14.067$

O	E	$O-E$	$(O-E)^2$	$\dfrac{(O-E)^2}{E}$
96	96.6214	−0.6214	0.3861	0.0040
98	93.7586	4.2414	17.9895	0.1919
72	69.4243	2.5757	6.6342	0.0956
80	78.0129	1.9871	3.9486	0.0506
74	71.5714	2.4286	5.8981	0.0824
44	46.5214	−2.5214	6.3575	0.1367
25	29.3443	−4.3443	18.8729	0.6432
12	15.7457	−3.7457	14.0303	0.8911
39	38.3786	0.6214	0.3861	0.0101
33	37.2414	−4.2414	17.9895	0.4831
25	27.5757	−2.5757	6.6342	0.2406
29	30.9871	−1.9871	3.9486	0.1274
26	28.4286	−2.4286	5.898	0.2075
21	18.4786	2.5214	6.3575	0.3440
16	11.6557	4.3443	18.8729	1.6192
10	6.2543	3.7457	14.0303	2.2433
				7.3705

$\chi^2 \approx 7.3705$

Because $\chi^2 < 14.067$, fail to reject H_0. There is not enough evidence at the 5% level of significance to reject the claim that the proportions of motor vehicle crash deaths involving males and females are the same for each group.

31. Right-tailed

33.

	Educational Attainment			
Status	**Not a High school Graduate**	**High School Graduate**	**Some College, No Degree**	**Associate's, Bachelor's, or Advanced Degree**
Employed	0.046	0.156	0.101	0.312
Unemployed	0.004	0.010	0.005	0.009
Not in the labor force	0.059	0.123	0.061	0.114

35. (a) 0.9%

(b) 6.1%

37.

	Educational Attainment			
Status	**Not a High School Graduate**	**High School Graduate**	**Some College, No Degree**	**Associate's, Bachelor's, or Advanced Degree**
Employed	0.076	0.253	0.164	0.507
Unemployed	0.150	0.350	0.183	0.317
Not in the labor force	0.164	0.344	0.172	0.320

39. 17.2%

41. 26.3%

10.3 COMPARING TWO VARIANCES

10.3 TRY IT YOURSELF SOLUTIONS

1. $\alpha = 0.05$ $\quad F_0 = 2.45$

2. $\alpha = 0.01$ $\quad F_0 = 18.31$

3. The claim is "the variance of the time required for nutrients to enter the bloodstream is less with the specially treated intravenous solution than the variance of the time without the solution."

$H_0: \sigma_1^2 \le \sigma_2^2$; $H_a: \sigma_1^2 > \sigma_2^2$ (claim)

$\alpha = 0.01$

$\text{d.f.}_\text{N} = n_1 - 1 = 24$

$\text{d.f.}_\text{D} = n_2 - 1 = 19$

$F_0 = 2.92$; Rejection region: $F > 2.92$

$$F = \frac{s_1^2}{s_2^2} = \frac{180}{56} \approx 3.21$$

Because $F > 2.92$, reject H_0.

There is enough evidence at the 1% level of significance to support the researcher's claim that a specially treated intravenous solution decreases the variance of the time required for nutrients to enter the bloodstream.

4. The claim is "the pH levels of the soil in two geographic regions have equal standard deviations."

$H_0: \sigma_1 = \sigma_2$ (claim); $H_a: \sigma_1 \ne \sigma_2$

$\alpha = .01$

$\text{d.f.}_\text{N} = n_1 - 1 = 15$

$\text{d.f.}_\text{D} = n_2 - 1 = 21$

$F_0 = 3.43$; Rejection region: $F > 3.43$

$$F = \frac{s_1^2}{s_2^2} = \frac{(0.95)^2}{(0.78)^2} \approx 1.48$$

Because $F < 3.43$. fail to reject H_0.

There is not enough evidence at the 1% level of significance to reject the biologist's claim that the pH levels of the soil in the two geographic regions have equal standard deviations.

10.3 EXERCISE SOLUTIONS

1. Specify the level of significance α. Determine the degrees of freedom for the numerator and denominator. Use Table 7 in Appendix B to find the critical value F.

3. (1) The samples must be randomly selected, (2) the samples must be independent, and (3) each population must have a normal distribution.

5. $F_0 = 2.54$ **7.** $F_0 = 2.06$ **9.** $F_0 = 9.16$ **11.** $F_0 = 1.80$

13. $H_0: \sigma_1^2 \le \sigma_2^2$; $H_a: \sigma_1^2 > \sigma_2^2$ (claim)

d.f.$_N = 4$, d.f.$_D = 5$

$F_0 = 3.52$; Rejection region: $F > 3.52$

$$F = \frac{s_1^2}{s_2^2} = \frac{773}{765} \approx 1.010$$

Because $F < 3.52$, fail to reject H_0. There is not enough evidence at the 10% level of significance to support the claim.

15. $H_0: \sigma_1^2 \le \sigma_2^2$ (claim); $H_a: \sigma_1^2 > \sigma_2^2$

d.f.$_N = 10$, d.f.$_D = 9$

$F_0 = 5.26$; Rejection region: $F > 5.26$

$$F = \frac{s_1^2}{s_2^2} = \frac{842}{836} \approx 1.007$$

Because $F < 5.26$, fail to reject H_0. There is not enough evidence at the 1% level of significance to reject the claim.

17. $H_0: \sigma_1^2 = \sigma_2^2$ (claim); $H_a: \sigma_1^2 \ne \sigma_2^2$

d.f.$_N = 12$, d.f.$_D = 19$

$F_0 = 3.76$; Rejection region: $F > 3.76$

$$F = \frac{s_1^2}{s_2^2} = \frac{9.8}{2.5} \approx 3.920$$

Because $F > 3.76$, reject H_0. There is enough evidence at the 1% level of significance to reject the claim.

19. Population 1: Company B; Population 2: Company A

(a) The claim is "the variance of the life of Company A's appliances is less than the variance of the life of Company B's appliances."
$H_0: \sigma_1^2 \le \sigma_2^2;\ H_a: \sigma_1^2 > \sigma_2^2$ (claim)

(b) $\text{d.f.}_N = 24$, $\text{d.f.}_D = 19$
$F_0 = 2.11$; Rejection region: $F > 2.11$

(c) $F = \dfrac{s_1^2}{s_2^2} = \dfrac{3.9}{1.8} \approx 2.167$

(d) Because $F > 2.11$, reject H_0.

(e) There is enough evidence at the 5% level of significance to support Company A's claim that the variance of life of its appliances is less than the variance of life of Company B appliances.

21. Population 1: Age group 35-49; Population 2: Age group 18-34

(a) The claim is "the variances of the waiting times differ between the two age groups."
$H_0: \sigma_1^2 = \sigma_2^2;\ H_a: \sigma_1^2 \ne \sigma_2^2$ (claim)

(b) $\text{d.f.}_N = 5$, $\text{d.f.}_D = 8$
$F_0 = 4.82$ Rejection region: $F > 4.82$

(c) $F = \dfrac{s_1^2}{s_2^2} = \dfrac{458.7}{177.7} \approx 2.58$

(d) Because $F < 4.82$, fail to reject H_0.

(e.) There is not enough evidence at the 5% level of significance to conclude that the variance of the waiting times differ between the two age groups.

23. Population 1: District 1; Population 2: District 2

(a) The claim is "the standard deviations of science achievement test scores for eighth grade students are the same in Districts 1 and 2."
$H_0: \sigma_1^2 = \sigma_2^2$(claim); $H_a: \sigma_1^2 \ne \sigma_2^2$

(b) $\text{d.f.}_N = 11$, $\text{d.f.}_D = 13$
$F_0 = 2.635$; Rejection region: $F > 2.635$

(c) $F = \dfrac{s_1^2}{s_2^2} = \dfrac{(36.8)^2}{(32.5)^2} \approx 1.282$

(d) Because $F < 2.635$, fail to reject H_0.

(e) There is not enough evidence at the 10% level of significance to reject the administrator's claim that the standard deviations of science assessment test scores for eighth grade students are the same in Districts 1 and 2.

25. Population 1: New York; Population 2: California

(a) The claim is "the standard deviation of the annual salaries for actuaries is less in California than in New York."
$H_0: \sigma_1^2 \le \sigma_2^2$; $H_a: \sigma_1^2 > \sigma_2^2$ (claim)

(b) $\text{d.f.}_N = 40$, $\text{d.f.}_D = 60$
$F_0 = 1.59$ Rejection region: $F > 1.59$

(c) $F = \dfrac{s_1^2}{s_2^2} = \dfrac{(37{,}100)^2}{(32{,}400)^2} \approx 1.31$

(d) Because $F < 1.59$, fail to reject H_0.

(e.) There is not enough evidence at the 5% level of significance to conclude that the standard deviation of the annual salaries for actuaries is less in California than in New York.

27. Right-tailed: $F_R = 14.73$
Left-tailed:
$\text{d.f.}_N = 3$ and $\text{d.f.}_D = 6$
$F = 6.60$
Critical value is $\dfrac{1}{F} = \dfrac{1}{6.60} \approx 0.15$.

29. $\dfrac{s_1^2}{s_2^2}\dfrac{1}{F_R} < \dfrac{\sigma_1^2}{\sigma_2^2} < \dfrac{s_1^2}{s_2^2}\dfrac{1}{F_L} \rightarrow \left(\dfrac{10.89}{9.61}\right)\left(\dfrac{1}{3.33}\right) < \dfrac{\sigma_1^2}{\sigma_2^2} < \left(\dfrac{10.89}{9.61}\right)(3.02) \rightarrow 0.340 < \dfrac{\sigma_1^2}{\sigma_2^2} < 3.422$

10.4 ANALYSIS OF VARIANCE

10.4 TRY IT YOURSELF SOLUTIONS

1. The claim is "there is a difference in the mean a monthly sales among the sales regions."
$H_0: \mu_1 = \mu_2 = \mu_3 = \mu_4$
H_a: At least one mean is different from the others. (claim)
$\alpha = 0.05$
$\text{d.f.}_N = 3$; $\text{d.f.}_D = 14$

$F_0 = 3.34$; Rejection region: $F > 3.34$

Variation	Sum of Squares	Degrees of Freedom	Mean Squares	F
Between	549.8	3	183.3	4.22
Within	608.0	14	43.4	

$F \approx 4.22$

Because $F > 3.34$, reject H_0.

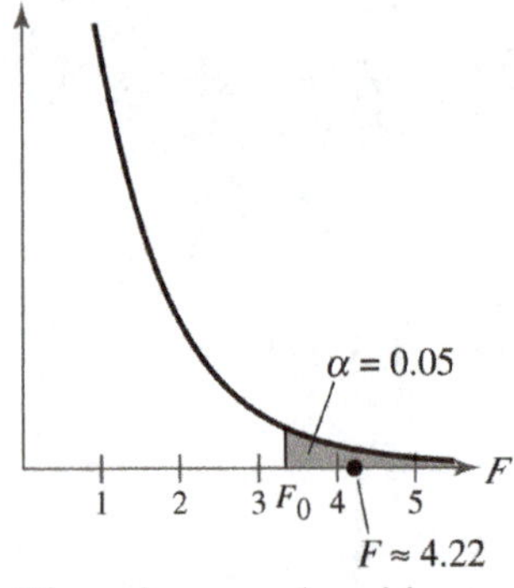

There is enough evidence at the 5% level of significance to conclude that there is a difference in the mean monthly sales among the sales regions.

2. The claim is "there is a difference in the means of the GPAs."
   $H_0: \mu_1 = \mu_2 = \mu_3 = \mu_4$
   H_a: At least one mean is different from the others. (claim)
   Enter the data.

Variation	Sum of Squares	Degrees of Freedom	Mean Squares	F
Between	0.584	3	0.195	1.34
Within	4.360	30	0.145	

$F = 1.34 \rightarrow P\text{-value} = 0.280$

$0.280 > 0.05$

Because P-value > 0.05, fail to reject H_0. There is not enough evidence to conclude that there is a difference in the means of the GPAs.

10.4 EXERCISE SOLUTIONS

1. $H_0: \mu_1 = \mu_2 = \ldots = \mu_k$
   H_a: At least one mean is different from the others.

3. The MS_B measures the differences related to the treatment given to each sample.
   The MS_W measures the differences related to entries within the same sample.

5. (a) The claim is that "the mean costs per ounce are different."
$H_0: \mu_1 = \mu_2 = \mu_3$
H_a: At least one mean is different from the others. (claim)

(b) $\text{d.f.}_N = k - 1 = 2$; $\text{d.f.}_D = N - k = 12$
$F_0 = 3.89$; Rejection region: $F > 3.89$

(c)

Variation	Sum of Squares	Degrees of Freedom	Mean Squares	*F*
Between	0.07406	2	0.037032	4.80
Within	0.09267	12	0.007723	

$F \approx 4.80$

(d) Because $F > 3.89$, reject H_0.

(e) There is enough evidence at the 5% level of significance to conclude that at least one mean cost per ounce is different from the others.

7. (a) The claim is that "at least one mean vacuum cleaner weight is different from the others."
$H_0: \mu_1 = \mu_2 = \mu_3$
H_a: At least one mean is different from the others. (claim)

(b) $\text{d.f.}_N = k - 1 = 2$; $\text{d.f.}_D = N - k = 15$
$F_0 = 6.36$; Rejection region: $F > 6.36$

(c)

Variation	Sum of Squares	Degrees of Freedom	Mean Squares	*F*
Between	122.1111	2	61.0556	16.11
Within	56.8333	15	3.7889	

$F \approx 16.11$

(d) Because $F > 6.36$, reject H_0.

(e) There is enough evidence at the 1% level of significance to conclude that at least one mean vacuum cleaner weight is different from the others

9. (a) The claim is "at least one mean age is different from the others."
$H_0: \mu_1 = \mu_2 = \mu_3 = \mu_4$
H_a: At least one mean is different from the others. (claim)

(b) $\text{d.f.}_N = k - 1 = 3$; $\text{d.f.}_D = N - k = 48$
$F_0 = 2.84$; Rejection region: $F > 2.84$

(c)

Variation	Sum of Squares	Degrees of Freedom	Mean Squares	*F*
Between	10.07692	3	3.358974	0.62
Within	262.1538	48	5.461538	

$F \approx 0.62$

(d) Because $F < 2.84$, fail to reject H_0.

(e) There is not enough evidence at the 5% level of significance to conclude that at least one mean age is different from the others.

11. (a) The claim is that "the mean scores are the same for all regions."
$H_0: \mu_1 = \mu_2 = \mu_3 = \mu_4$ (claim)
H_a: At least one mean is different from the others.

(b) $\text{d.f.}_N = k - 1 = 3$; $\text{d.f.}_D = N - k = 30$
$F_0 = 2.28$; Rejection region: $F > 2.28$

(c)

Variation	Sum of Squares	Degrees of Freedom	Mean Squares	*F*
Between	7.777	3	2.5923	3.67
Within	21.2124	30	0.7071	

$F \approx 3.67$

(d) Because $F > 2.28$, reject H_0.

(e) There is enough evidence at the 10% level of significance to reject the claim that the mean scores are the same for all regions.

13. (a)The claim is that "the mean salary is different in at least one of the areas."
$H_0: \mu_1 = \mu_2 = \mu_3 = \mu_4 = \mu_5 = \mu_6$
H_a: At least one mean is different from the others. (claim)

(b) $\text{d.f.}_N = k - 1 = 5$; $\text{d.f.}_D = N - k = 30$
$F_0 = 2.53$; Rejection region: $F > 2.53$

(c)

Variation	Sum of Squares	Degrees of Freedom	Mean Squares	*F*
Between	335,665,495	5	67,133,099	2.06
Within	975,795,651	30	32,526,522	

$F \approx 2.06$

(d) Because $F < 2.53$, fail to reject H_0.

(e) There is not enough evidence at the 5% level of significance to conclude that the mean salary is different in at least one of the areas.

15. H_0: Advertising medium has to effect on mean ratings.
H_a: Advertising medium has an effect on mean ratings.
H_0: Length of ad has no effect on mean ratings.

H_a : Length of ad has an effect on mean ratings.

H_0 : There is no interaction effect between advertising medium and length of ad on mean ratings.

H_a : There is an interaction effect between advertising medium and length of ad on mean ratings.

Source	d.f.	SS	MS	F	P
Ad medium	1	1.25	1.25	0.57	0.459
Length of ad	1	0.45	0.45	0.21	0.655
Interaction	1	0.45	0.45	0.21	0.655
Error	16	34.80	2.175		
Total	19	36.95			

Fail to reject all null hypotheses. The interaction between the advertising medium and length of the ad has no effect on the rating and therefore there is no significant difference in the means of the ratings.

17. H_0 : Age has no effect on mean GPA.

H_a : Age has an effect on mean GPA.

H_0 : Gender has no effect on mean GPA.

H_a :Gender has an effect on mean GPA.

H_0 : There is no interaction effect between age and gender on mean GPA.

H_a : There is no interaction effect between age and gender on mean GPA.

Source	d.f.	SS	MS	F	P
Age	3	0.4146	0.138	0.12	0.948
Gender	1	0.1838	0.184	0.16	0.697
Interaction	3	0.2912	0.097	0.08	0.968
Error	16	18.6600	1.166		
Total	23	19.5496			

Fail to reject all null hypotheses. The interaction between age and gender has no effect GPA and therefore there is no significant difference in the means of the GPAs.

19.

	Mean	Size
Pop 1	0.455	6
Pop2	0.606	5
Pop3	0.460	4

$SS_W = 0.09267$

$\sum(n_i - 1) = N - k = 12$

$F_0 = 3.89 \rightarrow \text{CV}_{\text{Scheffe}'} = 3.89(3-1) = 7.78$

$$\frac{(\bar{x}_1 - \bar{x}_2)^2}{\frac{SS_W}{\sum(n_i - 1)}\left[\frac{1}{n_1} + \frac{1}{n_2}\right]} \approx 8.05 \rightarrow \text{ Significant difference}$$

$$\frac{(\bar{x}_1 - \bar{x}_3)^2}{\frac{SS_W}{\sum(n_i - 1)}\left[\frac{1}{n_1} + \frac{1}{n_3}\right]} \approx 0.01 \rightarrow \text{No difference}$$

$$\frac{(\bar{x}_2 - \bar{x}_3)^2}{\frac{SS_W}{\sum(n_i - 1)}\left[\frac{1}{n_2} + \frac{1}{n_3}\right]} \approx 6.13 \rightarrow \text{No difference}$$

21. $CV_{Scheffe} = 10.98$

$(1,2) \rightarrow 34.18$ Significant difference between the means of groups 1 and 2 because the calculated statistic is greater than the critical value. (Test statistic calculated as $\frac{(77.86 - 55.82)^2}{\frac{1918.441}{27}\left(\frac{2}{10}\right)}$)

$(1,3) \rightarrow 64.14$ Significant difference between the means of groups 1 and 3 because the calculated statistic is greater than the critical value. (Test statistic calculated as $\frac{(77.86 - 47.67)^2}{\frac{1918.441}{27}\left(\frac{2}{10}\right)}$)

$(2,3) \rightarrow 4.67$ No significant difference between the means of groups 2 and 3 because the calculated statistic is less than the critical value. (Test statistic calculated as $\frac{(55.82 - 47.67)^2}{\frac{1918.441}{27}\left(\frac{2}{10}\right)}$)

CHAPTER 10 REVIEW EXERCISE SOLUTIONS

1. (a)

Response	Distribution
Less than 9	4%
10-12	24%
13-16	26%
17-20	22%
21-24	6%
25 or more	18%

H_0: The distribution of the lengths of office visits is 4% less than 9 minutes, 24% 10-12 minutes, 26% 13-16 minutes, 22% 17-20 minutes, 6% 21-24 minutes, and 18% 25 or more minutes.

H_a: The distribution of the lengths differ from the claimed or expected distribution. (claim)

(b) $\chi_0^2 = 15.086$; Rejection region: $\chi^2 > 15.086$

(c)

Response	Distribution	Observed	Expected	$\frac{(O-E)^2}{E}$
Less than 9	4%	20	16	1
10-12	24%	80	96	2.667
13-16	26%	113	104	0.779
17-20	22%	91	88	0.102
21-24	6%	40	24	10.667
25 or more	18%	56	72	3.556
				18.771

$\chi^2 \approx 18.771$

(d) Because $\chi^2 > 15.086$, reject H_0.

(e) There is not enough evidence at the 1% level of significance to conclude that the distribution of the lengths differs from the expected distribution.

3. (a)

Response	Distribution
Short-game shots	65%
Approach and swing	22%
Driver shots	9%
Putting	4%

H_0: The distribution of responses from golf students about what they need the most help with is 22% approach and swing, 9% driver shots, 4% putting and 65% short-game shots. (claim)
H_a: The distribution of responses differs from the claimed or expected distribution.

(b) $\chi_0^2 = 7.815$; Rejection region: $\chi^2 > 7.815$

(c)

Response	Distribution	Observed	Expected	$\frac{(O-E)^2}{E}$
Short-game shots	65%	276	282.75	0.161
Approach and swing	22%	99	95.70	0.114
Driver shots	9%	42	39.15	0.207
Putting	4%	18	17.40	0.021
				0.503

$\chi^2 \approx 0.503$

(d) Because $\chi^2 < 7.815$, fail to reject H_0.

(e) There is not enough evidence at the 5% level of significance to conclude that the distribution of golf students' responses is the same as the claimed or expected distribution.

5. (a) Expected frequencies:

	Years of full-time teaching experience				
Gender	Less than 3 years	3–9 years	10–20 years	20 years or more	Total
Male	95.4	349.2	383.4	222	1050
Female	222.6	814.8	894.6	518	2450
Total	318	1164	1278	740	3500

(b) The claim is "gender is related to the years of full-time teaching experience."
H_0: Years of full-time teaching experience is independent of gender.
H_a: Years of full-time teaching experience is dependent on gender. (claim)

(c) d.f. $=(r-1)(c-1)=3$; $\chi_0^2=11.345$; Rejection region: $\chi^2>11.345$

(d)

O	E	$O-E$	$(O-E)^2$	$\frac{(O-E)^2}{E}$
102	95.4	6.6	43.56	0.4566
339	349.2	−10.2	104.04	0.2979
402	383.4	18.6	345.96	0.9023
207	222.0	−15	225	1.0135
216	222.6	−6.6	43.56	0.1957
825	814.8	10.2	104.04	0.1277
876	894.6	−18.6	345.96	0.3867
533	518.0	15	225	0.4344
				3.8148

$\chi^2\approx 3.815$

(e) Because $\chi^2<11.345$, fail to reject H_0.

(f) There is not enough evidence at the 1% level of significance to conclude that the years of full-time teaching experience is dependent on gender.

7. (a) Expected frequencies:

	Vertebrate Group					
Status	Mammals	Birds	Reptiles	Amphibians	Fish	Total
Endangered	136.8	121	46.4	23.6	65.2	393
Threatened	37.2	33	12.6	6.4	17.8	107
Total	174	154	59	30	83	500

(b) The claim is "a species' status is not related to vertebrate group."
H_0: A species' status is independent of vertebrate group. (claim)
H_a: A species' status is dependent on vertebrate group.

(c) d.f. $=(r-1)(c-1)=4$; $\chi_0^2=9.488$; Rejection region: $\chi^2>9.488$

(d)

O	E	$O-E$	$(O-E)^2$	$\frac{(O-E)^2}{E}$
151	136.8	14.2	201.64	1.4740
137	121.0	16	256	2.1157
37	46.4	–9.4	88.36	1.9043
18	23.6	–5.6	31.36	1.3288
50	65.2	–15.2	231.04	3.5436
23	37.2	–14.2	201.64	5.4204
17	33.0	–16	256	7.7576
22	12.6	9.4	88.36	7.0127
12	6.4	5.6	31.36	4.9
33	17.8	15.2	231.04	12.9798
				48.4369

$\chi^2 \approx 48.437$

(e) Because $\chi^2 > 9.488$, reject H_0.

(f) There is enough evidence at the 5% level of significance to reject the claim that a species status is independent of vertebrate group.

9. $F_0 \approx 2.295$ **11.** $F_0 = 2.39$ **13.** $F_0 = 2.06$ **15.** $F_0 = 2.08$

17. (a) The claim is "the standard deviations of hotel room rates for San Francisco, CA and Sacramento, CA are the same."
$H_0: \sigma_1^2 = \sigma_2^2 \text{(claim)};\ H_a: \sigma_1^2 \neq \sigma_2^2$

(b) $\text{d.f.}_N = 35$; $\text{d.f.}_D = 30$
$F_0 = 2.575$; Rejection region: $F > 2.575$

(c) $F = \frac{s_1^2}{s_2^2} = \frac{(75)^2}{(44)^2} \approx 2.905$

(d) Because $F > 2.575$, reject H_0.

(e) There is enough evidence at the 1% level of significance to reject the claim that the standard deviations of hotel room rates for San Francisco, CA and Sacramento, CA are the same.

19. (a) The claim is "the variance of SAT critical reading scores for females is different than the variance of SAT critical reading scores for males"
$H_0: \sigma_1^2 = \sigma_2^2;\ H_a: \sigma_1^2 \neq \sigma_2^2$ (claim)

(b) $\text{d.f.}_N = 11$; $\text{d.f.}_D = 11$
$F_0 = 5.32$ Rejection region: $F > 5.32$

(c) $F = \frac{s_1^2}{s_2^2} = \frac{20790.15}{15169.7} \approx 1.37$

(d) Because $F < 5.32$, fail to reject H_0.

(e) There is not enough evidence at the 1% level of significance to support the claim that the test score variance for females is different than the test score variance for males.

21. (a) The claim is "the mean amount spent on energy in one year is different in at least one of the regions."
$H_0: \mu_1 = \mu_2 = \mu_3 = \mu_4$
H_a: At least one mean is different from the others. (claim)

(b) $\text{d.f.}_N = k - 1 = 3$; $\text{d.f.}_D = N - k = 28$
$F_0 = 2.29$; Rejection region: $F > 2.29$

(c)

Variation	Sum of Squares	Degrees of Freedom	Mean Squares	F
Between	2,207,334.3	3	735,778.08	6.19
Within	3,329,531.8	28	118,911.85	

$F \approx 6.19$

(d) Because $F > 2.29$, reject H_0.

(e) There is enough evidence at the 10% level of significance to conclude that at least one mean amount spent on energy is different from the others.

CHAPTER 10 QUIZ SOLUTIONS

1. (a) H_0: The distribution of educational attainment for people in the United States ages 30–34 is 4.7% none-8th grade, 6.9% 9th-11th grade, 29.5% high school graduates, 16.6% some college, no degree, 9.8% associate's degree, 20.5% bachelor's degree, 8.7% master's degree, and 3.3% professional/doctoral degree.
H_a: The distribution of educational attainment for people in the United States ages 30–34 differs from the distribution for people ages 25 and older. (claim)

(b) $\chi_0^2 = 14.067$; Rejection region: $\chi^2 > 14.067$

(c)

	Distribution	Observed	Expected	$\frac{(O-E)^2}{E}$
None-8th grade	4.7%	10	15.04	1.6889
9th-11th grade	6.9%	23	22.08	0.0383
High school graduate	29.5%	80	94.4	2.1966
Some college, no degree	16.6%	56	53.12	0.1561
Associate's degree	9.8%	34	31.36	0.2222
Bachelor's degree	20.5%	75	65.6	1.3470
Master's degree	8.7%	31	27.84	0.3587
Professional/Doctoral degree	3.3%	11	10.56	0.0183
		320		6.0261

$\chi^2 \approx 6.026$

(d) Because $\chi^2 < 14.067$, fail to reject H_0.

(e) There is not enough evidence at the 5% level of significance to conclude that the distribution for people in the United States ages 30–34 differs from the distribution for people ages 25 and older.

2. (a) H_0: Age and educational attainment are independent.

H_a: Age and educational attainment are dependent. (claim)

(b) $\text{d.f.} = (r-1)(c-1) = 7$; $\chi_0^2 = 18.475$; Rejection region: $\chi^2 > 18.475$

(c)

O	E	$O-E$	$(O-E)^2$	$\frac{(O-E)^2}{E}$
10	13.89	−3.89	15.1321	1.0894
23	20.63158	2.36842	5.609413	.271885
80	90.95	−10.95	119.9025	1.3183
56	56	0	0	0
34	31.58	2.42	5.8564	0.1854
75	64.42	10.58	111.9364	1.7376
31	30.32	0.68	0.4624	0.0153
11	12.21	−1.21	1.4641	0.1200
23	19.11	3.89	15.1321	0.7918
26	28.37	−2.37	5.6169	0.1980
136	125.05	10.95	119.9025	0.9588
77	77	0	0	0
41	43.42	−2.42	5.8564	0.1349
78	88.58	−10.58	111.9364	1.2637
41	41.68	−0.68	0.4624	0.0111
18	16.79	1.21	1.4641	0.0872
				8.183

$\chi^2 \approx 8.183$

(d) Because $\chi^2 < 18.475$, fail to reject H_0.

(e) There is not enough evidence at the 1% level of significance to conclude that educational attainment is dependent on age.

3. Population 1: Ithaca $\rightarrow s_1^2 = 72.59$

Population 2: Little Rock $\rightarrow s_2^2 = 52.48$

(a) $H_0: \sigma_1^2 = \sigma_2^2$; $H_a: \sigma_1^2 \neq \sigma_2^2$ (claim)

(b) $\text{d.f.}_N = 12$, $\text{d.f.}_D = 14$;
$F_0 = 4.43$; Rejection region: $F > 4.43$

(c) $F = \dfrac{s_1^2}{s_2^2} = \dfrac{72.59}{52.48} \approx 1.38$

(d) Because $F < 4.43$, fail to reject H_0.

(e) There is not enough evidence at the 1% level of significance to conclude the variances in annual wages for Ithaca, NY and Little Rock, AR are different.

4. (a) $H_0: \mu_1 = \mu_2 = \mu_3$ (claim)
H_a: At least one mean is different from the others.

(b) $\text{d.f.}_N = 2$, $\text{d.f.}_D = 40$
$F_0 = 2.44$; Rejection region: $F > 2.44$

(c)

Variation	Sum of Squares	Degrees of Freedom	Mean Squares	*F*
Between	699.2111	2	349.6055	6.18
Within	2264.2159	40	56.6054	

$F \approx 6.18$

(d) Because $F > 2.44$, reject H_o.

(e) There is enough evidence at the 10% level of significance to reject the claim that the mean annual wages are equal for all three cities.

CUMULATIVE REVIEW, CHAPTERS 9-10

1. (a) $r \approx 0.827$; strong positive linear correlation

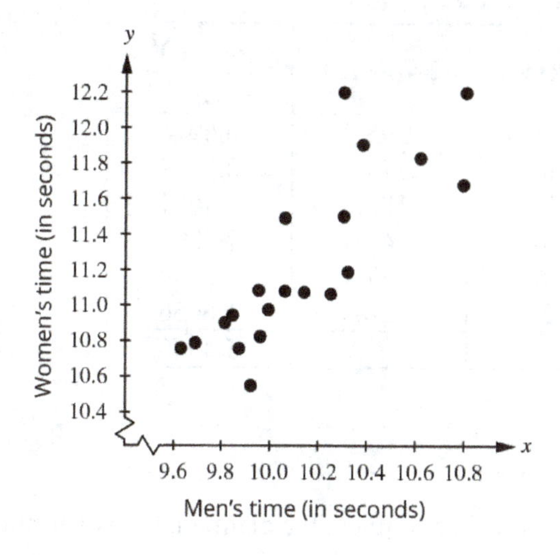

(b) There is enough evidence at the 5% level of significance to conclude that there is a significant linear correlation between the men's and women's winning 100-meter times.

(c) $\hat{y} = 1.216x - 1.088$

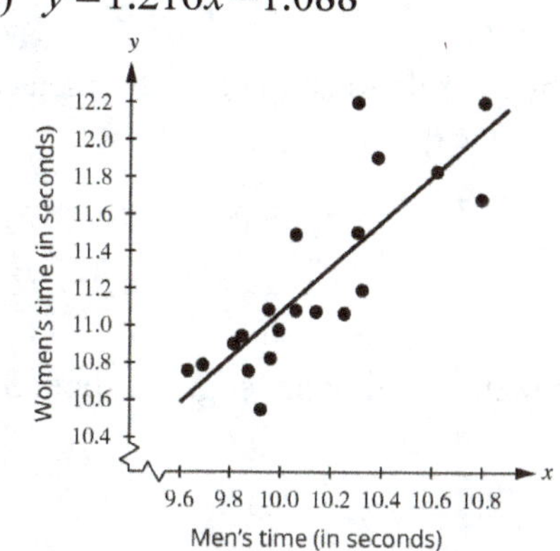

(d) 10.95 seconds

3. (a) $\hat{y} = 16,212 - 0.227(110,000) + 0.221(100,000)$
$= 12,442$

(b) $\hat{y} = 16,212 - 0.227(125,000) + 0.212(115,000)$
$= 12,217$

5. H_0: The distribution of credit card debts for college students are distributed as 32.8% \$0, 42.2% \$1-\$500, 11.4% \$501-\$1000, 5.1% \$1001-\$2000, 5.3% \$2001-\$4000, and 3.2% More than \$4000. (claim)

H_a: The distribution of credit card debts for college students differs from the claimed distribution.

$\chi_0^2 = 11.071$; Rejection region: $\chi^2 > 11.071$

	Distribution	Observed	Expected	$\frac{(O-E)^2}{E}$
\$0	.328	290	295.2	0.0916
\$1-\$500	.422	397	379.8	0.7789
\$501-\$1000	.114	97	102.6	0.3057
\$1001-\$2000	.051	54	45.9	1.4294
\$2001-\$4000	.053	40	47.7	1.2430
More than \$4000	.032	22	28.8	1.6056
				5.4541

$\chi^2 \approx 5.45$

Because $\chi^2 < 11.071$, fail to reject H_0.

There is not enough evidence at the 5% level of significance to reject the claim that the distributions are the same.

7. (a) $r^2 = \dfrac{\Sigma(\hat{y}_i - \bar{y})^2}{\Sigma(y_i - \bar{y})^2} \approx 0.751$

About 75.1% of the variation in height can be explained by the relationship between metacarpal bone length and height, and about 24.9% of the variation is unexplained.

(b) $s_e = \sqrt{\dfrac{\Sigma(y_i - \hat{y}_i)^2}{n-2}} = \sqrt{\dfrac{149.4}{10}} \approx 3.87$

The standard error of estimate of the height for a specific metacarpal bone length is about 3.876 centimeters.

(c) $n = 12$ d.f. $= 10$, $t_c = 2.228$, $s_e \approx 3.87$

$\hat{y} = 1.707x + 94.380 = 1.707(50) + 94.38 = 179.73$

$$E = t_c s_e \sqrt{1 + \frac{1}{n} + \frac{n(x-\bar{x})^2}{n(\Sigma x^2) - (\Sigma x)^2}}$$

$$\approx (2.228)(3.87)\sqrt{1 + \frac{1}{12} + \frac{12(50 - 44.5833)^2}{12(24007) - (535)^2}}$$

$$\approx 9.7$$

$\hat{y} \pm E \rightarrow 179.73 \pm 9.7 \rightarrow (170.03,\ 189.43)$

You can be 95% confident that the height will be between 170.03 centimeters and 189.43 centimeters when the metacarpal bone length is 50 centimeters.

CHAPTER 11

Nonparametric Tests

11.1 THE SIGN TEST

11.1 TRY IT YOURSELF SOLUTIONS

1. The claim is "the median number of days a home is on the market in its city is greater than 120."

H_0 :median$\leq$ 120; H_a : median > 120 (claim)

$\alpha = 0.025$

$n = 24$

The critical value is 6.

$x = 6$

Because $x \leq 6$, reject H_0.

There is enough evidence at the 2.5% level of significance to support the agency's claim that the median number of days a home is on the market in its city is greater than 120.

2. The claim is "the median age of museum workers in the United States is 46 years old."

H_0 : median = 46 (claim)

H_a : median $\neq$ 46

$\alpha = 0.10$

$n = 91$

The critical value is -1.645.

$x = 34$

$$z = \frac{(x+0.5)-0.5n}{\frac{\sqrt{n}}{2}} = \frac{(34+0.5)-0.5(91)}{\frac{\sqrt{91}}{2}} \approx \frac{-11}{4.77} \approx -2.306$$

Because $z < -1.645$, reject H_0.

There is enough evidence at the 10% level of significance to reject the claim that the median age of museum workers in the United States is 46 years old.

3. The claim is "a new vaccine will decrease the number of colds in adults."

H_0 : The number of colds will not decrease.

H_a : The number of colds will decrease. (claim)

$\alpha = 0.05$

$n = 11$

The critical value is 2.

$x = 2$

Because $x \leq 2$ R, reject H_0.

There is enough evidence at the 5% level of significance to support the researcher's claim that the new vaccine will decrease the number of colds in adults.

11.1 EXERCISE SOLUTIONS

1. A nonparametric test is a hypothesis test that does not require any specific conditions concerning the shapes of populations or the values of any population parameters.

A nonparametric test is usually easier to perform than its corresponding parametric test, but the nonparametric test is usually less efficient.

3. When n is less than or equal to 25, the test statistic is equal to x (the smaller number of + or – signs). When n is greater than 25, the test statistic is equal to $z = \dfrac{(x+0.5)-0.5n}{\dfrac{\sqrt{n}}{2}}$.

5. Verify that the sample is random. Identify the claim and state H_0 and H_a. Identify the level of significance and sample size. Find the critical value using Table 8 (if $n \le 25$) or Table 4 ($n > 25$). Calculate the test statistic. Make a decision and interpret it in the context of the problem.

7. (a) The claim is"the median credit card balance of college students is more than \$300."
H_0: median $\le$ \$300; H_a: median > \$300 (claim)

(b) The critical value is 1.

(c) $x = 5$

(d) Because $x > 1$, fail to reject H_0.

(e) There is not enough evidence at the 1% level of significance to support the accountant's claim that the median credit card balance of college students is more than \$300.

9. (a) The claim is "the median sales price of new privately owned one-family homes sold in a recent month is \$253,000 or less."
H_0: median $\le$ \$253,000 (claim); H_a: median > \$253,000

(b) The critical value is 1

(c) $x = 4$

(d) Because $x > 1$, fail to reject H_0.

(e) There is not enough evidence at the 5% level of significance to reject the agent's claim that the median sales price of new privately owned one-family homes sold in a recent month is \$253,000 or less.

11. (a) The claim is "the median amount of credit card debt for families holding such debt is at least \$2300."
H_0: median $\ge$ \$2300 (claim); H_a: median < \$2300

(b) The critical value is $z_0 = -2.05$

(c) $x = 44$

$$z = \frac{(x+0.5)-0.5n}{\frac{\sqrt{n}}{2}} = \frac{(44+0.5)-0.5(104)}{\frac{\sqrt{104}}{2}} \approx \frac{-7.5}{5.099} \approx -1.47$$

(d) Because $z > -2.05$, fail to reject H_0.

(e) There is not enough evidence at the 2% level of significance to reject the institution's claim that the median amount of credit card debt for families holding such debts is at least $2300.

13. (a) The claim is "the median age of the users of a social media website is greater than 30 years old."
H_0: median ≤ 30; H_a: median > 30 (claim)

(b) The critical value is 4.

(c) $x = 10$

(d) Because $x > 4$, fail to reject H_0.

(e) There is not enough evidence at the 1% level of significance to support the research group's claim that the median age of social media website users is greater than 30 years old.

15. (a) The claim is "the median number of rooms in renter-occupied units is four."
H_0: median $= 4$; (claim) H_a: median $\neq 4$

(b) The critical value is $z_0 = -1.96$.

(c) $x = 29$

$$z = \frac{(x+0.5)-0.5(n)}{\frac{\sqrt{n}}{2}} = \frac{(29+0.5)-0.5(82)}{\frac{\sqrt{82}}{2}} \approx \frac{-11.5}{4.53} \approx -2.54$$

(d) Because $z < -1.96$, reject H_0.

(e) There is enough evidence at the 5% level of significance to reject the organization's claim that the median number of rooms in renter-occupied units is four.

17. (a) The claim is "the median hourly wage of computer systems analysts is $41.93."
H_0: median = \$41.93 (claim); H_a: median ≠ \$41.93

(b) The critical value is –2.575

(c) $x = 18$

$$z = \frac{(x+0.5)-0.5(n)}{\frac{\sqrt{n}}{2}} = \frac{(18+0.5)-0.5(43)}{\frac{\sqrt{43}}{2}} \approx \frac{-3}{3.28} \approx -0.91$$

(d) Because $z > -2.575$, fail to reject H_0.

(e) There is not enough evidence at the 1% level of significance to reject the labor organization's claim that the median hourly wage of computer systems analysts is $41.93.

19. (a) The claim is "the lower back pain intensity scores will decrease after acupuncture treatment."
H_0: The lower back pain intensity scores will not decrease.
H_a: The lower back pain intensity scores will decrease. (claim)

(b) The critical value is 1.

(c) $x = 0$

(d) Because $x \leq 1$, reject H_0.

(e) There is enough evidence at the 5% level of significance to support the physician's claim that lower back pain intensity scores will decrease.

21. (a) The claim is "the student's math SAT scores will improve."
H_0: The SAT scores will not improve.
H_a: The SAT scores will improve. (claim)

(b) The critical value is 1.

(c) $x = 1$

(d) Because $x \leq 1$, reject H_0.

(e) There is enough evidence at the 5% level of significance to support the agency's claim that the SAT scores will improve.

23. (a) The claim is "the proportion of adults who feel older than their real age is equal to the proportion of adults who feel younger than their real age."
H_0: The proportion of adults who feel older than their real age is equal to the proportion of adults who feel younger than their real age. (claim)
H_a: The proportion of adults who feel older than their real age is different from the proportion of adults who feel younger than their real age.
The critical value is 3.
$x = 3$
Because $x \leq 3$, reject H_0.

(b) There is enough evidence at the 5% level of significance to reject the claim that the proportion of adults who feel older than their real age is equal to the proportion of adults who feel younger than their real age.

25. (a) The claim is "the median weekly earnings of female workers is less than or equal to \$765."
H_0 : median $\le \$765$ (claim); H_a : median $> \$765$

(b) The critical value is $z_0 = 2.33$.

(c) $x = 29$

$$z = \frac{(x-0.5)-0.5(n)}{\frac{\sqrt{n}}{2}} = \frac{(29-0.5)-0.5(47)}{\frac{\sqrt{47}}{2}} \approx \frac{5}{3.428} \approx 1.46$$

(d) Because $z < 2.33$, fail to reject H_0

(e) There is not enough evidence at the 1% level of significance to reject the organization's claim that the weekly earnings of female workers is less than or equal to \$765.

27. (a) The claim is "the median age of brides at the time of their first marriage is less than or equal to 27 years old."
H_0: median ≤ 27 (claim); H_a: median > 27

(b) The critical value is $z_0 = 1.645$.

(c) $x = 35$

$$z = \frac{(x-0.5)-0.5(n)}{\frac{\sqrt{n}}{2}} = \frac{(35-0.5)-0.5(59)}{\frac{\sqrt{59}}{2}} \approx \frac{5}{3.841} \approx 1.30$$

(d) Because $z < 1.645$, fail to reject H_0.

(e) There is not enough evidence at the 5% level of significance to reject the counselor's claim that the median age of brides at the time of their first marriage is less than or equal to 27 years old.

11.2 THE WILCOXON TESTS

11.2 TRY IT YOURSELF SOLUTIONS

1. The claim is "a spray-on water repellant is effective."
H_0: The water repellent does not increase the water repelled.
H_a: The water repellent increases the water repelled. (claim)
$\alpha = 0.01$

$n = 11$
The critical value is 7.

No repellent	Repellent applied	Difference	Absolute value	Rank	Signed rank
8	15	−7	7	11	−11
7	12	−5	5	9	−9
7	11	−4	4	7.5	−7.5
4	6	−2	2	3.5	−3.5
6	6	0	0	—	—
10	8	2	2	3.5	3.5
9	8	1	1	1.5	1.5
5	6	−1	1	1.5	−1.5
9	12	−3	3	5.5	−5.5
11	8	3	3	5.5	5.5
8	14	−6	6	10	−10
4	8	−4	4	7.5	−7.5

Sum of negative ranks = −55.5
Sum of positive ranks = 10.5
$w_s = 10.5$
Because $w_s > 7$, fail to reject H_0.
There is not enough evidence at the 1% level of significance to support the claim that the spray-on water repellent is effective.

2. The claim is "there is a difference in the claims paid by paid by the companies."
H_0: There is no difference in the claims paid by paid by the companies.
H_a: There is a difference in the claims paid by paid by the companies. (claim)
$\alpha = 0.05$
The critical values are $z_0 = \pm 1.96$.
$n_1 = 12$ and $n_2 = 12$

Ordered data	Sample	Rank
1.7	B	1
1.8	B	2
2.2	B	3
2.5	A	4
3.0	A	5.5
3.0	B	5.5
3.4	B	7
3.9	A	8
4.1	B	9
4.4	B	10
4.5	A	11
4.7	B	12

Ordered data	Sample	Rank
5.3	B	13
5.6	B	14
5.8	A	15
6.0	A	16
6.2	A	17
6.3	A	18
6.5	A	19
7.3	B	20
7.4	A	21
9.9	A	22
10.6	A	23
10.8	B	24

R = sum ranks of company B = 120.5 (*or* $R = 179.5$)

$$\mu_R = \frac{n_1(n_1 + n_2 + 1)}{2} = \frac{12(12 + 12 + 1)}{2} = 150$$

$$\sigma_R = \sqrt{\frac{n_1 n_2 (n_1 + n_2 + 1)}{12}} = \sqrt{\frac{(12)(12)(12+12+1)}{12}} \approx 17.321$$

$$z = \frac{R - \mu_R}{\sigma_R} \approx \frac{120.5 - 150}{17.321} \approx -1.703 \ (or\ z = 1.703)$$

$\frac{1}{2}\alpha = 0.025$ $\frac{1}{2}\alpha = 0.025$

-3 $-z_0$ -1 0 1 z_0 3 z

$z \approx -1.703$

Because $-1.96 < z < 1.96$, fail to reject H_0.

There is not enough evidence at the 5% level of significance to conclude that there is a difference in the claims paid by the companies.

11.2 EXERCISE SOLUTIONS

1. When the samples are dependent, use the Wilcoxon signed-rank test. When the samples are independent, use the Wilcoxon rank sum test.

3. (a) The claim is "there was no reduction in diastolic blood pressure."

H_0: There is no reduction in diastolic blood pressure. (claim)

H_a: There is a reduction in diastolic blood pressure.

(b) Wilcoxon signed-rank test

(c) The critical value is 10.

(d)

Before treatment	After treatment	Difference	Absolute difference	Rank	Signed rank
108	99	9	9	8	8
109	115	–6	6	4.5	–4.5
120	105	15	15	12	12
129	116	13	13	10.5	10.5
112	115	–3	3	2	–2
111	117	–6	6	4.5	–4.5
117	108	9	9	8	8
135	122	13	13	10.5	10.5
124	120	4	4	3	3
118	126	–8	8	6	–6
130	128	2	2	1	1
115	106	9	9	8	8

Sum of negative ranks = –17

Sum of positive ranks = 61

$w_s = 17$

(e) Because $w_s > 10$, fail to reject H_0.

(f) There is not enough evidence at the 1% level of significance to reject the claim that there was no reduction in diastolic blood pressure.

5. (a) The claim is "there is a difference in the earnings of people with bachelor's degrees and those with advanced degrees."
H_0: There is no difference in the earnings.
H_a: There is a difference in the earnings. (claim)

(b) Wilcoxon rank sum test

(c) The critical values are $z_0 = \pm 1.96$.

(d)

Ordered data	sample	Rank
52	B	1
58	B	2.5
58	B	2.5
60	B	4
62	B	5.5
62	B	5.5
64	B	7
68	B	8
71	B	9
78	B	10
84	B	11

Ordered data	sample	Rank
85	A	12
87	A	13
88	A	14
90	A	15
91	A	16.5
91	A	16.5
95	A	18
98	A	19.5
98	A	19.5
99	A	21

R = sum ranks of advanced degree = 165

$$\mu_R = \frac{n_1(n_1+n_2+1)}{2} = \frac{10(10+11+1)}{2} = 110$$

$$\sigma_R = \sqrt{\frac{n_1 n_2(n_1+n_2+1)}{12}} = \sqrt{\frac{(10)(11)(10+11+1)}{12}} \approx 14.201$$

$$z = \frac{R-\mu_R}{\sigma_R} \approx \frac{165-110}{14.201} \approx 3.87$$

(e) Because $z > 1.96$, reject H_0.

(f) There is enough evidence at the 5% level of significance to support the administrator's claim that there is a difference in the earnings of people with bachelor's degrees and those with advanced degrees.

7. (a) The claim is "there is a difference in the salaries earned by teachers in Wisconsin and Michigan."
H_0: There is no difference in salaries.
H_a: There is a difference in salaries. (claim)

(b) Wilcoxon rank sum test

(c) The critical values are $z_0 = \pm 1.96$.

(d)

Ordered data	sample	Rank
47	W	1
49	W	2
51	W	3
52	W	4
53	M	5.5
53	W	5.5
55	M	8
55	W	8
55	W	8
56	W	10
57	M	11
58	M	12
59	W	13.5
59	M	13.5
61	W	16.5
61	W	16.5
61	M	16.5
61	M	16.5
62	M	19
65	M	20
67	M	21.5
67	M	21.5
76	M	23

R = sum ranks of Wisconsin = 88

$$\mu_R = \frac{n_1(n_1+n_2+1)}{2} = \frac{11(11+12+1)}{2} = 132$$

$$\sigma_R = \sqrt{\frac{n_1 n_2(n_1+n_2+1)}{12}} = \sqrt{\frac{(11)(12)(11+12+1)}{12}} \approx 16.248$$

$$z = \frac{R-\mu_R}{\sigma_R} \approx \frac{88-132}{16.248} \approx -2.71$$

(e) Because $z < -1.96$, reject H_0.

(f) There is enough evidence at the 5% level of significance to support the representative's claim that there is a difference in the salaries earned by teachers in Wisconsin and Michigan.

9. The claim is "a certain fuel additive improves a car's gas mileage."

H_0: The fuel additive does not improve gas mileage.

H_a: The fuel additive does improve gas mileage. (claim)

The critical value is $z_0 = -1.28$.

Before	After	Difference	Absolute value	Rank	Signed rank
36.4	36.7	−0.3	0.3	4.5	−4.5
36.4	36.9	−0.5	0.5	11	−11
36.6	37.0	−0.4	0.4	7	−7
36.6	37.5	−0.9	0.9	17	−17
36.8	38.0	−1.2	1.2	19.5	−19.5
36.9	38.1	−1.2	1.2	19.5	−19.5
37.0	38.4	−1.4	1.4	25	−25
37.1	38.7	−1.6	1.6	30.5	−30.5
37.2	38.8	−1.6	1.6	30.5	−30.5
37.2	38.9	−1.7	1.7	32	−32
36.7	36.3	0.4	0.4	7	7
37.5	38.9	−1.4	1.4	25	−25
37.6	39.0	−1.4	1.4	25	−25
37.8	39.1	−1.3	1.3	21.5	−21.5
37.9	39.4	−1.5	1.5	28.5	−28.5
37.9	39.4	−1.5	1.5	28.5	−28.5
38.1	39.5	−1.4	1.4	25	−25
38.4	39.8	−1.4	1.4	25	−25
40.2	40.0	0.2	0.2	2.5	2.5
40.5	40.0	0.5	0.5	11	11
40.9	40.1	0.8	0.8	16	16
35.0	36.3	−1.3	1.3	21.5	−21.5
32.7	32.8	−0.1	0.1	1	−1
33.6	34.2	−0.6	0.6	14.5	−14.5
34.2	34.7	−0.5	0.5	11	−11
35.1	34.9	0.2	0.2	2.5	2.5
35.2	34.9	0.3	0.3	4.5	4.5
35.3	35.3	0	0	—	—
35.5	35.9	−0.4	0.4	7	−7
35.9	36.4	−0.5	0.5	11	−11
36.0	36.6	−0.6	0.6	14.5	−14.5
36.1	36.6	−0.5	0.5	11	−11
37.2	38.3	−1.1	1.1	18	−18

Sum of negative ranks = −484.5
Sum of positive ranks = 43.5
$w_s = 43.5$

$$z = \frac{w_s - \frac{n(n+1)}{4}}{\sqrt{\frac{n(n+1)(2n+1)}{24}}} = \frac{43.5 - \frac{32(32+1)}{4}}{\sqrt{\frac{32(32+1)[(2)32+1]}{24}}} = \frac{-220.5}{\sqrt{2860}} = -4.123$$

Note: $n = 32$ because one of the differences is zero and should be discarded.
Because $z < -1.28$, reject H_0. There is enough evidence at the 10% level of significance for the engineer to conclude that the gas mileage is improved.

11.3 THE KRUSKAL-WALLIS TEST

11.3 TRY IT YOURSELF SOLUTIONS

1. The claim is "the distribution of the veterinarians' salaries in at least one of the three states is different from the others."

   H_0: The distribution of the salaries is the same in all three states.

   H_a: The distribution of the salaries in at least one state is different from the others. (claim)

   $\alpha = 0.05$

   $\text{d.f.} = k - 1 = 2$

   $\chi_0^2 = 5.991$; Rejection region: $\chi^2 > 5.991$

Ordered data	State	Rank
57.6	CA	1
63.3	TX	2
74.8	TX	3
80.4	FL	4
83.4	CA	5
84.7	FL	6
85.3	TX	7
89.6	TX	8
91	TX	9
95	FL	10
97.9	TX	11
101.1	TX	12
105.3	FL	13
106.7	FL	14
111.3	CA	15
113.2	CA	16
116.6	FL	17
118.7	TX	18
121.6	FL	19
126.8	CA	20
131	CA	21
135.9	FL	22
143.3	FL	23
146.1	CA	24
147.7	TX	25
149.9	TX	26
154	CA	27
160.2	CA	28

$R_1 = 121$, $R_2 = 128$, $R_3 = 157$

$$H = \frac{12}{N(N+1)}\left(\frac{R_1^2}{n_1} + \frac{R_2^2}{n_2} + \frac{R_3^2}{n_3}\right) - 3(N+1)$$

$$= \frac{12}{28(28+1)}\left(\frac{(121)^2}{10} + \frac{(128)^2}{9} + \frac{(157)^2}{9}\right) - 3(28+1)$$

$$\approx 2.015$$

Because $H < 5.991$, fail to reject H_0.

There is not enough evidence at the 5% level of significance to support the claim that the distribution for at least one state is different from the others.

11.3 EXERCISE SOLUTIONS

1. The conditions for using a Kruskal-Wallis test are that the samples must be random and independent, and the size of each sample must be at least 5.

3. (a) The claim is "the distributions of the annual premiums in at least one state is different from the others."

H_0: The distribution of the annual premiums is the same in all three states.

H_a: The distribution of the annual premiums in at least one state is different from the others. (claim)

(b) $\chi_0^2 = 5.991$; Rejection region: $\chi^2 > 5.991$

(c)

Ordered data	Sample	Rank
755	VA	1
766	VA	2
838	VA	3
845	VA	4
950	VA	5
1034	MA	6
1035	VA	7
1098	CT	8
1132	VA	9
1166	MA	10
1179	CT	11

Ordered data	Sample	Rank
1257	MA	12
1263	CT	13
1302	MA	14
1303	CT	15
1320	CT	16
1382	MA	17
1387	MA	18
1413	CT	19
1538	CT	20
1572	MA	21

$R_1 = 102,\ R_2 = 98,\ R_3 = 31$

$$H = \frac{12}{N(N+1)}\left(\frac{R_1^2}{n_1} + \frac{R_2^2}{n_2} + \frac{R_3^2}{n_3}\right) - 3(N+1)$$
$$= \frac{12}{21(21+1)}\left(\frac{(102)^2}{7} + \frac{(98)^2}{7} + \frac{(31)^2}{7}\right) - 3(21+1)$$
$$\approx 11.807$$

(d) Because $H > 5.991$, reject H_0.

(e) There is enough evidence at the 5% level of significance to conclude that the distribution of annual premiums in at least one state is different from the others.

5. (a) The claim is "the distribution of the annual salaries of private industry workers in at least one state is different from the others."
H_0: The distribution of the annual salaries is the same in all four states.
H_a: The distribution of the annual salaries in at least one state is different from the others. (claim)

(b) $\chi_0^2 = 6.251$; Rejection region: $\chi^2 > 6.251$

(c)

Ordered data	Sample	Rank
32.9	KY	1
34.8	WV	2.5
34.8	WV	2.5
35.4	SC	4.5
35.4	SC	4.5
36.6	WV	6
38.1	WV	7
38.3	KY	8
39.9	KY	9.5
39.9	KY	9.5
40.3	SC	11
40.8	NC	12
41.6	KY	13
41.7	SC	14

Ordered data	Sample	Rank
41.9	NC	15
43	SC	16
44.9	NC	17
45.1	WV	18
45.9	WV	19
47.2	NC	20
48.5	SC	21
48.8	NC	22.5
48.8	NC	22.5
49.1	SC	24
50.3	WV	25
50.5	KY	26
59.6	NC	27
62.1	KY	28

$R_1 = 95,\ R_2 = 136,\ R_3 = 95,\ R_4 = 80$

$$H = \frac{12}{N(N+1)}\left(\frac{R_1^2}{n_1} + \frac{R_2^2}{n_2} + \frac{R_3^2}{n_3}\right) - 3(N+1)$$
$$H = \frac{12}{28(28+1)}\left(\frac{(95)^2}{7} + \frac{(136)^2}{7} + \frac{(95)^2}{7} + \frac{(80)^2}{7}\right) - 3(28+1)$$
$$\approx 3.667$$

(d) Because $H < 6.251$, fail to reject H_0.

(e) There is not enough evidence at the 10% level of significance to conclude that the distribution of annual salaries in at least one state is different from the others.

7. (a) The claim is "the number of days patients spend in the hospital is different in at least one region of the United States."

H_0: The number of days spent in the hospital is the same for all four regions.

H_a: The number of days spent in the hospital is different in at least one region. (claim)

$\chi_0^2 = 11.345$; Rejection region: $\chi^2 > 11.345$

Ordered data	Sample	Rank
1	NE	2.5
1	MW	2.5
1	S	2.5
1	S	2.5
2	W	5
3	NE	8.5
3	NE	8.5
3	MW	8.5
3	MW	8.5
3	W	8.5
3	W	8.5
4	MW	13.5
4	MW	13.5
4	MW	13.5
4	W	13.5
5	NE	19
5	MW	19
5	S	19
5	S	19

Ordered data	Sample	Rank
5	S	19
5	W	19
5	W	19
6	NE	26
6	NE	26
6	NE	26
6	MW	26
6	W	26
6	W	26
6	W	26
7	MW	30.5
7	S	30.5
8	NE	33.5
8	NE	33.5
8	S	33.5
8	S	33.5
9	MW	36
11	NE	37

$R_1 = 220.5,\ R_2 = 171.5,\ R_3 = 159.5,\ R_4 = 151.5$

$$H = \frac{12}{N(N+1)}\left(\frac{R_1^2}{n_1} + \frac{R_2^2}{n_2} + \frac{R_3^2}{n_3} + \frac{R_4^2}{n_4}\right) - 3(N+1)$$

$$= \frac{12}{37(37+1)}\left(\frac{(220.5)^2}{10} + \frac{(171.5)^2}{10} + \frac{(159.5)^2}{8} + \frac{(151.5)^2}{9}\right) - 3(37+1)$$

$$\approx 1.507$$

Because $H < 11.345$, fail to reject H_0. There is not enough evidence at the 1% level of significance to support the underwriter's claim that the number of days patients spend in the hospital is different in at least one region of the United States.

(b)

Variation	Sum of squares	Degrees of freedom	Mean squares	F	p
Between	9.17	3	3.06	0.52	0.673
Within	194.72	33	5.90		

Because $\alpha = 0.01$,the critical value is about 4.51. Because $F \approx 0.52$ is less than the critical value, the decision is to fail to reject H_0. There is not enough evidence at the 1% level of significance to support the underwriter's claim that the number of days patients spend in the hospital is different in at least one region of the United States.

(c) Both tests come to the same decision, which is that there is not enough evidence to support the claim that the number of days patients spend in the hospital is different in at least one region of the United States.

11.4 RANK CORRELATION

11.4 TRY IT YOURSELF SOLUTIONS

1. The claim is "there is a significant correlation between the oat and wheat prices."
H_0: $\rho_s = 0$; H_a: $\rho_s \neq 0$ (claim)
The critical value is 0.714.

Oat	Rank	Wheat	Rank	d	d^2
1.84	1	3.67	2	−1	1
1.97	2	3.49	1	1	1
2.03	3	3.68	3	0	0
2.25	4	3.88	4	0	0
2.35	6	3.91	5	1	1
2.31	5	4.02	6	−1	1
2.40	7	4.15	7	0	0
					$\sum d^2 = 4$

$$r_s \approx 1 - \frac{6\sum d^2}{n(n^2-1)} = 1 - \frac{6(4)}{7(7^2-1)} \approx 0.929$$

Because $r_s > 0.714$, reject H_0.
There is enough evidence at the 10% level of significance to conclude that there is a significant correlation between the oat and wheat prices.

11.4 EXERCISE SOLUTIONS

1. The Spearman rank correlation coefficient can be used to describe the relationship between linear and nonlinear data. Also, it can be used for data at the ordinal level and it is easier to calculate by hand than the Pearson correlation coefficient.

3. The ranks of corresponding data pairs are exactly identical when r_s is equal to 1.
The ranks are in "reverse" order when r_s is equal to −1.
The ranks of have no relationship when r_s is equal to 0.

5. (a) The claim is "there is a significant correlation between purchased seed expenses and fertilizer and lime expenses in the farming business."
H_0:: $\rho_s = 0$; H_a: $\rho_s \neq 0$ (claim)

(b) The critical value is 0.738

(c)

Seed expenses	Rank	Fertilizer/lime expense	Rank	*d*	d^2
490	4.5	480	3.5	1	1
1530	8	2060	8	0	0
490	4.5	480	3.5	1	1
266	1	402	1	0	0
741	6	642	6	0	0
380	3	470	2	1	1
879	7	858	7	0	0
360	2	560	5	−3	9
					$\sum d^2 = 12$

$$r_s \approx 1 - \frac{6\sum d^2}{n(n^2-1)} \approx 1 - \frac{6(12)}{8(8^2-1)} \approx 0.857$$

(d) Because $r_s > 0.738$, reject H_0

(e) There is enough evidence at the 5% level of significance to conclude that there is a significant correlation between purchased seed expenses and fertilizer and lime expenses.

7. (a) The claim is "there is a significant correlation between the barley and corn prices."
H_0: $\rho_s = 0$; H_a: $\rho_s \neq 0$ (claim)

(b) The critical value is 0.700

(c)

Barley	Rank	Corn	Rank	*d*	d^2
4.89	4	3.21	1	3	9
4.52	1	3.22	2	−1	1
4.85	2	3.29	4	−2	4
4.97	6	3.23	3	3	9
5.12	9	3.33	5	4	16
4.91	5	3.4	6	−1	1
5.08	8	3.44	8	0	0
4.98	7	3.49	9	−2	4
4.87	3	3.43	7	−4	16
					$\sum d^2 = 60$

$$r_s \approx 1 - \frac{6\sum d^2}{n(n^2-1)} \approx 1 - \frac{6(60)}{9(9^2-1)} \approx 0.5$$

(d) Because $r_s < 0.700$, fail to reject H_0.

(e) There is not enough evidence at the 5% level of significance to conclude that there is a significant correlation between the barley and corn prices.

9. The claim is "there is a significant correlation between science achievement scores and GNI."
H_0: $\rho_s = 0$; H_a: $\rho_s \neq 0$ (claim)

Science average	Rank	GNI	Rank	*d*	d^2
528	8	1529	4	4	16
495	5	2458	6	–1	1
509	7	3437	7	0	0
481	2	1815	5	–3	9
538	9	4549	8	–1	1
416	1	1143	2	–1	1
493	3.5	1192	3	0.5	0.25
493	3.5	503	1	2.5	6.25
496	6	18496	9	–3	9
					$\sum d^2 = 43.5$

$$r_s \approx 1 - \frac{6\sum d^2}{n(n^2-1)} \approx 1 - \frac{6(43.5)}{9(9^2-1)} \approx 0.638$$

Because $r_s > 0.600$, reject H_0.

There is enough evidence at the 10% level of significance to conclude that there is a significant correlation between science achievement scores and GNI.

11. The claim is "there is a significant correlation between science and mathematics achievement scores."
H_0: $\rho_s = 0$; H_a: $\rho_s \neq 0$ (claim)
The critical value is 0.600

Science average	Rank	Math average	Rank	*d*	d^2
528	8	516	8	0	0
495	5	493	5	0	0
509	7	506	7	0	0
481	2	490	4	–2	4
538	9	532	9	0	0
416	1	408	1	0	0
493	3.5	486	3	0.5	0.25
493	3.5	494	6	–2.5	6.25
496	6	470	2	4	16
					$\sum d^2 = 26.5$

$$r_s \approx 1 - \frac{6\sum d^2}{n(n^2-1)} \approx 1 - \frac{6(26.5)}{9(9^2-1)} \approx 0.779$$

Because $r_s > 0.600$, reject H_0.

There is enough evidence at the 10% level of significance to conclude that there is a significant correlation between science and mathematics achievement scores.

13. The claim is "there is a significant correlation between average hours worked and the number of on-the-job injuries."

H_0: $\rho_s = 0$; H_a: $\rho_s \neq 0$ (claim)

The critical values are $\frac{\pm z}{\sqrt{n-1}} = \frac{\pm 1.645}{\sqrt{33-1}} \approx \pm 0.291.$

Hours worked	Rank	Injuries	Rank	D	D^2
46	26.5	22	14	12.5	156.25
43	18	25	19.5	−1.5	2.25
41	10	18	8	2	4
40	5.5	17	5.5	0	0
41	10	20	11	−1	1
42	15.5	22	14	1.5	2.25
45	22.5	28	24	−1.5	2.25
45	22.5	29	27	−4.5	20.25
42	15.5	24	18	−2.5	6.25
45	22.5	26	21.5	1	1
44	19.5	26	21.5	−2	4
44	19.5	25	19.5	0	0
45	22.5	27	23	−0.5	0.25
46	26.5	29	27	−0.5	0.25
47	29.5	29	27	2.5	6.25
47	29.5	30	31.5	−2	4
46	26.5	29	27	−0.5	0.25
46	26.5	29	27	−0.5	0.25
49	31	30	31.5	−0.5	0.25
50	32.5	30	31.5	1	1
50	32.5	30	31.5	1	1
42	15.5	23	16.5	−1	1
41	10	22	14	−4	16
42	15.5	23	16.5	−1	1
41	10	21	12	−2	4
41	10	19	10	0	0
41	10	18	8	2	4
41	10	18	8	2	4
40	5.5	17	5.5	0	0
39	3	16	2.5	0.5	0.25
38	1	16	2.5	−1.5	2.25
39	3	16	2.5	0.5	0.25
39	3	16	2.5	0.5	0.25
					$\sum d^2 = 246$

$$r_s \approx 1-\frac{6\sum d^2}{n(n^2-1)} \approx 1-\frac{6(246)}{33(33^2-1)} \approx 0.959$$

Because $r_s > 0.291$, reject H_0. There is enough evidence at the 10% level of significance to conclude that there is a significant correlation between average hours worked and the number of on-the-job injuries.

11.5 THE RUNS TEST

11.5 TRY IT YOURSELF SOLUTIONS

1. *PPP F P F PPPP FF P F PP FFF PPP F PPP*
13 groups $\Rightarrow$ 13 runs
3, 1, 1, 1, 4, 2, 1, 1, 2, 3, 3, 1, 3

2. The claim is "the sequence of genders is not random."
H_0: The sequence of genders is random.
H_a: The sequence of genders is not random. (claim)
$\alpha = 0.05$
M FFF MM FF M F MM FFF
n_1 = number of F's = 9
n_2 = number of M's = 6
G = number of runs = 8
lower critical value = 4
upper critical value = 13
$G = 8$
Because $4 < G < 13$, fail to reject H_0.
There is not enough evidence at the 5% level of significance to support the claim that the sequence of genders is not random.

3 The claim is "the sequence of weather conditions is not random."
H_0: The sequence of weather conditions is random.
H_a: The sequence of weather conditions is not random. (claim)
$\alpha = 0.05$
n_1 = number of N's = 21
n_2 = number of S's = 10
G = number of runs = 17
The critical values are $z_0 = \pm 1.96$.

$$\mu_G = \frac{2n_1n_2}{n_1+n_2}+1 = \frac{2(21)(10)}{21+10}+1 = 14.55$$

$$\sigma_G = \sqrt{\frac{2n_1n_2(2n_1n_2-n_1-n_2)}{(n_1+n_2)^2(n_1+n_2-1)}} = \sqrt{\frac{2(21)(10)(2(21)(10)-21-10)}{(21+10)^2(21+10-1)}} \approx 2.38$$

$$z = \frac{G - \mu_G}{\sigma_G} \approx \frac{17 - 14.55}{2.38} \approx 1.03$$

Because $z < 1.96$, fail to reject H_0.

There is not enough evidence at the 5%level of significance to support the claim that the sequence of weather conditions is not random.

11.5 EXERCISE SOLUTIONS

1. Answers will vary. *Sample answer:* It is called the runs test because it considers the number of runs of data in a sample to determine whether the sequence of data was randomly selected.

3. Number of runs = 8
Run lengths = 1, 1, 1, 1, 3, 3, 1, 1

5. Number of runs = 9
Run lengths = 1, 1, 1, 1, 1, 6, 3, 2, 4

7. n_1 = number of *T*'s = 6
n_2 = number of *F*'s = 6

9. n_1 = number of *M*'s = 10
n_2 = number of *F*'s = 10

11. n_1 = number of *T*'s = 6
n_1 = number of *F*'s = 6
too high: 11; too low: 3

13. n_1 = number of *N*'s = 11
n_1 = number of *S*'s = 7
too high: 14; too low: 5

15. (a) The claim is "the tosses were not random."
H_0: The coin tosses were random.
H_a: The coin tosses were not random. (claim)

(b) n_1 = number of *H*'s = 7
n_2 = number of *T*'s = 9
lower critical value = 4
upper critical value = 14

(c) $G = 9$ runs

(d) Because $4 < G < 14$, fail to reject H_0.

(e) There is not enough evidence at the 5% level of significance to support the claim that the coin tosses were not random.

17. (a) The claim is "the sequence of leagues of World Series winning teams is not random."
H_0: The sequence of leagues of winning teams is random.
H_a: The sequence of leagues of winning teams is not random. (claim)

(b) n_1 = number of N's = 22
n_2 = number of A's = 25
The critical values are $z_0 = \pm 1.96$

(c) G = 31 runs

$$\mu_G = \frac{2n_1n_2}{n_1+n_2} + 1$$

$$\mu_G = \frac{2(22)(25)}{22+25} + 1 \approx 24.404$$

$$\sigma_G = \sqrt{\frac{2n_1n_2(2n_1n_2 - n_1 - n_2)}{(n_1+n_2)^2(n_1+n_2-1)}}$$

$$\sigma_G = \sqrt{\frac{2(22)(25)(2(22)(25) - 22 - 25)}{(22+25)^2(22+25-1)}} \approx 3.376$$

$$Z = \frac{G - \mu_G}{\sigma_G}$$

$$Z = \frac{31 - 24.404}{3.376} \approx 1.95$$

(d) Because $z > -1.96$, fail to reject H_0.

(e) There is not enough evidence at the 5% level of significance to conclude that the sequence of leagues of World Series winning teams is not random.

19. (a) The claim is "the microchips are random by gender."
H_0: The microchips are random by gender.
H_a: The microchips are not random by gender. (claim)

(b) n_1 = number of M's = 9
n_2 = number of F's = 20
lower critical value = 8; upper critical value = 18

(c) G = 12 runs

(d) Because $8 < G < 18$, fail to reject H_0.

(e) There is not enough evidence at the 5% level of significance to reject the claim that the microchips are random by gender.

21. The claim is "the daily high temperatures do not occur randomly."
H_0: Daily high temperatures occur randomly.
H_a: Daily high temperatures do not occur randomly. (claim)
median = 87
n_1 = number of above = 15
n_2 = number of below = 13
lower critical value = 9
upper critical value = 21
G = 11 runs

Because $9 < G < 21$, fail to reject H_0,
There is not enough evidence at the 5% level of significance to support the claim that the daily high temperatures do not occur randomly.

23. Answers will vary.

CHAPTER 11 REVIEW EXERCISE SOLUTIONS

1. (a) The claim is "the median number of customers per day is no more than 650."
H_0: median ≤ 650 (claim)
H_a: median > 650

(b) The critical value is 2.

(c) $x = 7$

(d) Because $x > 2$, fail to reject H_0.

(e) There is not enough evidence at the 1% level of significance to reject the store manager's claim that the median number of customers per day is no more than 650.

3. (a) The claim is "median sentence length for all federal prisoners is 2 years."
H_0: median = 2 (claim); H_a: median $\neq 2$

(b) The critical value is $z_0 = -1.645$.

(c) $x = 65$

$$z = \frac{(x+0.5)-0.5(n)}{\frac{\sqrt{n}}{2}} = \frac{(65+0.5)-0.5(174)}{\frac{\sqrt{174}}{2}} \approx \frac{-21.5}{6.595} \approx -3.26$$

(d) Because $z_0 < -1.645$, reject H_0.

(e) There is enough evidence at the 10% level of significance to reject the agency's claim that the median sentence length for all federal prisoners is 2 years.

5. (a) The claim is "there was no reduction in diastolic blood pressure."
H_0: There is no reduction in diastolic blood pressure. (claim)
H_a: There is a reduction in diastolic blood pressure.

(b) The critical value is 2.

(c) $x = 3$

(d) Because $x > 2$, fail to reject H_0.

(e) There is not enough evidence at the 5% level of significance to reject the claim that there was no reduction in diastolic blood pressure.

7. (a) The claim is "there is a difference in the total times required to earn a doctorate degree by female and male graduate students."
H_0: There is no difference in the total times required to earn a doctorate degree by female and male graduate students.
H_a: There is a difference in the total times required to earn a doctorate degree by female and male graduate students. (claim)

(b) Wilcoxon rank sum test

(c) The critical values are $z_0 = \pm 2.575$.

(d)

Ordered data	Sample	Rank
6	F	1.5
6	F	1.5
7	M	4.5
7	M	4.5
7	M	4.5
7	M	4.5
8	F	9
8	F	9
8	M	9
8	M	9
8	M	9
9	F	14.5
9	F	14.5
9	F	14.5
9	M	14.5
9	M	14.5
9	M	14.5
10	F	19
10	M	19
10	M	19
11	F	21.5
11	F	21.5
12	F	23
13	F	24

$$\mu_R = \frac{n_1(n_1+n_2+1)}{2} = \frac{12(12+12+1)}{2} = 150$$

$$\sigma_R = \sqrt{\frac{n_1 n_2(n_1+n_2+1)}{12}} \sqrt{\frac{(12)(12)(12+12+1)}{12}} \approx 17.32$$

$$z = \frac{R - \mu_R}{\sigma_R} \approx \frac{173.5 - 150}{17.32} \approx 1.357$$

$z = -1.357$ or 1.357.

(e) Because $z > -2.575$, fail to reject H_0.

(f) There is not enough evidence at the 1% level of significance to support the claim that there is a difference in the total times required to earn a doctorate degree by female and male graduate students.

9. (a) The claim is "the distributions of the ages of the doctorate recipients in at least one field of study is different from the others."

H_0: The distribution of the ages of doctorate recipients is the same in all three fields of study.

H_a: The distribution of the ages of doctorate recipients in at least one field of study is different from the others.(claim)

(b) $\chi_0^2 = 9.210$; Rejection region: $\chi^2 > 9.210$.

(c)

Ordered data	Sample	Rank
29	L	1.5
29	P	1.5
30	L	5.5
30	P	5.5
30	P	5.5
30	P	5.5
30	P	5.5
30	S	5.5
31	L	12.5
31	L	12.5
31	L	12.5
31	P	12.5
31	P	12.5
31	P	12.5
31	S	12.5
31	S	12.5
32	L	20

Ordered data	Sample	Rank
32	L	20
32	L	20
32	P	20
32	P	20
32	S	20
32	S	20
33	P	25
33	S	25
33	S	25
34	L	28
34	L	28
34	S	28
35	L	31
35	S	31
35	S	31
36	S	33

$R_1 = 191.5,\ R_2 = 126,\ R_3 = 243.5$

$$H = \frac{12}{N(N+1)}\left(\frac{R_1^2}{n_1} + \frac{R_2^2}{n_2} + \frac{R_3^2}{n_3}\right) - 3(N+1)$$

$$= \frac{12}{33(33+1)}\left(\frac{(191.5)^2}{11} + \frac{(126)^2}{11} + \frac{(243.5)^2}{11}\right) - 3(33+1)$$

$$\approx 6.741$$

(d) Because $H < 9.210$, fail to reject H_0.

(e) There is not enough evidence at the 1% level of significance to conclude that the distribution of ages of the doctorate recipients in at least one field of study is different from the others.

11. (a) The claim is "there is a correlation between the overall score and the price."
$H_0: \rho_s = 0$; $H_a: \rho_s \neq 0$ (claim)

(b) The critical value is 0.829.

(c)

Score	Rank	Price	Rank	d	d^2
93	6	500	5.5	0.5	0.25
91	5	300	4	1	1
90	4	500	5.5	−1.5	2.25
87	3	150	2	1	1
85	2	250	3	−1	1
69	1	130	1	0	0
					$\sum d^2 = 5.5$

$$r_s = 1 - \frac{6\sum d^2}{n(n^2-1)} = 1 - \frac{6(5.5)}{6(6^2-1)} = 0.8429$$

(d) Because $r_s > 0.829$, reject H_0.

(e) There is enough evidence at the 10% level of significance to conclude that there is a significant correlation between overall score and the price.

13. (a) The claim is "the stops were not random by gender."
H_0: The traffic stops were random by gender.
H_a: The traffic stops were not random by gender. (claim)

(b) n_1 = number of F's = 12
n_2 = number of M's = 13
lower critical value = 8
upper critical value = 19

(c) G = 14 runs

(d) Because $8 < G < 19$, fail to reject H_0.

(e) There is not enough evidence at the 5% level of significance to support the claim that the stops were not random by gender.

CHAPTER 11 QUIZ SOLUTIONS

1. (a) The claim is "the median number of annual volunteer hours is 52."
H_0: median = 52 (claim)
H_a: median ≠ 52

(b) Sign test

(c) The critical values are $z_0 = \pm 1.96$

(d) $x = 23$

$$z = \frac{(x+0.5)-0.5n}{\frac{\sqrt{n}}{2}} = \frac{(23+0.5)-0.5(70)}{\frac{\sqrt{70}}{2}} = -2.75$$

(e) Because $z < -1.96$, reject H_0.

(f) There is enough evidence at the 5% level of significance to reject the organization's claim that the median number of annual volunteer hours is 52.

2. (a) The claim is "there is a difference in the hourly earnings of union and nonunion workers in state and local governments."

H_0: There is no difference in the hourly earnings.

H_a: There is a difference in the hourly earnings. (claim)

(b) Wilcoxon rank sum test

(c) The critical values are $z_0 = \pm 1.645$.

(d)

Ordered data	Sample	Rank
20.45	N	1
21.20	N	2
21.40	N	3
22.05	N	4
22.25	N	5
22.50	N	6
23.10	N	7
24.75	N	8
26.15	N	9
26.75	U	10

Ordered data	Sample	Rank
26.95	N	11
27.35	U	12
27.60	U	13
27.85	U	14
28.15	U	15
29.05	U	16
29.75	U	17
32.30	U	18
32.88	U	19
35.52	U	20

R = sum ranks of nonunion workers = 56

$$\mu_R = \frac{n_1(n_1+n_2+1)}{2} = \frac{10(10+10+1)}{2} = 105$$

$$\sigma_R = \sqrt{\frac{n_1 n_2(n_1+n_2+1)}{12}} \sqrt{\frac{(10)(10)(10+10+1)}{12}} \approx 13.229$$

$$z = \frac{R-\mu_R}{\sigma_R} \approx \frac{56-105}{13.229} \approx -3.70 \ (or \ 3.70)$$

(e) Because $z < -1.645$, reject H_0.

(f) There is enough evidence at the 10% level of significance to support the claim that there is a difference in the hourly earnings of union and nonunion workers in state and local governments.

3. (a) The claim is "the distribution of the sales prices in at least one region is different from the others."
H_0: The distribution of the sales prices is the same in all four regions.
H_a: The distribution of the sales prices in at least one region is different from the others.(claim)

(b) Kruskal-Wallis test

(c) The critical value is $\chi^2 = 11.345$

(d)

Ordered data	Sample	Rank
148.5	MW	1
153.9	MW	2
155.6	S	3
156.7	S	4
163.3	MW	5
165.1	MW	6
166.8	S	7
166.9	MW	8
169.9	MW	9
170.4	S	10
175.3	S	11
178.4	S	12
178.9	MW	13
181.3	S	14
183.1	MW	15
196.3	S	16

Ordered data	Sample	Rank
242.7	NE	17
243.4	NE	18
250.3	NE	19
254.8	NE	20
257.3	NE	21
264.2	NE	22
270.7	NE	23
275	NE	24
291.6	W	25
303.6	W	26
308	W	27
320.2	W	28
321.7	W	29
327.4	W	30
331.7	W	31
357.4	W	32

$R_1 = 164,\ R_2 = 59,\ R_3 = 77,\ R_4 = 228$

$$H = \frac{12}{N(N+1)}\left(\frac{R_1^2}{n_1} + \frac{R_2^2}{n_2} + \frac{R_3^2}{n_3} + \frac{R_4^2}{n_4}\right) - 3(N+1)$$

$$= \frac{12}{32(32+1)}\left(\frac{(164)^2}{8} + \frac{(59)^2}{8} + \frac{(77)^2}{8} + \frac{(228)^2}{8}\right) - 3(32+1)$$

$$\approx 26.412$$

(e) Because $H > 11.345$, reject H_0.

(f) There is enough evidence at the 1% level of significance to conclude that the distribution of the sales prices in at least one region is different from the others.

4. (a) The claim is "there is a correlation between the number of emails sent and the number of emails received."
H_0: $\rho_s = 0$; H_a: $\rho_s \neq 0$ (claim)

(b) Spearman rank correlation coefficient.

(c) The critical value is 0.833.

Emails sent	Rank	Emails received	Rank	d	d^2
30	8.5	32	8	0.5	0.25
30	8.5	36	9	−0.5	0.25
25	4.5	21	3	1.5	2.25
26	6	22	4.5	1.5	2.25
24	3	20	1.5	1.5	2.25
18	1.5	20	1.5	0	0
18	1.5	22	4.5	−3	9
25	4.5	23	6.5	−2	4
28	7	23	6.5	0.5	0.25
					$\sum d^2 = 20.5$

$$r_s = 1 - \frac{6\sum d^2}{n(n^2-1)} = 1 - \frac{6(20.5)}{9(9^2-1)} \approx 0.829$$

(e) Because $r_s < 0.833$, fail to reject H_0.

(f) There is not enough evidence at the 1% level of significance to conclude that there is a significant correlation between the number of emails sent and the number of emails received.

5. (a) The claim is "days with rain are not random."
H_0: The days with rain are random.
H_a: The days with rain are not random. (claim)

(b) Runs test

(c) n_1 = number of N's = 15
n_2 = number of R's = 15
lower critical value = 10
upper critical value = 22

(d) $G = 16$ runs

(e) Because $10 < G < 22$, fail to reject H_0.

(f) There is not enough evidence at the 5% level of significance for the meteorologist to conclude that days with rain are not random.

Alternative Presentation of the Standard Normal Distribution

APPENDIX **A**

TRY IT YOURSELF SOLUTIONS

1. (1) 0.4857

(2) $z = \pm 2.17$

2.

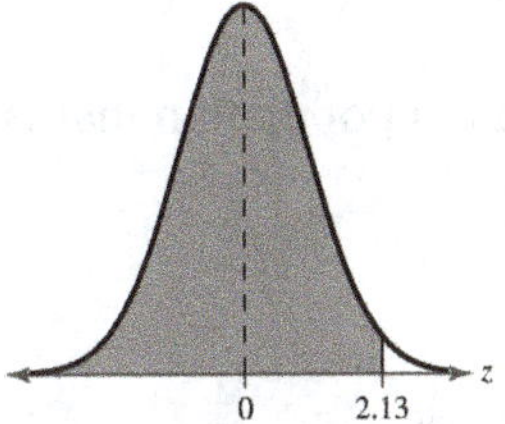

The area corresponding to $z = 2.13$ is 0.4834
Area = 0.5 + 0.4834 = 0.9834

3.

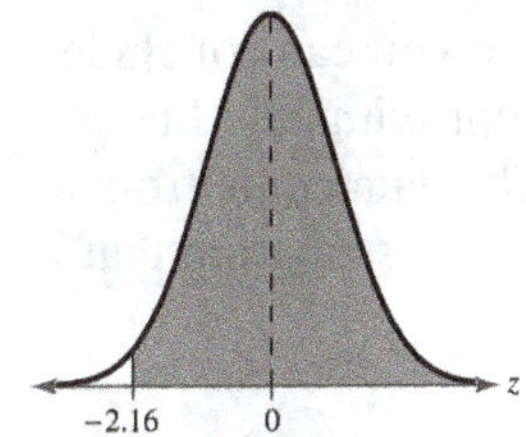

The area corresponding to $z = -2.16$ is 0.4846
Area = 0.50+00.4846 = 0.9846

4.

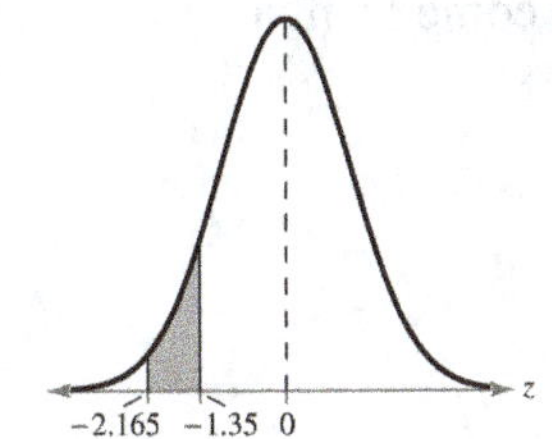

The area corresponding to $z = -1.35$ is 0.0885
The area corresponding to $z = -2.165$ is 0.0152
Area = $0.0885 - 0.0152 = 0.0733$

Normal Probability Plots

TRY IT YOURSELF SOLUTIONS

1. 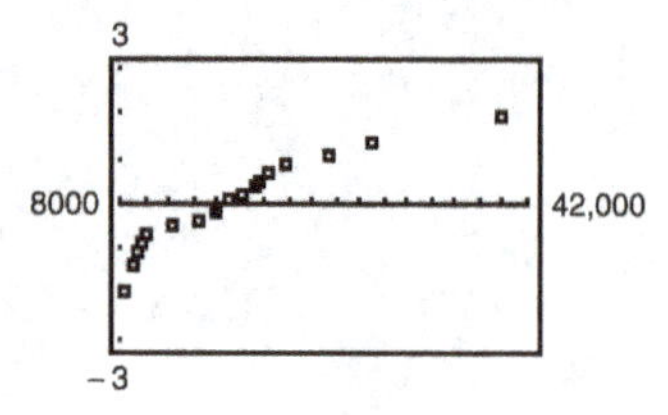

The points show a curve, so you can conclude that the sample data come from a population that is not normally distributed.

APPENDIX C EXERCISE SOLUTIONS

1. The observed values are usually plotted along the horizontal axis. The expected z-scores are plotted along the vertical axis.

2. If the plotted points in a normal probability plot are approximately linear, then you can conclude that the data come from a normal distribution. If the plotted points are not approximately linear or follow some type of pattern that is not linear, then you can conclude that the data come from a distribution that is not normal. Multiple outliers or clusters of points indicate a distribution that is not normal.

3. Because the points appear to follow a nonlinear pattern, you can conclude that the data do not come from a population that has a normal distribution.

4. Because the points are approximately linear, you can conclude that the data come from a population that has a normal distribution.

5. 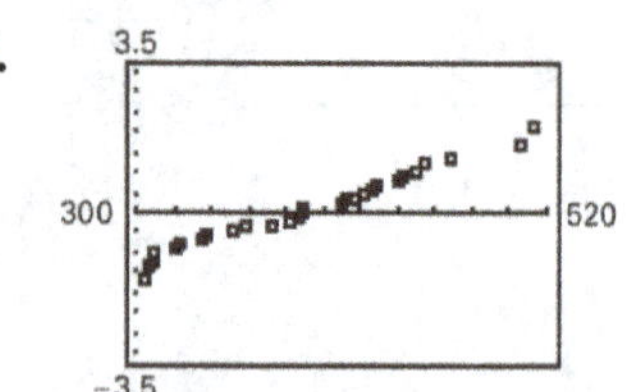

Because the points are approximately linear, you can conclude that the data come from a population that has a normal distribution.

6. 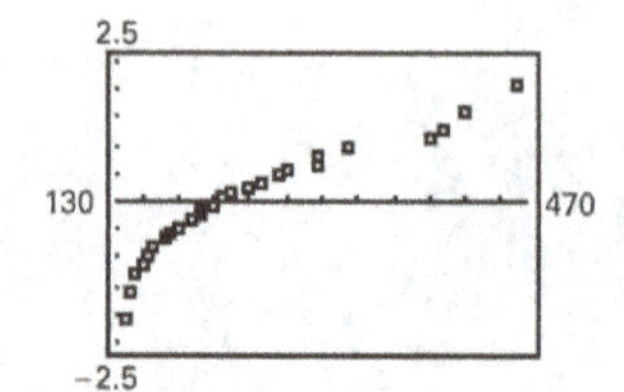

Because the points appear to follow a nonlinear pattern, you can conclude that the data do not come from a population that has a normal distribution.